METHODS IN GENOMIC NEUROSCIENCE

METHODS & NEW FRONTIERS IN NEUROSCIENCE

Series Editors
Sidney A. Simon, Ph.D.
Miguel A.L. Nicolelis, M.D., Ph.D.

Published Titles

Apoptosis in Neurobiology
Yusuf A. Hannun, M.D., Professor/Biomedical Research and Department Chairman/
    Biochemistry and Molecular Biology, Medical University of South Carolina
Rose-Mary Boustany, M.D., tenured Associate Professor/Pediatrics and Neurobiology,
    Duke University Medical Center

Methods for Neural Ensemble Recordings
Miguel A.L. Nicolelis, M.D., Ph.D., Associate Professor/Department of Neurobiology,
    Duke University Medical Center

Methods of Behavioral Analysis in Neuroscience
Jerry J. Buccafusco, Ph.D., Professor/Pharmacology and Toxicology,
    Professor/Psychiatry and Health Behavior, Medical College of Georgia

Neural Prostheses for Restoration of Sensory and Motor Function
John K. Chapin, Ph.D., MCP and Hahnemann School of Medicine
Karen A. Moxon, Ph.D., Department of Electrical and Computer Engineering,
    Drexel University

Computational Neuroscience: Realistic Modeling for Experimentalists
Eric DeSchutter, M.D., Ph.D., Department of Medicine, University of Antwerp

Methods in Pain Research
Lawrence Kruger, Ph.D., Professor Emeritus/Neurobiology, UCLA School of Medicine

Motor Neurobiology of the Spinal Cord
Timothy C. Cope, Ph.D., Department of Physiology, Emory University School of Medicine

Nicotinic Receptors in the Nervous System
Edward Levin, Ph.D., Associate Professor/Department of Pharmacology and Molecular
    Cancer Biology and Department of Psychiatry and Behavioral Sciences,
    Duke University School of Medicine

Methods in Genomic Neuroscience
Helmin R. Chin, Ph.D., NIMH, NIH Genetics Research
    Steven O. Moldin, Ph.D, NIMH, NIH Genetics Research

METHODS IN GENOMIC NEUROSCIENCE

Edited by
Hemin R. Chin
Steven O. Moldin

CRC PRESS

Boca Raton   London   New York   Washington, D.C.

QP356.22
M48

Library of Congress Cataloging-in-Publication Data

Methods in genomic neuroscience / edited by Hemin R. Chin and Steven O.
Moldin.
p. ; cm. -- (Methods and new frontiers in neuroscience)
Includes bibliographical references and index.
ISBN 0-8493-2397-5 (alk. paper)
1. Neurogenetics--Methodology.
[DNLM: 1. Genomics--methods.   2. Neurosciences--methods.   WL 100
M5916 2001]   I. Chin, Hemin R. II. Moldin, Steven O. III. Methods &
new frontiers in neuroscience series.
QP356.22   .M48 2001
612.8--dc21
2001002144

Visit the CRC Press Web site at www.crcpress.com

This book was written as part of the authors' official duties as U.S. Government employees.

International Standard Book Number 0-8493-2397-5
Library of Congress Card Number 2001002144
Printed in the United States of America 1 2 3 4 5 6 7 8 9 0
Printed on acid-free paper

Foreword

This latest volume in the CRC Press *Methods and New Frontiers in Neuroscience* series arrives at a most propitious moment, contemporaneously with the long-awaited, and much-ballyhooed releases of the first working drafts of the human genome. Given the enormous excitement generated by the not-quite-usable-yet inventories of probable genes, those ready to apply the discovery and definition methods of modern molecular genetics to the unsolved problems of the basic and clinical neurosciences should find here a new spectrum of strategies to add to their experimental armamentarium.

When modern cloning technology began to be applied in earnest to the genes expressed in the brain slightly more than 2 decades ago, the earliest success stories consisted of short circuiting peptide purification schemes, and the requirements for hundreds of micrograms of starting material, to getting a few amino acids of sequence and then synthesizing oligonucleotides to move the hunt from proteins to mRNA. Once it became clear that the brain's structural and functional complexity was also predicated on an unimaginably rich constitution of genes that were either nervous-system-specific or at least highly enriched in brain, the admirably innovative maneuvers crafted by our never-daunted practitioners have progressively sought new short cuts to pin down quickly the most "interesting" of the brain's long-held secret cache of gene products.

Not 10 years after the cloning of what were widely regarded as the last of the classical hypothalamic hypophysiotrophic hormones, new methods of molecular discovery using genetic manipulations have led to wholly unanticipated additions to the corticotropin-releasing hormone message family, and to a second member of the somatostatin family. Looking for mRNA activated by pharmacological manipulation has revealed an unanticipated neuropeptide that may well be related to appetite and reward; looking for genes that have no property other than being enriched in the hypothalamus and able to activate orphaned G-protein coupled receptors has revealed yet another peptide engaged in the regulation of appetite, blood pressure, and sleep, and quickly identified as the gene product mutated in narcolepsy. All that and more as the bountiful information of this volume will attest, by just clever hunting and a lot of hard work.

Now the need for clever experimental strategies of functional gene discovery may be giving way to the existence of the soon-to-be completed genomic sequences for man and mouse in which the size of all current gaps and the order and orientation of all sequences will be known and at least partially annotated as to known or presumed function. But that departure will only lead inevitably to the next series of much more profound puzzles — where are these genes expressed, and how in the intact individual do the cells expressing these genes adapt to the unpredicted demands of their environment to maintain the brain in a top-notch mode of healthful operation, and where do we spot those probably redundant points of vulnerability or resistance that can make the difference between health and disease, and which will likely form the next big evolution of individually tailored medications?

It is into this gap that the contributions of this volume should prove themselves of maximum benefit. Without attempting to overview the entire volume, let me highlight some of the chapters that I found most useful for my needs. Sokolowski and Wahlsten look specifically at the genetic underpinnings of complex behaviors, and how the new tools can be used most effectively. Frankel looks at what has been learned so far from the study of naturally occurring mutations, and Mayford and Kandel show what has already been accomplished with conditional and inducible gene manipulations that can sharpen the power and focus and greatly reduce covert effects of genetic perturbations on development. Lockhart and Barlow describe their initial uses of DNA arrays and the degree to which interstrain differences can be assessed, while Mirnics, Lewis, and Levitt describe their applications of this technology to the postmortem analysis of gene expression in schizophrenia — one of the very first studies to apply the technology to tissues as highly heterogeneous as brain.

What a great new world, and this volume can be your passport.

Floyd E. Bloom, M.D.
Department of Neuropharmacology
The Scripps Research Institute
LaJolla, California

Methods & New Frontiers in Neuroscience

Series Editors
Sidney A. Simon, Ph.D.
Miguel A. L. Nicolelis, M.D., Ph.D.

Our goal in creating the Methods & New Frontiers in Neuroscience Series is to present the insights of experts on emerging experimental techniques and theoretical concepts that are, or will be, at the vanguard of neuroscience. Books in the series cover topics ranging from methods to investigate apoptosis to modern techniques for neural ensemble recordings in behaving animals. The series also covers new and exciting multidisciplinary areas of brain research, such as computational neuroscience and neuroengineering, and describes breakthroughs in classical fields such as behavioral neuroscience. We want these to be the books every neuroscientist will use in order to get acquainted with new methodologies in brain research. These books can be given to graduate students and postdoctoral fellows when they are looking for guidance to start a new line of research.

Each book is edited by an expert and consists of chapters written by the leaders in a particular field. Books are richly illustrated and contain comprehensive bibliographies. Chapters provide substantial background material relevant to the particular subject. Hence, they are not just "methods books." They contain detailed "tricks of the trade" and information as to where these methods can be safely applied. In addition, they include information about where to buy equipment and Web sites helpful in solving both practical and theoretical problems.

We hope that as the volumes become available, the effort put in by us, the publisher, the book editors, and the individual authors will contribute to the further development of brain research. The extent to which we achieve this goal will be determined by the utility of these books.

Preface

This volume in the CRC Press *Methods and New Frontiers in Neuroscience* series explores the application of genetic and genomic methodologies to fundamental problems in neuroscience, and was stimulated by the recognition that understanding the genetic bases of the mammalian nervous system and its diseases will revolutionize clinical diagnosis, treatment, and prevention. The past few years have witnessed extraordinary advances in molecular genetic techniques and the accumulation of an immense amount of structural genomic information about both human and model organisms, culminating with the publication of the genome sequences of *Caenorhabditis elegans, Drosophila melanogaster, Saccharomyces cerevisiae, Homo sapiens*, and *Mus musculus*. The dramatic availability and increase in the amount of such rich genomic information and technologies for genetic analysis will usher in an exciting era in which the basic principles and methodologies of functional genomics will be successfully applied in neuroscience and basic behavioral science. With the full anatomy of the human genome at hand, researchers for the first time will be able to move beyond traditional gene-by-gene approaches and take a global view of neurobiological processes, in which the function and expression of all approximately 30,000 mammalian genes are considered simultaneously.

The application of high-precision tools and technologies to understand individual variation in the entire genome and the expression and function of all genes in all cell types in the nervous system will usher in a new era of exploration and discovery. This newly emerging field of genomic neuroscience will provide an innovative paradigm by which problems of a previously unthinkable scope and scale can be approached and solved as they have been in many other areas of biology.

We anticipate that this volume will benefit newcomers and experienced neuroscientists alike, and will serve as a reference for applying powerful, cutting-edge genetic techniques to experimental studies of nervous system functions and complex behaviors. Our goal is to provide a unified resource that catalogs cutting-edge genetic and genomic tools and technologies. The first section (Chapters 1, 2, and 3) describes approaches and methodologies that are currently available for analyzing the genetic underpinnings of naturally occurring neurological mutations and complex behaviors, and for the genetic manipulation of specific genes important in nervous system functioning. The second section (Chapters 4, 5, and 6) covers state-of-the-art experimental strategies and techniques for genome-wide mutation screens, both phenotype and genotype based, that aim to identify and characterize mutations that provide deep insights into many neurodevelopmental and physiological processes. The third section consists of two chapters on microarray technologies, which have become a standard tool for molecular biology. Chapter 7 focuses on DNA chips and their utility in determining gene expression profiles in the brain of various mouse strains. Chapter 8 describes the use of cDNA arrays for the analysis of postmortem human brain specimens, and highlights the enormous potential of this methodology for both basic and clinical neuroscience applications. Strategies and methods for generating much needed full-length cDNA libraries (Chapter 9), new developments in viral vector systems for the effective delivery and sustained expression of targeted genes to neural tissue (Chapter 10), and emerging stem cell technologies for preparing specific neural cell populations as an innovative means of repairing damage in neural tissue (Chapter 11) are covered in the next section. Finally, this volume concludes with two chapters dealing with genome-wide and population-based statistical methodologies for the identification of disease vulnerability genes (Chapter 12) and the analysis of common genetic sequence variation in humans (Chapter 13). Clear understanding of the genetic bases for individual genetic variation in complex traits will provide a foundation for developing therapeutic drug targets and interventions suited for individual genotypes.

Our expectation is that the strategies and experimental designs described in this volume ultimately will be useful for neuroscientists wishing to obtain broad perspectives on the emerging field of genomic neuroscience, as well as for those wishing to manipulate and analyze genomes as a means of gaining insight into the genetic bases of the mammalian nervous system and its complex disorders. We anticipate that this volume makes a useful contribution by providing a firm foundation for using data on gene expression and function to develop new therapeutics for mental disorders.

We express our deep gratitude to each of the contributors who worked so diligently on their chapters. We also thank the staff of CRC Press, especially Barbara Norwitz and Tiffany Lane, for their help, patience, and encouragement. The enthusiastic support of Sidney A. Simon and especially Miguel A. L. Nicolelis, the editors for the CRC Press *Methods and New Frontiers in Neuroscience* series, was invaluable. We are also greatly appreciative of the encourgement of our colleague Dennis L. Glanzman, who encouraged us to develop a compendium of cutting-edge genetics and genomics methods for a broad neuroscience audience.

Hemin R. Chin, Ph.D.
Steven O. Moldin, Ph.D.
Bethesda, Maryland
March 23, 2001

The Editors

Hemin R. Chin, Ph.D. is chief of Genetics Basis of Neural Function Program in the Genetics Research Branch in the Division of Neuroscience and Basic Behavioral Science at the National Institute of Mental Health (NIMH), National Institutes of Health (NIH).

Dr. Chin received a B.S. in biology and chemistry from the University of Illinois, and his Ph.D. in neurobiology and physiology from Northwestern University in 1983. He received 2 years of training under a National Research Service Award in molecular neurobiology at the Laboratory of Biochemical Genetics, National Heart, Lung, and Blood Institute, NIH, and continued his independent research in the Laboratory of Molecular Biology and the Laboratory of Neurochemistry, National Institute of Neurological Disorders and Stroke, NIH until he took his current position at NIMH in 1998. During this period, Dr. Chin had pursued his research interests in the molecular and genetic mechanisms of calcium channel function in the nervous system.

Dr. Chin serves on a number of trans-NIH genomics and genetic initiatives coordinating committees including the Brain Molecular Anatomy Project Coordinating Committee, Mammalian Gene Collection Inter-Institute Coordinating Committee, Mouse Sequencing Consortium Oversight Committee, Pharmacogenetics Research Network Steering Committee, Rat Genome Working Group, Coordinating Committee for Non-Mammalian Models, Trans-NIH Zebrafish Coordinating, and the Committee on Gene Transfer Research.

Dr. Chin received an NIH Merit Award in 1994 for his contribution to the field of molecular neuroscience, and an NIH Director's Award in 1999. Dr. Chin has organized international and national symposia and meetings and published over 40 papers and book chapters.

Steven O. Moldin, Ph.D. is chief of the Genetics Research Branch in the Division of Neuroscience and Basic Behavioral Science at the National Institute of Mental Health (NIMH), National Institutes of Health (NIH). He current oversees the NIMH extramural research portfolio on the human genetics of mental disorders, the genetic basis of complex behavior, and the genetic basis of neural function.

Dr. Moldin received his B.A. (magna cum laude with distinction and Phi Beta Kappa) in psychology from Boston University in 1983, his M.A. in psychology from Yeshiva University, New York, NY in 1985, and his Ph.D. in clinical psychology from Yeshiva University in 1988. After working as an assistant research scientist in the Department of Medical Genetics at the New York State Psychiatric Institute, New York, and completing a clinical internship at Hillside Hospital–Long Island Jewish Medical Center, Glen Oaks, NY, from 1988 to 1991, he received postdoctoral training in quantitative genetics at Washington University School of Medicine in

St. Louis, MO. From 1991 to 1995 he was an assistant professor in the Department of Psychiatry at Washington University School of Medicine, and director of the Center for Psychiatric Genetic Counseling at the Washington University Medical Center from 1993 to 1995.

Dr. Moldin joined NIMH in 1995 as chief of the Genetics Research Program in the Schizophrenia Research Branch of the Division of Clinical and Treatment Research. After an administrative re-organization, he was first acting chief and then chief of the newly established Genetics Research Branch in 1998. He has served as a consultant to the Institute of Medicine's Committee on Health and Behavior and currently serves on numerous trans-NIH committees, including the Trans-NIH Mouse Genomics and Genetics Resources Committee and the Center for Inherited Disease Research's Board of Governors. His current editorial responsibilities include serving as associate editor of *Schizophrenia Bulletin* and of *Genes, Brain, and Behavior,* and serving on the editorial board of the *American Journal of Medical Genetics (Neuropsychiatric Genetics).*

Dr. Moldin has published over 40 papers and book chapters, and has presented over a dozen invited lectures at national and international meetings. Dr. Moldin received fellowships from the New England Psychological Association and the Leopold Schepp Foundation, as well as awards from the International Congress on Schizophrenia Research and the Society for Life History Research. His research on statistical genetic methods and schizophrenia was previously supported by the Scottish Rite Schizophrenia Research Program–Northern Masonic Jurisdiction, NIMH, and the National Alliance for Research on Schizophrenia and Depression.

Editor and Contributor List*

EDITORS

Hemin R. Chin, Ph.D.
Genetic Basis of Neural Function
  Program
Genetics Research Branch
Division of Neuroscience and Basic
  Behavioral Science
National Institute of Mental Health
Bethesda, MD
E-mail: hchin@mail.nih.gov

Steven O. Moldin, Ph.D.
Genetics Research Branch
Division of Neuroscience and Basic
  Behavioral Science
National Institute of Mental Health
Bethesda, MD
E-mail: smoldin@mail.nih.gov

CONTRIBUTORS

Carrolee Barlow, M.D., Ph.D. (7)
Laboratory of Genetics
The Salk Institute for Biological Studies
La Jolla, CA
E-mail: barlow@salk.edu

Floyd E. Bloom, M.D. (Forward)
Department of Neuropharmacology
The Scripps Research Institute
La Jolla, CA
E-Mail: fbloom@scripps.edu

Xandra O. Breakefield, Ph.D. (10)
Department of Neurology
Molecular Neurogenetics Unit, Room
  6203
Massachusetts General Hospital
Harvard Medical School
Charlestown, MA
E-mail:
  breakefield@helix.mgh.harvard.edu

Alice B. Brown, M.D., Ph.D. (10)
Department of Neurology
Molecular Neurogenetics Unit
Massachusetts General Hospital
Harvard Medical School
Charlestown, MA
E-mail: brown@helix.mgh.harvard.edu

Maja Bucan, Ph.D. (6)
Department of Psychiatry
University of Pennsylvania
Philadelphia, PA
E-mail: bucan@pobox.upenn.edu

Aravinda Chakravarti, Ph.D. (13)
McKusick–Nathans Institute of Genetic
  Medicine
Johns Hopkins University School of
  Medicine
Baltimore, MD
E-mail: aravinda@jhmi.edu

---

* Contributor chapter number given in parentheses.

David J. Cutler, Ph.D. (13)
McKusick–Nathans Institute of Genetic
  Medicine
Johns Hopkins University School of
  Medicine
Baltimore, MD
E-mail: dcutler@jhmi.edu

Wayne N. Frankel, Ph.D. (2)
The Jackson Laboratory
Bar Harbor, ME
E-mail: wnf@jax.org

Lisa V. Goodrich, Ph.D. (5)
Departments of Anatomy and of
  Biochemistry and Biophysics
Howard Hughes Medical Institute
University of California
San Francisco, CA
E-mail: goodric@itsa.ucsf.edu

Jürgen A. Hampl, M.D. (10)
Department of Neurology
Molecular Neurogenetics Unit
Massachusetts General Hospital
Harvard Medical School
Bethesda, MD
E-mail: jhampl@helix. mgh. harvard.edu

Munetomo Hida, Ph.D. (9)
Department of Genome Structure
  Analysis
Institute of Medical Science
University of Tokyo
Minato-Ku, Tokyo, Japan
E-mail: ss77233@ims.u-tokyo.ac.jp

Eric Kandel, M.D. (3)
Howard Hughes Medical Institute
Center for Neurobiology and Behavior
Columbia University
New York, NY
E-mail: erk5@columbia.edu

Stephen J. Kanes, M.D. Ph.D. (6)
Center for Neurobiology and Behavior
Department of Psychiatry
University of Pennsylvania
Philadelphia, PA
E-mail:
  Kanes@bblmail.psycha.upenn.edu

Olivia G. Kelly, Ph.D. (5)
Department of Molecular and Cell
  Biology
University of California
Berkeley, CA
E-mail: ogkelly@uclink.berkeley.edu

Gillian R. Leach (6)
Center for Neurobiology and Behavior
Department of Psychiatry
University of Pennsylvania
Philadelphia, PA
E-mail: gleach@mail.med.upenn.edu

Philip A. Leighton, Ph.D. (5)
Departments of Anatomy and of
  Biochemistry and Biophysics
Howard Hughes Medical Institute
University of California
San Francisco, CA
E-mail: leighto@itsa.ucsf.edu

Pat Levitt, Ph.D. (8)
Department of Neurobiology
University of Pittsburgh School of
  Medicine
Pittsburgh, PA
E-mail: plevitt@pitt.edu

David A. Lewis, M.D. (8)
University of Pittsburgh
Western Psychiatric Institute and Clinic
Pittsburgh, PA
E-mail: lewisda@msx.upmc.edu

David J. Lockhart, Ph.D. (7)
Laboratory of Genetics
The Salk Institute for Biological Studies
La Jolla, CA
E-mail: Lockhart@salk.edu

Xiaowei Lu, Ph.D. (5)
Departments of Anatomy and of
  Biochemistry and Biophysics
Howard Hughes Medical Institute
University of California
San Francisco, CA
E-mail: xwlu@itsa.ucsf.edu

Judy Mak (5)
Departments of Anatomy and of
  Biochemistry and Biophysics
Howard Hughes Medical Institute
University of California
San Francisco, CA
E-mail: jsmak@itsa.ucsf.edu

Mark Mayford, Ph.D. (3)
Department of Cell Biology
The Scripps Research Institute
La Jolla, CA
E-mail: mmayford@scripps.edu

Ron McKay, Ph.D. (11)
Laboratory of Molecular Biology
National Institute of Neurological
  Disorders and Stroke
National Institutes of Health
Bethesda, MD
E-mail: mckayr@ninds.nih.gov

Károly Mirnics, M.D. (8)
Department of Psychiatry
University of Pittsburgh School of
  Medicine
Biomedical Science Tower
Pittsburgh, PA
E-mail: karoly@pitt.edu

Kevin J. Mitchell, Ph.D. (5)
Howard Hughes Medical Institute
Department of Anatomy
University of California
San Francisco, CA
E-mail: kjmtchl@itsa.ucsf.edu

Jurg Ott, Ph.D. (12)
Laboratory of Statistical Genetics
Rockefeller University
New York, NY
E-mail: ott@rockvax.rockefeller.edu

Edivinia Pangilinan (5)
Department of Molecular and Cell
  Biology
University of California
Berkeley, CA
E-mail: dspangil@yahoo.com

Kathy Pinson, Ph.D. (5)
Department of Molecular and Cell
  Biology
University of California
Berkeley, CA
E-mail: pinsonk@hotmail.com

Lawrence H. Pinto, Ph.D. (4)
Department of Neurobiology and
  Physiology
Northwestern University
Evanston, IL
E-mail: Larry-Pinto@northwestern.edu

Nikolai G. Rainov, M.D. (10)
The University of Liverpool
Department of Neurological Science
Clinical Sciences Centre for Research
  and Education
Halle, Germany
E-mail: rainov@helix.mgh.harvard.edu

Helen Rayburn, Ph.D. (5)
Departments of Anatomy and of
  Biochemistry and Biophysics
Howard Hughes Medical Institute
University of California
San Francisco, CA
E-mail: rayburn@itsa.ucsf.edu

Danielle Rottkamp (5)
Departments of Anatomy and of
  Biochemistry and Biophysics
Howard Hughes Medical Institute
University of California
San Francisco, CA
E-mail: daniele@itsa.ucsf.edu

Paul Scherz (5)
Department of Molecular and Cell
  Biology
University of California
Berkeley, CA
E-mail: pscherz@uclink4.berkeley.edu

William C. Skarnes, Ph.D. (5)
Department of Molecular and Cell
  Biology
University of California
Berkeley, CA
E-mail: skarnes@socrates.berkeley.edu

Marla B. Sokolowski, Ph.D., F.R.S.C. (1)
Department of Zoology
University of Toronto
Mississauga, ON, Canada
E-mail:
  msokolow@credit.erin.utoronto.ca

Lorenz Studer, M.D. (11)
Laboratory of Stem Cell and Tumor
  Biology
Neurosurgery and Cellular Biochemistry
  and Biophysics
Memorial Sloan Kettering Cancer
  Center
New York, NY
E mail: studer1@mskcc.org

Sumio Sugano, M.D. (9)
Department of Genome Structure
  Analysis
Institute of Medical Science
University of Tokyo
Minato-Ku, Tokyo, Japan
E-mail: ssugano@ims.u-tokyo.ac.jp

Yutaka Suzuki, Ph.D. (9)
Department of Genome Structure
  Analysis
Institute of Medical Science
University of Tokyo
Minato-Ku, Tokyo, Japan
E-mail:
  ysuzuki@manage.ims.u-tokyo.ac.jp

Joseph S. Takahashi, Ph.D. (4)
Howard Hughes Medical Institute and
  Department of Neurobiology and
  Physiology
Northwestern University
Evanston, IL
E-mail: j-takahashi@northwestern.edu

Peri Tate, Ph.D. (5)
Department of Molecular and Cell
  Biology
University of California
Berkeley, CA

Marc Tessier-Lavigne, Ph.D. (5)
Howard Hughes Medical Institute
Department of Anatomy
University of California
San Francisco, CA
E-mail: marctl@itsa.ucsf.edu

Douglas Wahlsten, Ph.D. (1)
Department of Psychology
Biological Sciences P220
University of Alberta
Edmonton, AB, Canada
E-mail: wahlsten@ualberta.ca

Paul Wakenight (5)
Department of Molecular and Cell
  Biology
University of California
Berkeley, CA
E-mail: pw@cub.bio.columbia.edu

Joe Zhong (5)
Departments of Anatomy and of
  Biochemistry and Biophysics
Howard Hughes Medical Institute
University of California
San Francisco, CA
E-mail: jwzhong@itsa.ucsf.edu

Joel Zupicich (5)
Department of Molecular and Cell
  Biology
University of California
Berkeley, CA
E-mail: joelz@socrates.berkeley.edu

Michael E. Zwick, Ph.D. (13)
McKusick–Nathans Institute
  of Genetic Medicine
Johns Hopkins University School
  of Medicine
Baltimore, MD
E-mail: mzwick@jhmi.edu

Table of Contents

SECTION 3

SECTION 4

SECTION 5

Section 1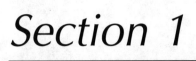

Section 7

1 Gene–Environment Interaction and Complex Behavior

Marla B. Sokolowski and Douglas Wahlsten

CONTENTS

1.1 INTRODUCTION

During the early decades of the last century, the statistical laws but not the molecular mechanisms of heredity were well understood, whereas the last two decades witnessed an explosive growth in knowledge of genes and their functions at the molecular level. Psychologists and developmental biologists have always been interested in interactions between heredity and environment, but only recently have we acquired the tools needed to understand how interactions actually work. This chapter focuses on both statistical and molecular biological methods applicable to the detection and analysis of gene–environment (or gene by environment, G×E) interactions. We highlight theories, techniques, and experimental strategies relevant to studies of individual differences in behavior, including natural behavioral variants and genetic mutants. Although the use of molecular approaches to study G×E interactions is still in its infancy, the advent of the genome projects combined with some of the latest technologies for analyses of genome-wide responses to the environment make investigations of this type timely and promising.

1.2 THE MEANING AND IMPORTANCE OF INTERACTION

Well before the dawn of modern genetics, the theory that preformed characters undergo quantitative enlargement or unfolding was rejected by embryologists in favor of epigenetic development, the notion that the parts of an organism emerge through qualitative transformation.[1] Early Mendelism, especially the doctrine of unit characters championed by Bateson,[2] revived preformationism by asserting that genes specify the properties of adult organisms in a one gene–one character fashion. Gottlieb et al.[3] characterize this view as predetermined epigenesis, in contrast with probabilistic epigenesis that allows many possible outcomes from the same set of genes. As pointed out by Strohman,[4] exclusively genetic determination of adult characters is still a widely held opinion, as expressed in mosaic theories of development.[5]

The concept we now term genotype–environment interaction was formulated early in the last century as an alternative to unit characters. Johannsen[6] proposed that the genotype, the set of all the individual's genes, is inherited from the parents, whereas the observable phenotype develops and may have many values. He observed that "some strains of wheat yield relatively much better than others on rich soil, while the reverse is realized on poorer soils." Woltereck[7] proposed that the organism inherits not the character but the "Norm of Reaction with all its numberless specific relations" to all conceivable conditions, a lawful norm that can lead to many different "biotypes." He equated the Norm of Reaction with Johannsen's genotype concept. In a more recent expression, Lewontin[8] refers to the norm of reaction as the graph of the phenotype values of "a particular genotype as a function of the environment," and emphasizes, as did Woltereck and Johannsen, that these graph lines may take rather different forms and even intersect. Nijhout[9] uses the term "reaction norm" to denote continuous variation in phenotypes in response to graded changes in environment and "polyphenic development" for situations where the phenotype is expressed in qualitatively different forms as a consequence of environmental conditions.

Hogben[10] set forth a view that is widely held today. "Characteristics of organisms are the result of interaction between a certain genetic equipment inherent in the fertilized egg and a certain configuration of extrinsic agencies." He cited many instances where the quantitative effects of changing the environment depended strongly on the genotype. Concerning the issue of how much of a difference is due to heredity and how much to the environment, Hogben argued: "The question is easily seen to be devoid of a definite meaning." He also said, "When we understand the *modus operandi* of the gene, we can state the kind of knowledge we need in order to control the conditions in which its presence will be recognized." This doctrine has now become the principal rationale for the Human Genome Project as a source of new therapies for medical disorders.

G×E interaction asserts that the response of an organism to an environmental treatment depends on its genotype, and the manifestation of genetic differences between individuals depends on the environment. This concept is presented graphically in Figure 1.1A. On the other hand, the reaction range concept claims that environmental effects are essentially the same for all genotypes, such that rank orders of genotypes are maintained over a wide range of environments, consistent with simple addition of genetic and environmental effects, and there is a gene-imposed upper limit on phenotypic development (Figure 1.1B; see References 11 and 12).

Interactionism as a doctrine assumes several forms. In psychology, Hull[13] maintained that behavior is governed by mathematical laws and that "the forms of the equation" are the same for all species and individuals, emphasizing, "Innate individual and species differences find expression in the 'empirical constants' which are essential constituents of the equation expressing the primary and secondary laws of behavior." For the behaviorism of Watson, Hull, and Skinner, the forms of the laws were identified with the biological structure of a nervous system that determined how environmental stimuli are sensed and associated. In the study of animal learning, this view led psychologists away from genetic research and justified the almost universal use of albino rats in the lab because the functional laws themselves were thought to be the same for all, including humans (see previous critique[14]).

In biology, Waddington[15] proposed a complex epigenetic landscape that governed how an individual would develop under different environmental conditions. The topography of the landscape, however, was said to be genetically specified. Like the notions of Hull, this theory of passive gene-related responsiveness to environment was reductionist, being based on strict genetic determination of an earlier phase of structural development and/or a lower level of organization (see Gottlieb[16]); genetic effects were held to be unidirectional from molecule upwards to morphology and behavior.

Interdependence or interpenetration of heredity and environment, on the other hand, holds that environment is an essential factor at all levels of organization; interactions are bidirectional, and the organism is an active agent in constructing and transforming its own environment.[8,17-19] According to Oyama,[20] a developmental system comes into being "…not as the reading off of a preexisting code, but as a complex of interacting influences, some inside the organism's skin, some external to it … It is in this ontogenetic crucible that form appears and is transformed, not because it is immanent in some interactants and nourished by others … but because any form is created by the precise activity of the system."

Thus, the fact that different genotypes evidence different shapes of reaction norms should be taken only as the starting point for investigation of development.[21] While the mathematical shapes of functions provide important clues, a deeper theory of development cannot be encapsulated in a few formulas.[20]

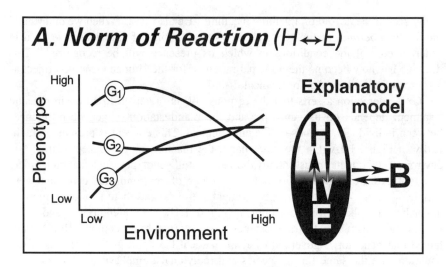

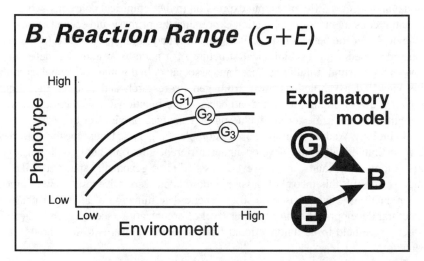

FIGURE 1.1 Two conceptual models of the functional relations between 3 genotypes and environment. A. The norm of reaction expects that the shape of the function will depend on genotype, and it allows for reversals of rank orders in different environments. This is possible because genotype and environment are interdependent causes whose interaction produces development. B. The reaction range expects essentially the same shape of the function for all genotypes. It asserts the developmental separation and hence additivity of genetic and environmental causes.

Ongoing disputes about appropriate models and credible assumptions for the analysis of human behavior continue to generate interest in G×E interaction. The prevalent view in quantitative genetic analysis in psychology denies the existence or importance of G×E interaction and instead asserts that genetic differences and environmental variations have additive effects, as expressed in the equation Phenotype = Genotype + Environment (P + G + E; see Plomin et al.[22]). When two variables are statistically independent and additive, their phenotypic variance (V_P) can be partitioned into two components attributable to genetic (V_G) and environmental (V_E) variance. The broad-sense heritability ratio is then $h^2 = V_G/V_P$. The meaning and magnitude of h^2 for IQ in particular has been hotly debated for many years and continues to be contested.[23-25]

Additivity implies that the manifestation of genetic differences is unaffected by the rearing environment and the consequences of environmental change should be the same for all genotypes (Figure 1.1B). Only when factors act separately in the process of development will their effects generally be separable statistically. Some behavior geneticists, while not outright denying the existence of G×E interaction, argue that interaction effects pertinent to human psychology are generally so small that an additive model is a good approximation of reality.[22] In this respect, there is a stark contrast between prevailing conceptions in genetic studies of human and nonhuman animals.[26,27]

Three kinds of criticisms have been directed at heritability analysis of human behavior: (1) available research designs with humans are incapable of cleanly separating genetic and environmental effects;[28-31] (2) statistical methods of analyzing variance into components are relatively insensitive to the presence of real interactions[26] and yield a false impression of additivity; and (3) enough is known about the regulation of gene expression during development to warrant rejection of additivity as a general principle.[8,11,27,32]

Research on human behavior does not require an assumption of additivity. In psychiatric genetics, for example, models involving G×E interaction have been prominent for many years and published in recent reviews.[33] The diathesis-stress theory of Gottesman and Shields,[34] the model-fitting methods of Kendler and Eaves,[35] and the interaction hypothesis of Wahlberg et al.[36] provide clear examples. Cloninger[37] concludes that schizophrenia involves the "nonlinear interaction of multiple genetic and environmental factors."

In essence, two questions are raised by this discussion. First, is research with animals in the laboratory pertinent for models of human behavior, or can a theory of human exceptionalism be defended? At the molecular level, there is so much in common between human and mouse that broad generality of basic principles is expected. For further discussion of this question see Skuse.[38] Second, could it be that medically significant disorders involve interaction while variations within the normal range that are of central interest to psychologists do not? Little has been published on this latter question; however, in the few cases where a genetic and molecular basis for normal individual differences in behavior is understood, G×E interaction cannot be excluded.[39]

1.3 GLOBAL APPROACHES TO STUDYING INTERACTION

1.3.1 EXPERIMENTAL DESIGNS

To ascertain the impact of an environmental variable, independent groups of animals having equivalent genotypes must be reared in different environments. This kind of one-way research design (Figure 1.2A) is easily arranged with an inbred mouse

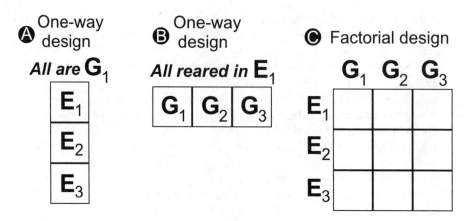

FIGURE 1.2 Experimental designs involving three genotypes and three environments. A. Raising genetically identical individuals in different environments. B. Raising different genotypes in the same environment. C. Raising each genotype in three different environments. Only the factorial design can reveal the presence of genotype × environment interaction, even though the two factors interact in the developmental sense even in the one-way designs.

strain, an isofemale line of *Drosophila*, or a clone of *Daphnia*. Likewise, genetic variation can be studied with a one-way design (Figure 1.2B) wherein different genotypes are reared in the same environment, as was done carefully by Mendel. Heredity and environment may be strongly interacting factors in the developmental sense in either kind of experiment, yet in neither case will the statistical interaction be apparent. Interaction is visible only when animals with different heredities are reared in different environments using a factorial design (Figure 1.2C). The smallest experiment conceivable is the 2 × 2 design, but the generality of findings will be greater when several strains are subjected to a wide variety of environments so that genotype-specific norms of reaction may be observed.

Following in Mendel's footsteps, modern geneticists usually strive to rear their subjects in a uniform environment in the lab or a carefully cultivated field so that data for different genotypes will not be confounded with environmental differences. Hence, results of many superb genetic analyses tell us nothing at all about the presence or magnitude of G×E interaction. A clue that interaction lurks in the background is sometimes seen when different labs fail to replicate effects of the same mutation. For example, three recent studies published simultaneously in *Nature Genetics* reported discrepant results of effects of a null mutation in the corticotrophin-releasing hormone (*Crh*) and its receptors (*Crhr1* and *Crhr2*) on mouse anxiety.[40-42] Because the genetic backgrounds of their strains as well as the details of the tests for measuring anxiety also differed between labs, it was not possible to attribute discrepancies to the rearing environments in the three labs.

Crabbe et al.[43] addressed this problem by testing the same eight genetic strains with identical test apparatus and protocols simultaneously in three labs. For certain phenotypes, ethanol preference in particular, the three labs observed essentially the same results, whereas measures of activity, anxiety, and activating effects of a cocaine injection yielded different patterns of data for certain strains in the three labs. The study was explicitly designed to yield large genetic effects by choosing strains known to differ greatly on several phenotypes, and great efforts were made to equate many aspects of the lab environment. Nevertheless, substantial G×E interaction was seen. This study contradicts the contention that interaction is to be expected only when extreme differences in environment are employed.[44]

A few large studies in behavioral genetics have replicated a complete genetic crossing experiment in two different environments. Henderson[45] conducted a diallel cross of four inbred strains to create 12 F_1 hybrids, and all 16 groups were reared in either standard lab cages or larger, enriched environments. He found that evidence of genetic influences was markedly suppressed by rearing in the small, impoverished lab cages. Carlier et al.[46] repeated an entire reciprocal crossing study with eight genetic groups derived from ovaries that were grafted into a hybrid female, and they found that an effect of the Y chromosome on fighting behavior of male mice having inbred mothers was not evident in the F_1 maternal environment.

Interaction also occupies an important place in the laboratory as a research tool for analyzing mechanisms of development. In *Drosophila*, for example, temperature-sensitive mutations make it possible to delineate critical periods for genetic effects and to study interactions among gene products; wild-type and mutant transgenes can also be engineered to be expressed at certain times during development or targeted to specific tissue.[47] Inducible mutations in mice, whereby production of a specific protein is shut down when an animal drinks water containing an antibiotic, make it possible to assess the role of the gene in formation of memories in the adult without the confounding developmental effects that are typical for most targeted mutations.[48] By inserting special regulatory sequences near a gene, its expression may also be limited to a particular kind of tissue in the brain.[49]

1.3.2 SINGLE GENES AND PLEIOTROPY

How can we meaningfully analyze the effect of alterations in a single gene on the performance of a behavior in several environments when we know that most genes have pleiotropic functions? When a gene known to influence behavior is knocked out or inactivated, severe disruptions in a number of phenotypes are often observed. This usually reflects a role for this gene in both development and behavior. Indeed many genes that alter behavior are vital genes that cause lethality when inactivated (e.g., in fly food search — *foraging*,[50] *scribbler*;[51] courtship- *fruitless*;[52] learning — *latheo*[53]). It is of interest that more subtle alterations in the gene, for example hypomorphic mutations that cause a small reduction in the amount of gene product, often exhibit the behavioral alteration but not the other pleiotropic phenotypes. Greenspan[54] in his review entitled "A Kinder, Gentler Genetic Analysis of Behavior: Dissection Gives Way to Modulation," argues for the importance of studying milder mutations because they are more similar to the subtler genetic influences on behavior found in nature. These milder mutations and the ability to target the expression of a gene to certain times in development and to certain tissues in the organisms may allow us to disentangle a gene's role in development from its role in behavioral functioning. As an example, variants with partial loss of function of the major serine/threonine protein kinases — cAMP-dependent protein kinase (PKA), calcium/calmodulin-dependent protein kinase type II

(CaMKII), or protein kinase C (PKC) — all cause effects in behavioral plasticity specific to learning and memory, while severe mutations in these genes are lethal.[54] These effects on learning and memory are seen when the level of kinase is reduced by only 10–20%. A 12% difference in the expression of the *foraging* gene which encodes a cGMP-dependent protein kinase (PKG) explains rover compared to sitter natural foraging behavior variants.[50] All of these kinases are involved in a wide range of biological processes; however, a subtle shift in kinase activity exerts a potent effect on the phenotype. It is likely that natural variants that have been selected under natural conditions involve these types of subtle mutations, because more severe mutations with their prevalent pleiotropic effects would be selected against. Future research on the molecular basis of natural behavioral variants will enable us to test this prediction.

One conclusion from the discussion above is that studies of the molecular mechanisms underlying G×E interactions on behavior should be done using natural variants, mutations, or transgenes that have subtle effects on the behavioral phenotype. If mutants with large effects are used, then one is more likely to identify genes and processes important to the many pleiotropic functions of the gene rather than to the behavior specifically. The task of teasing apart which specific mechanisms are associated with the behavioral function would then be overwhelming. In some cases, however, the developmental alterations in mutants may be the cause of the behavioral variation; for example, the presence of an altered level of a specific kinase or a second messenger such as cAMP during nervous system development may cause alterations in the morphology of the neurons and on their functioning.[55] In this case the connection between the developmental and behavioral phenomena can be determined using inducible transgenes prior to the molecular analyses of G×E interactions.

1.3.3 RESEARCH OUTSIDE THE LAB

Research with wild populations indicates that G×E interaction is not merely some oddity confined in a laboratory. On the contrary, genotype-dependent responsiveness to environment is crucial as a means of adapting organisms to a wide range of circumstances. Certain reptiles lack sex chromosomes, and sexual differentiation depends on clutches of eggs being laid in soils having different temperatures,[56] whereas many other species are strongly buffered against temperature effects. The specific kind of food, oak catkins or leaves, on which larvae of the geometrid moth *Nemoria arizonaria* dine leads to a remarkable development of morphology that matches the caterpillar to the texture and color of its host.[57] Many other examples are cited by Nijhout.[9] G×E interaction is seen in all parts of the animal and plant kingdoms. Knowledge of which aspects of development are sensitive to which features of the environment can teach us a great deal about the mode of life and evolution of a species.

1.3.4 STUDIES OF HUMANS

Research on G×E interaction is particularly difficult with humans because replicate genotypes simply do not exist. Monozygotic twins provide only two copies of a genotype, and the environments of the pair also tend to be correlated. In the study of schizophrenia, there is evidence for G×E interaction in the Finnish adoption studies that show elevated psychopathology in adopted-away offspring of schizophrenic mothers only when they are reared in psychologically inferior homes.[36] The data are consistent with the hypothesis of gene-related vulnerability to stressful or confusing environments, but the case is weak because there is no identification of genotype per se. Instead, the probands and matched controls are selected on the basis of a maternal phenotype that is not a reliable proxy for a genetic abnormality.

When a specific, major gene effect on human development is established, evidence for G×E interaction may be obtained. The classic case is the phenylalanine hydroxylase (PAH) mutation that leads to phenylketonuria.[28] Children homozygous for the recessive allele are much more sensitive to the level of phenylalanine in the diet and can thrive only with rearing on a low phenylalanine diet and careful monitoring of blood levels of the amino acid. In work on genetic diseases, the studies are not as well controlled as lab experiments with mice or flies, but large effects of a mutation nevertheless permit conclusions about interactions in many instances.

In psychology, strong claims have been made that G×E interaction effects involving intelligence in particular have been sought but cannot be detected.[58,59] A closer look at the nature of IQ tests reveals a very large interaction, however, one that is obscured by the manner in which test items are chosen and the raw test score is transformed into an IQ score. The rationale for intelligence test interpretation was stated clearly by Goodenough:[60] "...the intelligence tests in present use are indirect rather than direct measures. They deal with the results of learning, from which capacity to learn is inferred. When opportunity and incentives have been reasonably similar, the inference is sound, but its validity may be questioned when a comparison is to be made between two or more groups for whom these factors have been markedly different." In other words, given similar exposure to relevant educational material, if one child learns faster and therefore more than another, psychologists infer this must be because of differences in an inherent property of the nervous system termed intelligence (see Wahlsten[61]). Intelligence is believed to cause *differences in the ease or rate of learning*, as expressed in the slope of the function relating amount of acquired knowledge to cumulative experience (see Figure 1.3). This is an example of interaction *par excellence*. It has become customary to avoid discussion of raw intelligence test scores and instead convert the raw scores to standardized scores based on large, representative samples of children of different ages. The desired scaling results in a mean IQ of 100 and a standard deviation of 15 at every age, no matter what kind of items are on the test. This practice effectively obscures the real rates of growth of intelligence and conceals the interaction. It gives rise to perplexing facts, such as different brands of IQ tests that yield the same mean and variance of IQ but are far from perfectly correlated with each other. It also tends to minimize the indications of a dramatic increase in intelligence test score over a period of one or two decades in a society, because most IQ tests are altered and re-standardized every few years, which forces the mean back to 100.[62] Insisting that critics of heritability analysis should be able to show G×E interaction in IQ scores requires a large, gene-related difference in the *second derivative* of the function relating knowledge to experience.

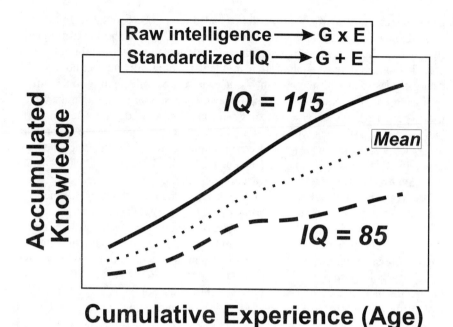

FIGURE 1.3 Accumulated knowledge, as expressed on an intelligence test, as a function of cumulative experience, for which age is a proxy variable. Hypothetical profiles are shown for two genetically unique individuals who have substantially different slopes of the experience–knowledge relation at certain ages, but nevertheless have stable IQ test scores across age. Converting the test score to a standard score tends to obscure the presence of interaction and make the factors appear to be additive.

1.4 STATISTICAL ANALYSIS OF INTERACTION

1.4.1 FACTORIAL DESIGNS

When J genotypes are reared in K different environments, the experiment with JK groups can be appraised with the analysis of variance (ANOVA) for fixed factors. Fisher and Mackenzie[63] devised this method to evaluate yield of 12 potato varieties under six conditions of manure. They divided the variance between the 72 groups into three portions, the two main effects and a third term, the "deviations from summation formula," a quantity we now assign the appellation "interaction." Statistically, interaction is defined as the variation among the JK group means that cannot be accounted for by the addition of the separate main effects of genotype and environment.

Whereas execution and interpretation of ANOVA are now quite routine, one crucial aspect of this methodology is not widely appreciated. For many interesting kinds of interaction that may exist in the real world, the ANOVA tends to be much less sensitive to presence of interaction than to the main effects.[26,64] That is, the statistical power of the test of interaction is often pathetically low and Type II errors (failure to reject a false null hypothesis that G and E are additive) are probably very

TABLE 1.1
Hypothetical Means for an Experiment with and without Strain by Lab Interaction

	Additive Main Effects				Model with Substantial Interaction			
Strain	Lab A	Lab B	Difference	Mean	Lab A	Lab B	Difference	Mean
A	30	40	10	35	30	50	20	40
B	35	45	10	40	35	35	0	35
C	40	50	10	45	40	50	10	45
D	45	55	10	50	45	65	20	55
E	50	60	10	55	50	50	0	50
F	55	65	10	60	55	65	10	60
G	60	70	10	65	60	80	20	70
H	65	75	10	70	65	65	0	65
I	70	80	10	75	70	80	10	75
Mean	50	60	10	55	50	60	10	55

Lab effect: $\sigma_M = 5.0$ $f = 0.2$ $n = 22$
Strain effect: $\sigma_M = 12.9$ $f = 0.5$ $n = 7$
Lab × Strain interaction: $\sigma_M = 5.8$ $f = 0.23$ $n = 31$

Note: Standard deviation within each group is set at 25 units. Sample size is calculated in order to yield power of 90% when Type I error probability is set at $\alpha = 0.01$. Values obtained from Tables 8.4.1 and 8.4.2 in Cohen[66] must be adjusted with the formula on page 396 in order to adapt tables computed for a one-way design for use with a factorial design.

common. Great attention is usually devoted to the proper choice of a criterion for Type I error (rejection of a true null hypothesis), especially in linkage studies, and this concern is appropriate because we expect that most genetic markers in a genome scan are not linked to a gene causing individual differences in a particular behavior.[65] On the other hand, in a study of inbred strains in different environments, we have good reason to suspect that the factors really do interact, and the null hypothesis lacks credibility; hence the central issue is the sensitivity of the test of interaction.

An effective remedy for low power of a test of interaction is readily prescribed. Larger samples are required to confer adequate power on an assessment of what, to the ANOVA procedure, appears to be a relatively small effect. Power and sample size calculations should be done before the data are collected, and we must propose credible but hypothetical values of group means, guided by previous studies. The method of Cohen[66] is convenient when working with effects having more than one degree of freedom.

An example is provided in Table 1.1 for a study where nine inbred strains are reared and tested with the same apparatus in two labs. Suppose that in Lab A the strain means on a test range evenly from 30 to 70, and in Lab B each strain scores 10 units higher, which is an instance of additive effects. Next we must propose a model of group means that expresses the kind of interaction we would like to be able to detect. It would be silly to suggest that in Lab B there will be no strain differences at all; this would be a huge interaction effect but not one we could plausibly expect to find, given decades of research with inbred strains. The model of interaction in Table 1.1 entails three strains that have identical means in both labs, three strains that differ by 10 points, and three strains that differ by 20 points. Note that the distributions of strain and lab means are the same under both models. The method of Cohen[66] requires that we find the standard deviation between group means (σ_M), and the effect size f is the ratio σ_M/σ, where σ is the standard deviation within groups (set at 25 in this example). For the strain main effect, σ_M is based on nine means, whereas it is based on only two for the lab main effect. For the interaction having eight degrees of freedom, we must take the average squared difference between all 18 group means expected under the hypothesis of interaction and the means expected from simple additivity. When criteria for Type I (α) and II (β) errors are set at 0.01 and 0.10 (90% power), respectively, only 7 mice per group and a total of 126 in the study would be needed to detect the large strain main effect and 22 would be needed to detect the medium-sized lab main effect, but one must test *31 per group* and 558 in the whole study in order to be able to detect the moderate interaction effect. Precisely how many more observations are needed to detect the interaction vs. main effects depends strongly on the specific kind of interaction that is likely to occur.[26]

In the specific case of a small factorial study of two strains in two labs, a general guideline can be proposed if we can agree on a criterion for the size of an interaction that would be considered noteworthy in our field of study. Wahlsten[61,67] proposes that we should certainly want to detect the interaction if the treatment effect on one genotype is *twice as large* as the effect on the other genotype. In this case, one must test *at least six times as many mice* in order to detect the interaction compared with the number needed to detect a substantial main effect. Considering the sample sizes commonly employed in neurobehavioral genetics, many researchers appear to be satisfied with studying main effects and rarely employ sample sizes that are adequate for the evaluation of substantial interactions. The problem is particularly severe for a simple 2×2 design where each effect in the ANOVA has only one degree of freedom.

1.4.2 CONTRAST ANALYSIS

Some of the more elegant experimental designs in behavioral and neural genetics cannot be evaluated with the usual ANOVA. Consider the reciprocal crossing and backcrossing experiment that can be used to study maternal environment, cytoplasmic, and Y chromosome effects.[46,68,69] As illustrated in Table 1.2, the 16 groups may be arranged conceptually as a 4×4 factorial design, but the main effects and the global interaction term are almost impossible to interpret scientifically. Clarity emerges, however, when specific pairs of groups or linear combinations of group means are compared with each other in a logical series of biologically informative questions, each embodied in a one degree of freedom contrast (see Wahlsten[5,64]).

The challenge of achieving sufficient power for tests of interaction is present for contrast analyses as well as factorial ANOVA methods. The required sample size to detect a particular kind of interaction effect can be determined conveniently with a formula that is a good approximation for the noncentral t distribution.[64] When an experiment is to be analyzed with several orthogonal contrasts, it is inevitable that

TABLE 1.2
Factorial Design that Is Better Analyzed with Logical Contrasts

Origin of Mother	Origin of Father			
	Strain A	Strain B	A × B Hybrid	B × A Hybrid
Strain A	1. Inbred	3. F_1 hybrid	5. Backcross to A	6. Reciprocal of cross 5
Strain B	4. F_1 hybrid	2. Inbred	9. Backcross to B	10. Reciprocal of cross 9
A × B Hybrid	7. Backcross to A	11. Backcross to B	13. F_2 hybrid	14. F_2 hybrid
B × A Hybrid	8. Reciprocal of cross 7	12. Reciprocal of cross 11	15. F_2 hybrid	16. F_2 hybrid

Note: Abbreviated contrast analysis; see Sokolowski[68] or Wahlsten[69] for a more complete presentation.

   i. Do inbred strains differ? (1 vs. 2)
   ii. Is there an effect of genes in groups with an inbred mother? Note that the question whether there is hybrid vigor is logically equivalent to this question. ([1 vs. 3] and [2 vs. 4])
   iii. Is there a Y effect in backcrosses with inbred mothers? ([5 vs. 6] and [9 vs. 10])
   iv. Is there a Y chromosome effect in F_2 hybrids? ([13 vs. 14] and [15 vs. 16])
   v. Is the magnitude of the Y effect different with inbred and hybrid mothers? {([5 vs. 6] and [9 vs. 10]) vs. ([13 vs. 14] and [15 vs. 16])}
   vi. Is there an effect of cytoplasmic organelles in backcrosses and F_2 hybrids? ([7 vs. 8] and [11 vs. 12] and [13 vs. 15] and [14 vs. 16])
   vii. Is there an effect of autosomal genes? ([3 vs. 5] and [4 vs. 10] and [7 vs. 14] and [11 vs. 13] and [8 vs. 16] and [12 vs. 15])
   viii. Is the autosomal gene effect larger when the mother is inbred? {([3 and 4] vs. [5 and 10]) vs. ([7 and 11] vs. [14 and 13])}

a larger sample size will be required to detect some effects than for others. In such a case, the experimenter should choose a sample size for the entire experiment that is adequate to allow detection of the smallest effect that he or she is seriously interested in evaluating. An example of the application of this method to the reciprocal cross breeding design in Table 1.2 is provided by Wahlsten,[64] and other examples are given by Wahlsten.[27,70]

1.4.3 MULTIPLE-REGRESSION ANALYSIS

Factorial ANOVA and contrast analysis are best employed when the study involves carefully controlled treatment conditions given to independent groups of subjects. These kinds of analyses can also be performed using multiple-regression methods. Multiple regression offers the added advantage of being able to incorporate continuous variables in the list of predictors in order to account for the influence of covariates. A model can even include terms to assess group differences in the slopes of response to a covariate or nonlinear trend of response. Along with the elegance of the method come many hazards that can undermine the credibility of an analysis.[71,72] Only one aspect of this very large topic will be discussed here.

In multiple regression, an equation is computed that gives the best prediction or expected value of a dependent variable (Y) from several predictors (X) using the method of least squares: $E(Y) = b_0 + \Sigma b_j X_j$, where b_0 is the Y-intercept when all predictors are zero and b_j is the regression coefficient for the jth predictor. A predictor X may be a "dummy" variable to code the difference between a particular strain and a reference group or an orthogonal contrast in a contrast analysis. One of the most valuable pieces of information disgorged from a computer analysis is the "tolerance" that shows the extent to which the predictors are independent from one another. When the predictors are indeed independent, tolerance is 1.0, the standard errors of the regression coefficients tend to be low, and the multiple R^2 for the entire equation can be decomposed into fractions, each of which is attributable to a single predictor; that is, the model is perfectly additive. When a contrast effect and a covariate are themselves correlated, on the other hand, tolerance will be less than 1.0, sometimes much less, and effects will be confounded, so that the variance can no longer be partitioned into non-overlapping portions. As discussed in detail by Aiken and West,[71] interaction effects in a multiple-regression model will usually have very low tolerance unless each variable is "centered" by expressing it as the deviation from the mean. Instead of coding the interaction term as $X_1 * X_2$, one should use $(X_1 - \text{Mean of } X_1) * (X_2 - \text{Mean of } X_2)$.

1.5 MOLECULAR TECHNIQUES FOR QUANTITATIVE ANALYSIS

Many examples of GxE interaction have been well documented in laboratory research with strains and mutations in flies, worms (*C. elegans*), and mice, and examples with humans are also well established for several mutations. One lesson from this body of research is that the specific features of the environment that are most influential in altering the consequences of a genetic variant depend strongly on the gene in question. The exquisite specificity of the gene–environment interaction is related to the nature of gene expression at the molecular level. It is therefore necessary that we gain a deeper understanding of this relation through molecular analysis. Perhaps in this way we can also discover more effective means to alter the course of development and devise better therapies for a wide range of mental and behavioral disorders. Thus, the demonstration of GxE interaction with classical methods for studying global effects of differences in heredity forms the foundation for a new direction of research in neurobehavioral genetics.

1.5.1 THE REGULATION OF GENE EXPRESSION

Genetic and molecular biological approaches using model organisms such as the fruit fly, *Drosophila melanogaster*, and the mouse have provided a basis for unraveling the complex hierarchical interactions between genes, their RNA, and proteins in certain aspects of development.[73] More recently, nervous system development and function have also become the subjects of genetic and molecular analyses.[49] To make this chapter accessible to a broad audience, we include a brief summary of how genes work and illustrate how the environment may modulate the action of genes (taken in part from Kandel[74]). Genes are comprised of long strands of DNA and every cell in the body (aside from germ cells) has the same complement of DNA. What makes cells different from each other is that only a small (<20%) subset of genes is expressed in a given cell type. The actual DNA sequences that are transmitted intact from parents to offspring through the generations are not directly responsive to environmental regulation. Rather it is the expression of these genes that is regulated. Gene expression can be regulated by transcriptional control that determines (1) whether or not a gene is transcribed and, if so, (2) the rate at which it is transcribed. Transcription involves the synthesis of RNA from DNA. It is initiated when RNA polymerase binds to the DNA in the promoter region so that nuclear RNA can be made from the DNA. This RNA is then processed and modified into cytoplasmic messenger RNA (mRNA), which is then translated into a protein. Differential protein modification (a posttranslational process) determines which proteins will be retained and function (via activation) in the cell.

Transcriptional control occurs through transcription factors that bind sites in the promoter called promoter elements. Transcription factors can be cell specific or ubiquitous. In some cases transcriptional regulation is thought to proceed when transcription factors form a hierarchy. This results in a cascade of expression of hierarchically arranged transcription factors. For example, studies in development have shown that only a few genes that code for transcription factors can have crucial effects on the expression of many other genes in development.[73]

Other regulatory elements or sequences in the genome are enhancer and response (silencer) elements. These elements can be found either upstream or downstream of the promoter. They contain sequences that bind specific proteins and they are involved in the tissue-specific control of gene expression. When an enhancer-protein complex is formed, it then interacts with the promoter. As a result, proteins involved in multiple signaling pathways can act on a transcription factor bound to a promoter. Signals such as hormones can act on these regulatory elements when, for example, an enhancer binds a hormone responsive transcription factor. Both intracellular and extracellular signals can be environmentally responsive and join with enhancer or response elements to act on the gene's promoter.

The level of transcription of a gene results from the net effect of the factors described above: enhancers, response elements, tissue-specific proteins, and extracellular regulators. This system of gene regulation provides organisms with a versatile approach that enables gene transcription to be superbly sensitive to environmental stimuli.[74] These environmental stimuli can include such complex factors as different learning paradigms and social experiences as well as more easily defined environmental factors such as the pattern of light/dark cycles in circadian rhythms.

1.5.2 DETECTING AND LOCALIZING RNA

In this section, we focus on techniques that can be used to quantify RNA abundance with particular focus on techniques that can measure differences in RNA expression. Northern Blot Analysis has been the molecular workhorse in providing measures of RNA abundance. It is the only method that provides information about mRNA size and alternative splicing. In Northern analysis, similar levels of total RNA or mRNA are loaded on a gel, the RNA is transferred to a membrane, and a labeled probe from the gene of interest is applied to the membrane. The abundance of the RNA in each lane is then visualized on film or on a phosphoimaging device. To obtain an estimate of the total RNA loaded and transferred in each lane, the membrane is also treated with a control probe usually taken from a ubiquitous, "housekeeping" type gene (e.g., a ribosomal protein such as *rp49* in *Drosophila*[75]). Good control probes are best obtained from genes expressed at a constant level during development and throughout the organism. The RNA abundance in the sample of interest is then adjusted by its loading control.

Northern blots have been extensively used in analyses of the cycling in RNA of genes involved in circadian rhythms in *Drosophila* and other organisms (for review see Dunlap[76]). The sensitivity limit of Northern hybridization is 1–5 pg of RNA target molecule,[77] and in some instances the sensitivity of Northern blots is not sufficient to detect RNA. This occurs when the amount of tissue sampled is limited and/or the RNA abundance of a particular gene is very low. This might occur when a small subset of tissue such as a brain region is used or in the case of organisms carrying null mutants of a vital gene where early mortality limits the number of samples available.

The localization of RNA transcripts in tissue (whole mount or sections) is done using *in situ* hybridization.[78] However, it is not always useful for quantification of differences in RNA levels between samples. The relative difference in the level of a signal between mutant and wild-type or treated and untreated animals can sometimes be visualized using this technique, but differences in abundance of RNA must be relatively large to be able to quantify these differences. Specifically, problems arise with the insensitive and inaccurate quantification of mRNA expressed at low levels. RNase protection assays[79] enable one to map the transcript initiation and termination sites and intron/exon boundaries and to discriminate among related mRNA of similar size that migrate to similar places on the Northern blot. All of these techniques suffer from low sensitivity.

1.5.3 REAL-TIME RT-PCR

The reverse-transcription polymerase chain reaction (RT-PCR) has been used to overcome many of the aforementioned problems because RNA of low abundance can be detected in small amounts of tissue. However, RT-PCR is a complex process and as a quantitative technique it suffers from the problems inherent to PCR. These problems include questions about the technique's true sensitivity, its reproducibility, and its specificity. The reproducibility problems that result are difficult to interpret because it is not possible to process controls for every PCR reaction.

A promising technology was recently developed to overcome these difficulties. It is a fluorescence-based kinetic RT-PCR procedure known as Quantitative Real Time PCR. The principle of TaqMan real-time detection is based on the fluorogenic 5′ nuclease assay that allows simple and rapid quantification of a target sequence during the extension phase of PCR amplification. The web page (http://www.applied-biosystems.com/techsupp/tools.html) provides detailed protocols and advice on probe design for this technology. Advantages of this technique are that (a) little tissue is required, (b) controls (often a housekeeping gene) can be run for each reaction, (c) optimization of the reaction is relatively easy, (d) the technique uses two-gene specific primers and a gene-specific probe that lies within the primers, making the technique highly sequence specific, and (e) with some technologies (the Roche thermal cycler — http://biochem.boehringer-mannheim.com/lightcycler/), the ongoing reaction kinetics can be visualized graphically. Bustin[80] provides an excellent review of the technical aspects of this technique, comparing the conventional and real-time RT-PCR approaches for quantifying gene expression and comparing the different systems commercially available for real-time PCR. The disadvantage of real time RT-PCR has been the high cost, but it is decreasing. In addition, RT-PCR cannot be used to identify differences in expression patterns in unknown genes because it is done using primers from known genes. The Molecular Tools web page at http://www.nlv.ch/Molbiotoolsrtpcr.html#PE compares the various technologies available for real-time PCR.

To our knowledge, analyses of GxE interactions on complex behavior have not yet been published using real-time PCR. In the last year, several studies in a variety of systems have used this technique successfully. These include the analysis of brain homogenates of adult Wistar rats for mRNA expression of the genes *bcl-2* and *bax*, both involved in chemical preconditioning in ischemia,[81] quantification of multiple human potassium-channel genes at the single-cell level,[82] gene expression of neuronal nitric oxide synthase and adrenomedullin in human neuroblastoma,[83] and analysis of gene expression of the D2 receptor in regions of the human brain.[84]

Proper experimental design including replication is crucial for accurately quantifying the relative differences in RNA using real-time RT-PCR. The experiments are designed as in Figure 1.2. It is important to run all GxE treatments and their replicates simultaneously in one randomized block representing one full replication of the experiment. Four independent mRNA extractions for all treatments and replicates comprise the four experimental blocks. This design produces highly reproducible results amenable to statistical analysis. This design mimics our behavioral analyses that test GxE interactions (see Figure 1.2) and enables both sets of data (behavioral and RNA expression data) to be analyzed statistically with analysis of variance.

1.5.4 DNA Microarray Technology

A microarray contains DNA sequences (full or partial cDNA) from both known and unknown genes. This DNA is spotted onto a solid support, usually nylon membranes or glass slides. The array is then hybridized with RNA isolated from different experimental conditions (e.g., mutant vs. wild-type; an environmental treatment vs. a control; drug treatment vs. placebo; experience vs. no experience; immature vs. mature). The expression of large numbers of genes (thousands of genes and in some cases entire genomes) is simultaneously analyzed for each experimental condition so that the expression of each gene in both conditions can be compared. Some genes will be up regulated, others will be down regulated, and still others will not be affected by the treatment. The data are visualized using a reader to detect many fluorescent spots in a grid pattern. Each spot represents one of the DNA clones initially put on the chip. The brightness of the spot gives an indication of the magnitude of the change in expression and the color of the spot, usually red or green, gives an idea about whether the expression of that gene has been up or down regulated by the experimental treatment. It is important to design microarray experiments and replicate them so that the number of false positives can be minimized, because it can take an inordinate amount of time to sift through these false positives. All positive clones (and often there are hundreds of them) need to be confirmed using an independent technique such as Northern analysis or real time RT-PCR. The sensitivity of the microarray technique is similar to that of Northern Blot Analysis; it is difficult to reliably detect gene expression changes less than 2- to 3-fold on average. This limitation should change as the technology improves. The technique is still very expensive and requires good knowledge of the technology. On the other hand, the DNA microarray technology provides us with the possibility of finding many of the genes and processes involved in the phenomenon of interest.

1.5.5 Microarrays: Experimental Design Issues

The particular design chosen for the microarray experiment is crucial to its success. Advice given in Section 1.3.1 and Figure 1.2 are directly applicable here. If genes are being manipulated, then the genetic background of the strains to be compared should be identical or else many differences in expression will be detected that are not related to the phenotype of interest. For example, if mutant and wild-type are to be compared, the strains should be co-isogenic; this means that allelic variation between the strains should only be in the locus of interest. Similarly, if a transgenic strain is being compared to a mutant or wild-type strain, then the transgene should be on an identical genetic background to the strain of interest. Strains should be reared in an identical fashion, and animals of the same sex and age should be compared so as not to cause gene expression to vary due to uncontrolled environmental factors. Dissections and RNA extraction must also be done under identical conditions. If the design involves an environmental treatment, then it is critical that there be no genetic variation within and between the strains used (as described in Section 1.3.1 above). The ideal situation is to treat the same clone (or group of highly inbred isogenic animals) with the environmental or pharmacological treatment. G×E interactions could be tested on microarrays by using for example the two-way design shown in Figure 1.2.

For instance, in one laboratory we could choose two natural strains of Drosophila flies called rover and sitter that differ only in their allelic composition at the *foraging* gene.[50] We could give each strain one of two treatments (food and water vs. water only) 3 h prior to their RNA extraction. This would give us four groups: rover fed, rover unfed, sitter fed, and sitter unfed. This experiment would be replicated several times so that there are at least three replicates for each array for a total of 12 arrays. The pattern that the four arrays produced could be analyzed for a strain effect, a feeding effect, or an interaction. The interaction would suggest that different strains (rover or sitter) respond differently to the feeding treatment. The response is measured as changes in the patterns of gene expression. For example, rovers may significantly up regulate genes a, b, and d, whereas sitters may down regulate c and d but upregulate b. This approach would uncover the molecular underpinnings of

GxE interactions on food search behavior. It is important to note that in this design DNA microarrays only examine short-term changes in gene expression. It is conceivable that a gene is important for the development of a structure or system that is crucial to the performance of the adult behavior but that this gene is not expressed in the adult stage of development. The role of such a gene in the development of adult behavior would remain undiscovered in the microarray experiment.

Microarray experiments designed to measure gene–environment interactions and changes in gene expression during development require statistical analysis which can handle this type and quantity of data. One decision to be made is what constitutes a significant change in gene expression — a 0.5-fold, 1-fold, or 2-fold change? Obviously, a lower cut-off yields more false positives. On the other hand, some genes that play crucial roles in the process of interest may only show a relatively small fold change, and by setting the cut-off too high, these genes would be missed. Another problem with analysis of microarray data stems from the newness of this technique. Software that enables exploration and statistical analysis of microarray data has been lacking (see Chapters 7 and 8 of this volume). Tools are required that can analyze the expression of individual genes, gene families, and gene clusters, compare expression patterns, and directly access other genomic databases for clones of interest.

A number of very recent studies successfully used DNA microarray analysis to identify changes in gene expression of known and novel genes. As was the case for real-time PCR, there is a paucity of studies that use microarrays to address issues of complex behavior and GxE interactions. The first comprehensive genome scan examined the response of the yeast genome to aerobic and anerobic fermentation conditions.[85] High-density DNA microarrays containing several thousand *Drosophila melanogaster* gene sequences were used to study changes in gene expression during a developmental stage called metamorphosis known to involve an integrated set of developmental processes controlled by a transcriptional hierarchy that affects hundreds of genes.[86] Of the differentially expressed genes found in this study, many could be assigned to developmental pathways known to play a role in metamorphosis, while others were involved in pathways not previously known to play a role in metamorphosis. Still other genes that were identified were novel and had previously unknown functions. Another study found that brains of aging mice showed parallels with human neurodegenerative disorders at the transcriptional level and that caloric restriction, which retards the aging process in mammals, selectively diminished the age-associated induction of genes encoding inflammatory and stress responses.[87] DNA microarrays have also been used to identify differentially expressed genes in purified follicle cells, demonstrating that the technique can be used for cell type-specific developmental analyses.[88] Changes in the expression patterns of >2,000 *Arabidopsis* genes after inoculation with or without a fungal pathogen or after treatment with plant-defense signaling molecules resulted in molecular evidence for coordinated defense responses,[89] suggesting multiple overlapping signal transduction pathways in plant defense mechanisms. The ability to detect interactions between different expression patterns in plant defense mechanisms shows promise for analysis of pathways involved in complex behavior patterns. DNA microarrays have also been used to study expression profiles in multiple sclerosis lesions and in Alzheimer's disease tangle-bearing CA1 neurons.[90,91] The technologies available for high throughput analysis of gene expression in the human brain are reviewed by Colantuoni et al.[92]

Dubnau and Tully (unpublished data) are using microarrays to unravel changes in gene expression associated with learning in *Drosophila*. They use (a) a genetic manipulation — comparing gene expression in isogenic populations which differ at a single gene that affects learning, (b) an environmental manipulation — comparing gene expression in one homozygous population which has been trained using different learning paradigms, and (c) a pharmacological manipulation — comparing gene expression in one homozygous population where half of the individuals have been treated with a chemical known to alter learning scores. The expectations from their experiment are that: (1) some of the genes and signal transduction pathways identified will be shared in common between all of the treatments whereas others will differ, (2) changes in the expression of genes known to be involved in learning will be identified along with known genes and pathways not previously thought to be involved in learning, and (3) previously unidentified novel genes will be associated with one or several of the treatments. This type of experimental design could in theory be applied to any behavior of interest using a genetically malleable organism.

1.6 SUMMARY

The relations between genes and behavior currently are studied in two ways: differences in behavior between (conspecific) individuals are associated with genotypic differences, and changes in the behavior of an individual are associated with changes in gene expression in the brain. Because these two approaches have historically proceeded independently, there is a major gap in our knowledge of precisely how genes and the environment interact to regulate behavior. Our challenge is to use the new technologies along with the data from the genome projects to unravel the molecular mechanisms underlying G×E interactions involved in the development and functioning of complex behavior.

The abundance, developmental timing, and localization of gene products can influence the probability of a behavior being performed. A predisposition to perform a behavior can be thought of as giving the adult organism a certain probability of performing a behavior under a certain set of environmental circumstances. However, there is a subtle interplay during development between predisposition and experience. Hence, one needs to consider the environment during development that influences gene expression and the environment during adulthood that affects the expression of the behavior of interest. We have discussed statistical and molecular techniques that enable the analysis of G×E interactions. For gene-brain-behavior relationships, however, ongoing feedback from the interaction of the organism with the environment often affects how the brain develops and functions. Performing the behavior itself can cause changes in gene expression and the function of nerve cells.[93] For example, when free-ranging sparrows hear a conspecific's song, this changes the level of ZENK, a transcriptional regulator thought to play a role in song learning.[94] Interactions between mothers and their infants are reflected in changes in brain neurochemistry during development and across generations.[95] Social modulation of amine responsiveness at particular synaptic sites occurs during lobster aggressive interactions.[96] Species-specific patterns of oxytocin and vasopressin receptor expression in the brain are associated with monogamous vs. nonmonogamous social structure in voles.[97] These complex environmental effects combined with the complexity of the genetic millieu contribute to the tremendous challenge ahead in addressing questions of the molecular underpinnings of gene–environment interactions during the development and functioning of complex behaviors.

ACKNOWLEDGMENTS

We thank Y. Ben Shahar for technical discussions and research grants from the Medical Research Council of Canada to MBS and the Natural Sciences and Engineering Council of Canada to MBS and DW. MBS is a CRCP Chairholder.

REFERENCES

1. Jacob, F., *The Logic of Life. A History of Heredity*, Vintage Books, New York, 1976.
2. Bateson, W., *Mendel's Principles of Heredity*, Cambridge University Press, Cambridge, 1913.
3. Gottlieb, G., Wahlsten, D., and Lickliter, R., The significance of biology for human development: A developmental psychobiological systems view, in *Handbook of Child Psychology, Vol. 1, Theoretical Models of Human Development*, 5th ed., R. M. Lerner, Ed., Wiley, New York, 1998, 233–273.
4. Strohman, R. C., The coming Kuhnian Revolution in biology, *Nat. Biotechnol.*, 15, 194–200, 1997.
5. Wahlsten, D., Genetics and the development of brain and behavior, in *Handbook of Developmental Psychology*, Valsiner, J. and Connolly, K., Eds., in press.
6. Johannsen, W., The genotype conception of heredity, *Am. Naturalist*, 45, 129–159, 1911.
7. Woltereck, R., Weitere experimentelle Untersuchungen über das Wesen quantitativer Artunterschieder bei Daphniden, *Verh. Dtsch. Zool. Ges.*, 19, 110–173, 1909.
8. Lewontin, R., *The Triple Helix. Gene, Organism, Environment*, Harvard University Press, Cambridge, MA, 2000.
9. Nijhout, H. F., Control mechanisms of polyphenic development in insects, *BioScience*, 49, 181–192, 1999.
10. Hogben, L., *Nature and Nurture*, Williams & Norgate, London, 1933.
11. Wahlsten, D. and Gottlieb, G., The invalid separation of effects of nature and nurture: Lessons from animal experimentation, in *Intelligence, Heredity, Environment*, R. J. Sternberg and E. L. Grigorenko, Eds., Cambridge University Press, New York, 1997, 163–192.
12. Platt, S. A. and Sanislow, C. A., Norm-of-reaction: Definition and misinterpretation of animal research, *J. Comp. Psychol.*, 102, 254–261, 1988.
13. Hull, C. L., The place of innate individual and species differences in a natural-science theory of behavior, *Psychol. Rev.*, 52, 55–60, 1945.
14. Beach, F., The snark was a boojum, *Am. Psychol.*, 5, 115–124, 1950.
15. Waddington, C. H., *The Strategy of the Genes*, Allen & Unwin, London, 1957.

16. Gottlieb, G., Experiential canalization of behavioral development: Theory, *Dev. Psychol.*, 27, 39–42, 1991.

17. Gottlieb, G., *Individual Development and Evolution. The Genesis of Novel Behavior,* Oxford University Press, New York, 1992.

18. Rose, S., *Lifelines. Biology beyond Determinism*, Oxford University Press, Oxford, 1997.

19. Rose, S., Kamin, L. J., and Lewontin, R. C., *Not in Our Genes*, Penguin, New York, 1984.

20. Oyama, S., *The Ontogeny of Information. Developmental Systems and Evolution*, Cambridge University Press, Cambridge, 1985.

21. Bateson, P., Biological approaches to the study of behavioral development, *Int. J. Behav. Dev.*, 10, 1–10, 1987.

22. Plomin, R., DeFries, J. C., McClearn, G. E., and Rutter, M., *Behavioral Genetics.* 3rd ed., Freeman, New York, 1997.

23. Devlin, B., Daniels, M., and Roeder, K., The heritability of IQ, *Nature*, 388, 468–470, 1997.

24. Neisser, U., Boodoo, G., Bouchard, T. J., Jr., Boykin, A. W., Brody, N., Ceci, S. J., Halpern, D. F., Loehlin, J. C., Perloff, R., Sternberg, R. J., and Urbina, S., Intelligence: Knowns and unknowns, *Am. Psychol.*, 51, 77–101, 1996.

25. Wahlsten, D., Single-gene influences on brain and behavior, *Annu. Rev. Psychol.*, 50, 599–624, 1999.

26. Wahlsten, D., Insensitivity of the analysis of variance to heredity–environment interaction, *Behav. Brain Sci.*, 13, 109–161, 1990.

27. Wahlsten, D., The intelligence of heritability, *Can. Psychol.*, 35, 244–258, 1994.

28. Scriver, C. R. and Waters, P. J., Monogenic traits are not simple — lessons from phenylketonuria, *Trends Genet.*, 15, 267–272, 1999.

29. Guo, S.-W., The behaviors of some heritability estimators in the complete absence of genetic factors, *Hum. Hered.*, 49, 215–228, 1999.

30. Kempthorne, O., Logical, epistemological and statistical aspects of nature–nurture data interpretation, *Biometrics*, 34, 1–23, 1978.

31. Kempthorne, O., How does one apply statistical analysis to our understanding of the development of human relationships, *Behav. Brain Sci.*, 13, 138–139, 1990.

32. Gottlieb, G., Normally occurring environmental and behavioral influences on gene activity: From central dogma to probabilistic epigenesis, *Psychol. Rev.*, 105, 792–802, 1998.

33. Moldin, S. O. and Gottesman, I. I., At issue: Genes, experience, and chance in schizophrenia — positioning for the 21st century, *Schizophr. Bull.*, 23, 547–561, 1997.

34. Gottesman, I. I., and Shields, J., *Schizophrenia and Genetics. A Twin Study Vantage Point*, Academic Press, New York, 1972.

35. Kendler, K. S. and Eaves, L., Models for the joint effect of genotype and environment on liability to psychiatric illness, *Am. J. Psychiatry*, 143, 279–289, 1986.

36. Wahlberg, K.-E., Wynne, L. C., Oja, H., Keskitalo, P., Pykäläinen, Lahti, I., Moring, J., Naarla, M., Sorri, A., Seitamaa, M., Läksy, K., Kolassa, J., and Tienari, P., Gene–environment interaction in vulnerability to schizophrenia: Findings from the Finnish adoptive family study of schizophrenia, *Am. J. Psychiatry*, 154, 355–362, 1997.

37. Cloninger, C. R., Multilocus genetics of schizophrenia, *Curr. Opin. Psychiatry*, 10, 5–10, 1997.

38. Skuse, D. H., Behavioural neuroscience and child psychopathology: Insights from model systems, *J. Child Psychol. Psychiatry*, 41, 3–31, 2000.

39. Sokolowski, M. B., Genes for normal behavioral variation: Recent clues from flies and worms, *Neuron*, 21, 1, 1998.

40. Bale, T. L., Contarino, A., Smith, G. W., Chan, R., Gold, L. H., Sawchenko, P. E., Koob, G. F., Vale, W. W., and Lee, K.-F., Mice deficient for corticotropin-releasing hormone receptor-2 display anxiety-like behaviour and are hypersensitive to stress, *Nat. Genet.*, 24, 410–414, 2000.

41. Coste, S. C., Kesterson, R. A., Heldwein, K. A., Stevens, S. L., Heard, A. D., Hollis, J. H., Murray, S. E., Hill, J. K., Pantely, G. A., Hohimer, A. R., Hatton, D. C., Phillips, T. J., Finn, D. A., Low, J. J., Rittenberg, M. B., Stenzel, P., and Stenzel-Poore, M. P., Abnormal adaptations to stress and impaired cardiovascular function in mice lacking corticotropin-releasing hormone receptor-2, *Nat. Genet.*, 24, 403–409, 2000.

42. Kishimoto, T., Radulovic, J., Radulovic, M., Lin, C. R., Schrick, C., Hooshmand, F., Hermanson, O., Rosenfeld, M. G., and Spiess, J., Deletion of Chrhr2 reveals and anxiolytic role for corticotropin-releasing hormone receptor-2, *Nat. Genet.*, 24, 415–419, 2000.

43. Crabbe, J. C., Wahlsten, D., and Dudek, B. C., Genetics of mouse behavior: Interactions with laboratory environment, *Science*, 284, 1670–1672, 1999.

44. Scarr, S., Developmental theories for the 1990s: Development and individual differences, *Child Dev.*, 63, 1–19, 1992.

45. Henderson, N. D., Genetic influences on behavior of mice can be obscured by laboratory rearing, *J. Comp. Physiol. Psychol.*, 73, 505–511, 1970.

46. Carlier, M., Roubertoux, P. L., and Wahlsten, D., Maternal effects in behavior genetic analysis, in *Neurobehavioral Genetics: Methods Applications*, P. Mormede and B. Jones, Eds., CRC Press, Boca Raton, FL, 1999, pp. 187–197.

47. Brand, A. H. and Perrimon, N., Targeted gene expression as a means of altering cell fates and generating dominant phenotypes, *Development,* 118, 401–415, 1993.

48. Mayford, M., Bach, M. E., Huang, Y., Wang, L., Hawkins, R. D., and Kandel, E. R., Control of memory formation through regulated expression of a CaMKII transgene, *Science,* 274, 1678–1683, 1996.

49. Crusio, W. E. and Gerlai, R. T., Eds., *Handbook of Molecular-Genetic Techniques for Brain Behavior Res.*, Elsevier, Amsterdam, 1999.

50. Osborne, K. A., Robichon, A., Burgess, E., Butland, S., Shaw, R. A., Coulthard, A., Pereira, H. S., Greenspan, R. J., and Sokolowski, M. B., Natural behavior polymorphism due to a cGMP-dependent protein kinase of Drosophila, *Science*, 277, 834–836, 1997.

51. Yang, P., Shaver, S. A., Hilliker, A. J., and Sokolowski, M. B., Abnormal turning behavior in *Drosophila* larvae: Identification and molecular analysis of *scribbler (sbb)*, *Genetics*, 155, 1161, 2000.

52. Ryner, L. C., Goodwin, S. F., Castrillon, D. H., Anand, A., Villelia, A., Baker, B. S., Hall, J. C., and Wasserman, S. A., Control of male sexual behavior and sexual orientation in Drosophila by the fruitless gene, *Cell*, 87, 1079–1089, 1996.

53. Boynton, S. and Tully, T., *latheo*, a new gene involved in associative learning and memory in Drosophila melanogaster, identified from P element mutagenesis, *Genetics*, 131, 655, 1992.

54. Greenspan, R. J., A kinder, gentler genetic analysis of behavior: Dissection gives way to modulation, *Curr. Op. Neurobiol.,* 7, 805–811, 1997.

55. Cheung, U.S., Shayan, A. J., Boulianne, G. L., and Atwood, H. L., Drosophila larval neuromuscular junction's responses to reduction of cAMP in the nervous system, *J. Neurobiol.*, 40, 1, 1999.

56. Bull., J. J., *Evolution of Sex Determining Mechanisms*, Benjamin/Cummings, Menlo Park, CA, 1983.
57. Greene, E., A diet-induced developmental polymorphism in a caterpillar, *Science*, 243, 643–646, 1989.
58. Detterman, D. K., Don't kill the ANOVA messenger for bearing bad interaction news, *Behav. Brain Sci.*, 13, 131–132, 1990.
59. Van Den Oord, E. J. C. G. and Rowe, D. C., An examination of genotype-environment interactions for academic achievement in an U.S. national longitudinal survey, *Intelligence*, 25, 205–228, 1998.
60. Goodenough, F. L., The measurement of mental growth in childhood, in *Manual of Child Psychology*, L. Carmichael, Ed., Wiley, New York, 1954, 459–491.
61. Wahlsten, D., The theory of biological intelligence: History and critical appraisal, in *The General Factor in Intelligence: How General Is It?* R. Sternberg and E. Grigorenko, Eds., in press.
62. Neisser, U., Rising scores on intelligence tests, *Sci. Am.*, 85, 440–447, 1997.
63. Fisher, R. A. and Mackenzie, W. A., Studies in crop variation. II. The manurial responses of different potato varieties, *J. Agric. Sci.*, 13, 311–320, 1923.
64. Wahlsten, D., Sample size to detect a planned contrast and a one degree-of-freedom interaction effect, *Psychol. Bull.*, 110, 587–595, 1991.
65. Lander, E. and Kruglyak, L., Genetic dissection of complex traits: Guidelines for interpreting and reporting linkage results, *Nat. Genet.*, 11, 241–247, 1995.
66. Cohen, J., *Statistical Power Analysis for the Behavioral Sciences*, Erlbaum, Hillsdale, NJ, 1988.
67. Wahlsten, D., Standardization of test of mouse behaviour: Reasons, recommendations, and reality, *Physiol. Behav.*, in press.
68. Sokolowski, M. B., Genetic analysis of behavior in the fruit fly, Drosophila melanogaster, in *Techniques for the Genetic Analysis of Brain Behavior. Focus on the Mouse*, D. Goldowitz, D. Wahlsten, and R. E. Wimer, Eds., Elsevier, Amsterdam, 1992, 497–512.
69. Wahlsten D., A critique of the concepts of heritability and heredity in behavior genetics, in *Theoretical Advances in Behavioral Genetics*, J. R. Royce and L. Mos, Eds., Sijthoff and Noorhoff, Alphen aan den Rijn, Netherlands, 1979, 425–481.
70. Wahlsten, D., Experimental design and statistical inference, in *Handbook of Molecular-Genetic Techniques for Brain and Behavior Research*, W. E. Crusio and R. T. Gerlai, Eds., Elsevier, Amsterdam, 1999, 41–57.
71. Aiken, L. S. and West, S. G., *Multiple Regression. Testing and Interpreting Interactions*, Sage, Thousand Oaks, CA, 1991.
72. Marascuilo, L. A. and Serlin, R. C., *Statistical Methods for the Social and Behavioral Sciences,* Freeman, New York, 1988.
73. Nusslein-Volhard, H. G., Frohnhöfer, H., and Lehmann, R., Determinants of anteroposterior polarity in Drosophila, *Science,* 238, 1675–1681, 1987.
74. Kandel, E. R., A new intellectual framework for psychiatry, *Am. J. Psychiatry,* 155, 457–469, 1998.
75. O'Connell, P. O. and Rosbash, M., Sequence, structure, and codon preference of the *Drosophila* ribosomal protein 49 gene, *Nucl. Acids Res.,* 12, 5495–5513, 1984.
76. Dunlap, J. C., Molecular bases for circadian clocks, *Cell,* 96, 271–290, 1999.
77. Sabelli, P. A., Northern Blot Analysis, in *Molecular Biomethods Handbook,* R. Rapley and J. M. Walker, Eds., Humana Press Inc., Totowa, NJ, 1998, 90.
78. Parker, R. M. and Barnes, N. M., mRNA: Detection by *in situ* and northern hybridization, *Meth. Mol. Biol.*, 106, 247–283, 1999.
79. Hod, Y., A simplified ribonuclease protection assay, *Biotechniques,* 13, 852–854, 1992.

80. Bustin, S. A., Absolute quantification of mRNA using real-time reverse transcription polymerase chain reaction assays, *J. Mol. Endocrinol.,* 25, 169–193, 2000.

81. Brambrink, T. M., Schneider, T., Noga, H., Astheimer, A., Gotz, T., Korner, T., Heimann, A., Welschof, M., and Kempski, O., Tolerance-inducing dose of 3-nitropropionic acid modulates bcl-2 and bax balance in the rat brain: A potential mechanism of chemical preconditioning, *Cere. J. Blood Flow Metab.,* 20, 1425, 2000.

82. Al-Taher, A., Bashein, A., Nolan, T., Hollingsworth, M., and Brady, G., Global cDNA amplification combined with real-time RT-PCR: Accurate quantification of multiple human potassium channel genes at the single cell level, *Yeast,* 17, 201–210, 2000.

83. Dotsch, J., Harmjanz, A., Christiansen, H., Hanze, J., Lampert, F., and Rascher, W., Gene expression of neuronal nitric oxide synthase and adrenomedullin in human neuroblastoma using real-time PCR, *Int. J. Cancer,* 88, 172–175, 2000.

84. Medhurst, A. D., Harrison, D. C., Read, S. J., Campbell, C. A., Robbins, M. J., and Pangalos, M. N., The use of TaqMan RT-PCR assays for semiquantitative analysis of gene expression in CNS tissues and disease models, *J. Neurosci. Meth.* 15, 9–20, 2000.

85. de Risi, J. L., Iyer, V. R., and Brown, P. O., Exploring the metabolic and genetic control of gene expression on a genomic scale, *Science,* 278, 680–686, 1997.

86. White, K. P., Rifkin, S. A., Hurban, P., and Hogness, D. S., Microarray analysis of Drosophila development during metamorphosis, *Science,* 286, 2179, 2000.

87. Lee, C.-K., Weindruch, R., and Prolla, T. A., Gene-expression profile of the ageing brain in mice, *Nat. Genet.,* 25, 294–297, 2000.

88. Bryant, Z., Subrahmanyan, L., Tworoger, M., LaTray, L., Liu, C. R., Li, M. J., van den Engh, G., and Ruohola-Baker, H., Characterization of differentially expressed genes in purified Drosophila follicle cells: Toward a general strategy for cell type-specific developmental analysis, *Proc. Natl. Acad. Sci. USA,* 96, 5559–5564, 1999.

89. Schenk, P. M., Kazan, K., Wilson, I., Anderson, J. P., Richmond, T., Somerville, S. C., and Manners J. M., Coordinated plant defense responses in arabidopsis revealed by microarray analysis, *Proc. Natl. Acad. Sci. USA,* 97, 11655–11660, 2000.

90. Whitney, L. W., Becker, K. G., Tresser, N. J., Caballero-Ramos, C. I., Munson, P. J., Prabhu, V. V., Trent, J. M., McFarland, H. F., and Biddison, W. E., Analysis of gene expression in mutiple sclerosis lesions using cDNA microarrays, *Ann. Neurol.,* 46, 425, 1999.

91. Ginsberg, S. D., Hemby, S. E., Lee, V. M., Eberwine, J. H., and Trojanowski, J. Q., Expression profile of transcripts in Alzheimer's disease tangle-bearing CA1 neurons, *Ann. Neurol.,* 48, 77–87, 2000.

92. Colantuoni, C., Purcell, A. E., Bouton, C. M., and Pevsner, J., High throughput analysis of gene expression in the human brain, *J. Neurosci. Res.,* 59, 1–10, 2000.

93. Stork, O. and Welzl, H., Memory formation and the regulation of gene expression, *Cell. Mol. Life Sci.,* 55, 575–592, 1999.

94. Jarvis, E. D., Schwabl, H., Ribeiro, S., and Mello, C. V., Brain gene regulation by territorial singing behavior in freely ranging songbirds, *NeuroReport,* 8, 2073–2077, 1997.

95. Fleming, A. S., O'Day, D. H., and Kraemer, G. W., Neurobiology of mother-infant interactions: Experience and central nervous system plasticity across development and generations, *Neurosci. Biobehav. Rev.,* 23, 673–685, 1999.

96. Kravitz, E. A., Serotonin and aggression: Insights gained from a lobster model system and speculations on the role of amine neurons in a complex behavior, *J. Comp. Physiol. A,* 186, 221–238, 2000.

97. Young, L. J., Wang, Z. and Insel, T. R., Neuroendocrine bases of monogamy, *Trends Neurosci.,* 21, 71–75, 1998.

2 Current Perspectives on the Genetic Analysis of Naturally Occurring Neurological Mutations in Mice

Wayne N. Frankel

CONTENTS

2.1 INTRODUCTION

In the past few decades, the biological research community has relied more and more on genetic variation in mice to gain insight into disease susceptibility and basic physiological processes. Although much of this work has focused historically on differences among common inbred strains, and the "interval mapping" paper of Lander and Botstein[1] made the genetic mapping of natural variants as quantitative trait loci (QTL) extremely popular, the most progressive use of natural variation has come from positional gene identification of spontaneous single-gene mutations. In many cases, spontaneous mutants have given us instant insight into gene function by mere correlation of a known gene with aspects of cellular neurobiology and disease, including neuronal migration (rostral cerebellar malformation[2], scrambler[3]), cell death (lurcher,[4] staggerer[5]), developmental patterning (swaying[6]), epilepsy (stargazer,[7] tottering,[8] lethargic,[9] slow-wave epilepsy[10]), deafness (shaker-1,[11] shaker-2,[12] Snell's waltzer[13]), retinal disorders (retinal degeneration,[14] ocular retardation[15]), energy homeostasis (fat[16]), and others. In a few cases, completely novel genes were discovered, thus starting inquiry into the basis for their existence (reeler,[17] obese,[18] diabetes,[19] tubby[20]). As the mouse genomics community approaches newer phenotype-driven gene discovery approaches on a daunting scale, such as random mutagenesis and proactive screening for traits of interest, it is appropriate to review lessons learned from its predecessor — the spontaneous mutations — and to consider where the future lies for this once groundbreaking resource.

2.2 ORIGIN OF MUTATIONS

What are the types of spontaneous mutations in mice and how often do they arise? Spontaneous mutations come in all shapes and sizes, ranging from single nucleotide substitutions, small intragenic deletions, larger deletions which remove multiple genes, DNA insertions (usually, but not always, of retroelements such as proviruses and transposons), larger chromosomal rearrangements such as inversions, translocations, or more complex events. The rate of spontaneous mutation has been estimated at several loci in various ways, and ranges from the very high 1.7×10^{-4} apparent mutations per gene per generation for the $H2K^b$ gene,[21] to about 2×10^{-6} mutations per gene per generation at recessive coat color mutations.[22] However, because the $H2$ complex mutations seem not to be *de novo* mutations at all but rather gene conversion-like events,[23,24] it is more likely the latter, lower rates are more representative of the average gene.

The types of *de novo* spontaneous mutation also vary and do not necessarily correspond to rate. For example, at the tottering locus, each of the six solved spontaneous recessive alleles of the calcium channel subunit alpha-1A (*Cacna1a*) are point mutations.[8,25,26,26a] In contrast, at least two of the three alleles of the stargazer, or *Cacng2* calcium channel gene, are insertions of one type of transposable element — the ETn.[7,26b] Likewise, each of the three spontaneous alleles at the reeler locus, *Reln*, encoding the cell–cell interaction protein reelin, also involve a major genetic rearrangement (deletion or insertion) and not a point mutation.[17,27,28] It is unclear, however, whether these differences in mutation type between loci such as *Cacna1a* compared to *Cacng2* or *Reln* relate more to inherent mutability *per se* rather than the selection for phenotype. That is, one could argue that perhaps some genes (e.g., *Cacna1a*) have more distinct and critical functional domains for their size than others.

Once having appreciated the frequency, type, and variation of spontaneous mutations that occur in mice, we can better understand the consequences of mutation on a gene's phenotypic expression. Many mutations seem to affect mRNA levels primarily; this is not surprising since there are a multitude of ways in which transcription, splicing, or mRNA stability can be altered by mutation. Unfortunately, the specific type of mutation is not predictive of severity: deletions,[27] insertions,[7] nucleotide substitutions at splice acceptor or donor sequences,[29] or single nucleotide substitutions in coding sequence[10] can all affect the amount of normal mRNA present in the cell. The remaining mutations, like the *Cacna1a* alleles, do not alter the amount of mRNA but rather the amount of protein or its activity. These are mostly smaller events resulting in alteration of one or a few codons, although analogous to the ENU-induced mutation *Clock*,[30] it is possible that some spontaneous splice-site mutations could result in "exon skipping," resulting in proteins with missing domains and thus altered function rather than reduced activity.

2.3 ASCERTAINMENT OF MUTANT MICE

The other key issue regarding spontaneous mutant mice is of their ascertainment: what makes them discoverable? The vast majority of mouse mutants were detected because animal caretakers, while changing the bedding or cages, observed a mouse's unusual feature or attribute. In the case of neurological mutants, these features have to be rather severe (from a mouse's viewpoint) for even the most careful of humans to notice. Thus, most of the extant spontaneous neurological mutants show very obvious abnormalities, from the characteristic circling or head-bobbing of inner ear mutants, to the wobbly gait of cerebellar mutants, to shivers and tremors in myelin mutants, to overt seizures in epileptic mutants.[31] It has often required rather detailed follow-up examination of these mutants to uncover some of their more important attributes, such as was accomplished in the discovery that a subset of wobbly mutants not only had abnormal cerebellar function but also had spike-wave seizures — a form of absence, or petit-mal epilepsy.[32] These four mutations, which, interestingly, occurred in different subunits of the same neuronal ion channel, would not have been discovered as epilepsy models if it were not for dedicated encephalographic screening. This contrast between obvious, severe neurological mutants and more subtle ones, e.g., psychiatric disorder models, will only increase as the dedicated neurological examinations are undertaken in large-scale mutation screening programs — programs in which both obvious and subtle phenotypes will be detected.

2.4 FOLLOW-UP GENETICS OF SPONTANEOUS MUTANT MICE: SEGREGATION AND LINKAGE ANALYSIS

Perhaps the most intimidating aspect of classical genetics for the average non-geneticist (e.g., neuroscientist) to grasp are segregation and linkage analysis, the variety of ways in which it is carried out (e.g., backcross, intercross, etc.), and its analysis and interpretation. In actuality, the underlying principle of segregation analysis is deceptively simple: genetics is ultimately a plus–minus, binary system (an individual either inherits the gene which causes an abnormal phenotype, or it does not). The confusing aspects arise with the following caveats (listed in increasing complexity):

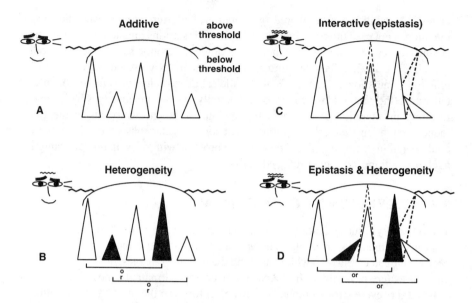

FIGURE 2.1 Typical models for multigene action in a threshold trait — one in which literal, subclinical phenotypes lead to a visible, unitary trait which the observer can recognize and contemplate. (A) In an additive model each of five genes contributes independently (not by interacting with each another directly) to the organism surpassing an inherent biological threshold. (B) In the heterogeneity model, either one or another set of genes (but not both) is required to surpass threshold. (C) In the interactive (epistatic) model, two genes directly interact with one another to surpass threshold. (D) In the combined model, probably the closest to real life, both epistasis and heterogeneity are involved.

1. For recessive phenotypes, both copies of the "bad" allele must be inherited for the abnormality to show; if dominant (or semidominant), one allele will suffice.
2. Encoding the full susceptibility at a locus does not always guarantee that a phenotypic abnormality will show. Some traits are fully penetrant, others are incompletely penetrant. In general, the most common neurobehavioral disorders are thought to be comprised of loci that, by themselves, are not fully penetrant.
3. Although penetrance can be the result of environmental factors, for complex traits the penetrance of a susceptibility locus segregating in a population usually depends on the presence of other susceptibility or modifier loci which are also segregating. The nature of this dependence among multiple loci may vary, and can range from relatively simple (Figure 2.1A) to very complex (Figure 2.1D).

   In the context of the last caveat, it is also important to recognize that the major difference between monogenic traits and polygenic traits — and the main practical advantage of the former — is not that monogenic traits do not depend on such complex interactions (because they do), but rather that the interacting loci do not vary in their allelic forms.

2.5 APPLICATIONS OF SEGREGATION AND LINKAGE ANALYSIS

Segregation and linkage analyses are important for (a) stock maintenance, (b) gross chromosomal mapping, (c) strain background effects, and (d) high-resolution mapping. Rather than review the detailed logistics of segregation mapping and linkage analysis of single-locus neurological mutations in mice which can be found elsewhere,[33] it is more useful here to review the role of segregation and linkage analysis in these applications.

2.5.1 STOCK MAINTENANCE

Careful segregation analysis during maintenance of a mutant allows one to determine efficiently whether a phenotype is purely recessive, semidominant, or dominant, and also to assess the degree of penetrance. Although the approach itself is quite simple, when initially establishing a mode of inheritance, it is extremely helpful to utilize mating partners that are of known genotype. Thus, consider a mutant mouse arising from a mating between normal parents (each of whom presumably carries a recessive allele). One is tempted to mate it initially with either the opposite sex parent or an unaffected sibling, in order to determine whether the mutation is heritable (a parent would, by definition be a carrier — as would two thirds of the unaffected siblings). However, if there is other evidence for heritability (e.g., other affected siblings), it is more efficient to assume heritability for the moment and to instead mate it to a completely unrelated member of the same strain (i.e., wild-type, or +/+). Thus the resultant progeny will be, at best, heterozygous for the mutation and if affected, the phenotype may be unambiguously declared semidominant or dominant. Testing penetrance becomes a simple matter of examining the number of affected F_1 progeny over total; the accuracy of this test being directly related to sample size. If the progeny are unaffected, recessivity may be tested by crossing them *inter se*.

2.5.2 GROSS LINKAGE MAPPING

Typically, the next most useful segregation analysis is some type of linkage mapping. As discussed by Frankel and Taylor,[33] the choice of type of cross (intercross or backcross) depends partly on the mode of inheritance and partly whether affected mutant mice can reproduce. Indeed, often an intercross is required since mice that are homozygous for many of the classical neurological mutations are not capable of mating. These mice must be produced by crosses between heterozygotes or by some type of specialized gamete transfer, e.g., ovary transplantation or *in vitro* fertilization. Regardless, a population of anywhere from 20 to 100 progeny are bred and examined for phenotype, and a small amount of DNA is prepared (usually excised tail tips) for a genome-wide scan using a genetic marker system of choice — the logistics of which have been considered previously.[33] When an unambiguous linkage is identified, markers can be selected that are on each side (centromeric, telomeric) of the mutation. A good set of flanking markers can serve as surrogates for many further genetic manipulations, including crossing the mutation to a given strain background, e.g., to look for phenotypic modifier loci, and for higher resolution mapping. Such uses of surrogate markers make a bigger difference for recessive phenotypes or for dominant phenotypes with incomplete penetrance, than for fully penetrant dominant, visible phenotypes which are evident in every generation.

2.5.3 STRAIN BACKGROUND MODIFIERS

One of the trendiest facets of modern mouse neurogenetics is the pursuit of strain background effects on phenotypes conferred by mutations. Thus, mutations that are fully penetrant or show a certain form of expression in one mouse strain may have modified phenotypes when transferred to other strains. This modification is the result of interactions between the mutation and natural variants among mouse strains. Whether these modifications are the result of one, a few, or many modifier genes depends greatly on the mutation and the phenotype. Some of these modifier genes may themselves be susceptibility alleles. A good example of this is the hearing loss of heterozygous deaf-waddler mice, the age-dependent susceptibility to which is modified by a locus, *Mdfw*, on chromosome 10.[34] Interestingly, in an independent study, the major quantitative trait locus (QTL) influencing strain susceptibility to age-dependent hearing loss, *Ahl1*, was found to map to the same small region of the genome and its strain distribution was consistent with it being allelic with *Mdfw*.[35] Approaches developed to pursue strain background effects of spontaneous mutations will be invaluable tools for gaining insight into the function of new mutants such as those derived from large-scale mutagenesis studies.

2.5.4 HIGH-RESOLUTION MAPPING OF MUTATIONS

The ultimate goal of linkage mapping is to identify the underlying gene and its defect in a given mutant mouse. From there, the so-called "real biology" takes over — the genocentric premise being that its value is maximized with the actual gene in hand. Increasingly, the "game" in trait-locus identification, whether spontaneous or chemically induced or polygenic mutations, becomes one of candidate gene-testing. Thus, with the full mouse genome sequence and annotation imminent and gene expression patterns not far behind, there is less of a burden on high-resolution mapping to incriminate a candidate gene and more on gaining as much incriminating information as possible from a low-resolution map to make "educated guesses" about which candidate genes to test. Although the ideal level of resolution is still under debate, it is difficult to argue against high-resolution mapping when there are sufficient animal breeding resources. The procedure to get the map location down to a few cM or less is a standard one and can best be summarized by "recombinants first, then phenotype." Thus, once a mutation has been mapped-down to 10% recombination or less ($\approx$10 cM), using a conventional cross with all samples phenotyped and genotyped, to refine map position further, it is more efficient to perform selective phenotyping of mice carrying recombinant chromosomes. Mice that carry crossovers which flank a mutation are found when the genotype of two flanking markers is of the recombinant (and not parental) haplotype (Figure 2.2). The phenotype is ascertained from further testing the progeny of mice carrying only recombinant haplotypes — resulting in a 90% or greater savings (i.e., for markers that are 10 cM apart, 90% of the haplotypes are nonrecombinant). This form of selective phenotyping is used frequently in F_2 progeny that carry recessive mutations (thus, in principle three fourths of the progeny are potentially informative — the homozygotes and the heterozygotes, taking full advantage of the ability to detect recombinations inherited from both F_1 parents), and also for N_2 progeny carrying mutations that are not fully penetrant. Thus, the phenotype of the mice initially ascertained to have a recombination is less important than that of the progeny.

2.5.5 WHERE BAC RESCUE COMES IN

High-resolution mapping need not depend only upon meiotic recombinants. With advances in large-insert recombinant DNA vectors, such as bacterial artificial chromosomes (BAC), a new form of physical mapping has emerged in mouse whereby BAC transgenic complementation can localize a gene to a region of less than 200 kb (Figure 2.3). Thus, one might map a mutation to within 2 Mb which is spanned by 12 overlapping BAC total (200 kb each) from a wild-type mouse strain. One would then make transgenic lines containing these BAC (one transgenic can easily harbor 2–3 independent clones); the line which complements the mutation must have the BAC that contains the gene of interest. This approach is most useful for some recessive mutations where it is difficult to achieve a high resolution map in short time but a physical map of the region already exists, or to accompany mapping crosses where there are "cold spots" of recombination. This type of approach, either using recombinant BAC or bacteriophage P1 clones, has now been used in the positional cloning of several mouse mutants (Clock,[36] vibrator,[37] shaker-2,[12] ashen,[38] pudgy[39] — also, see Camper and Saunders[40] for a review chapter on the subject of large DNA fragment transgenesis).

2.5.6 CANDIDATE GENE TESTING

The last stage in this analysis is actually finding a mutation in a candidate gene from a critical interval and being certain that this mutation is causally associated with the phenotypic abnormality. Although biochemical or physiological information about a gene's function together with the mutant mouse's abnormality can often be enough to "convince" an eager researcher of causality, it should be recognized that some physiological processes are so complex that plausible explanations can be made for a very large variety of molecules; if the critical interval is large enough to contain many genes, false leads may arise by chance alone. Therefore, ideally one would like to indict a candidate gene with overwhelming genetic evidence. Furthermore, one would like to have this evidence from existing resources, to spare the expense of new experiments (e.g., gene targeting) designed merely to prove this point. The following list represents a view of the strength of certain kinds of circumstantial evidence — ranked in descending order of their effectiveness:

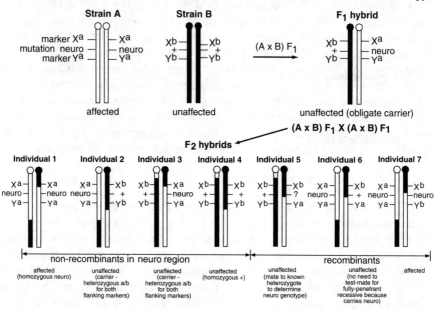

FIGURE 2.2 Use of flanking markers as surrogates to infer a recessive mutation's genotype. Shown is a single pair of chromosomes, whereby a recessive mutation ("*neuro*") is carried on Strain A (white chromosomes) and Strain B has only the wild-type ("+") allele. Flanking markers "*X*" and "*Y*" have distinctive "*a*" or "*b*" alleles from the respective strain parent. The F_1 hybrid is unaffected, but is an obligate carrier of the mutation. In this example, two F_1 hybrids are mated *inter se* to produce a collection of F_2 individuals in which the neuro mutation is segregating. These F_2 may also carry recombinations (occurring in the F_1) on one or the other homologue. The first set are F_2 that are "nonrecombinant" in the *neuro* locus region. One is affected (Individual 1), and thus homozygous *neuro/neuro*, and while the others are unaffected their *neuro* locus genotypes can be inferred with good reliability by examining the genotypes of markers X and Y and assuming that double-crossovers have not occurred. The second set contains recombinations within the interval defining the location of *neuro*. For fully penetrant recessive mutations, the genotype of Individual 7 is intuitively obvious (*neuro/neuro*) because it is affected (if the phenotype were not fully penetrant, a test-mating would be required). Furthermore, while the *neuro* genotype of Individual 6 is not immediately obvious, from the flanking marker genotypes we infer that it must have one copy of *neuro* and therefore if the phenotype is fully penetrant but the individual is unaffected, the other copy must be the + allele. However, the genotype of Individual 5 is unknown. We know that it must carry one copy of the + allele because of flanking marker genotypes, but to determine the genotype of the other homologue, it needs to be tested to a known carrier. If there are affected individuals in the next generation, it must have been heterozygous for *neuro*; if there are no affected individuals after screening an appreciable number to ensure transmission of the test chromosome, it must have been +/+ genotype.

Use of BAC Transgenic Complementation to Refine a Genetic Map

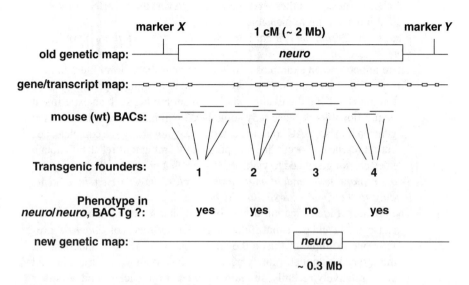

FIGURE 2.3 Use of BAC transgenic complementation to refine a genetic map. The critical interval of a recessive trait locus "*neuro*" is shown — first, after a high-resolution genetic map, and second, after the map has been refined based on a transgenic complementation with BAC that contain the wild-type allele of the underlying gene. Thus, in genotypically *neuro/neuro* homozygous mice, the only mice showing a phenotype will be those carrying BAC that do not have the correct gene.

1. *Only a single mutant allele exists, but the wild-type chromosome (mouse strain background) upon which this mutation arose is known and lacks the mutation.*

   An example of such a *coisogenic* mutation is the slow-wave epilepsy mutant mouse model for childhood absence epilepsy and cerebellar ataxia. The mutation arose in 1993 in the SJL/J strain at JAX and is a single nucleotide substitution in the ubiquitous sodium–hydrogen exchanger gene (*Slc9a1*, a.k.a. *Nhe1*) creating a premature stop codon and effectively a null phenotype.[10]

   Caveat: Coisogenic mutations are tough ones to beat, although we must remember the evidence is circumstantial. That is, it is formally possible (albeit very improbable because of the low spontaneous mutation rate) that a second "dummy" mutation with nothing to do with the phenotype exists in close linkage to the real mutation. Thus together with anything else that links the gene to the phenotype (e.g., expression pattern) and the severity of the mutation (e.g., predictive of a null phenotype), it can be quite convincing. In the case of *Nhe1*, the ubiquitous expression pattern

was not particularly helpful (although the relatively normal nonneuro-
logical phenotype of these mutants can be explained by redundancy in
other tissues, i.e., other NHE genes), however, the severity of the mutation made it more plausible.

2. *A single BAC from the critical interval rescues the mutation and only one gene resides on the BAC.*

   There are no known examples of this in the neurobehavioral literature.

   Caveat: Complementation by BAC containing a single expressed gene is
   tough to beat, but the likelihood of even attempting such an experiment
   is questionable, because only some BAC will have only one expressed
   gene on it. Also, BAC complementation may not always be complete, i.e.,
   partial rescue can occur, leaving potentially ambiguous results if enough
   lines are not generated to yield at least one that provides full rescue.

3. *Two or more independent mutations (alleles) have arisen in a gene, leading to the same or a similar phenotype.*

   The positional identification of the tottering and leaner mutations, which
   cause ataxia and petit-mal epilepsy in mice, as alleles of *Cacna1a*[8] provide one example of this, although probably having two independent
   mutations was not absolutely necessary since both were intragenic and
   arose relatively recently on known inbred strain background. Another
   example is of the shaker-1 missense mutation in the *Myo7a* gene,[11]
   causing deafness and imbalance, which occurred in the Bagg albino
   strain many years ago — a strain which is no longer in existence but
   which was one of the founders of modern strains including BALB/c. To
   be completely certain of the gene identity, it was very important to have
   independent mutations in the same gene — one of these was another
   spontaneous missense mutation which occurred recently in a known
   strain.[41]

   Caveat: Multiple, independent alleles also provide very good evidence,
   although for genes that are frequent targets for mutations that are not
   necessarily intragenic, i.e., large deletions or retro-transposon insertions, the altered expression of neighboring genes could formerly have
   been responsible for the phenotype.

4. *The high-resolution map for a mutation is so refined that only a single gene remains in the critical interval.*

   Again there are not yet any known examples of this in the neurobehavioral
   literature, although there are several examples of mapping crosses
   yielding intragenic recombinants on one side or the other of a locus
   (e.g., *Cacng2*,[7] *Fign*[42]); in such cases other alleles are usually brought
   in to ensure the gene's identity.

   Caveat: Completely exclusionary genetic mapping would seem tough to
   beat, but some mutations (large deletions, insertions) can have long-
   range action and might affect the expression of neighboring genes. It
   should be remembered that the genetic map is of the mutation itself, not
   necessarily the gene in which the mutation resides. Thus, a critical interval

produced in a high-resolution map could, in principle, exclude all coding sequence if the mutation was regulatory and resided in a large intron or in intergenic space.

5. *The protein or mRNA expression is altered in mutant mice but a mutation has not yet been found in the underlying gene.*

   There are many examples of this in the literature, as this type of result often provides the first clue for a candidate mutation — remembering that the ultimate effect of many causal mutations is on the amount or form of protein or mRNA. However, few cases appear in the literature with only this type of evidence.

   Caveat: Sufficient evidence exists for effects of gene defects on elements "downstream" in a pathway to say that a difference in gene or protein expression is hardly convincing evidence on its own. Another caveat is the fact that apparently downstream gene expression differences may arise when certain mutations which may affect large numbers of a cell type from a tissue (e.g., cerebellar Purkinje cells in a mouse with degenerative ataxia), would obviously reduce the expression of cell-type specific molecules.

2.6 FUTURE DIRECTIONS FOR SPONTANEOUS NEUROLOGICAL MUTATIONS: PASSING THE TORCH

Spontaneous mouse mutations have certainly played a critical historical role in the appreciation of how genetic defects can influence the development, integrity, and normal function of the mammalian nervous system and associated disorders in animal models of human disease. They allowed the community to move from an era in which the role of genes in neurology and behavior was questioned outright, into one where most neuroscientists now understand that a gene's role is either central to their interests, or at the very least provides a building block upon which neuronal plasticity and response to environment would later serve to fine-tune nervous system function. In recent years, positional gene identification of spontaneous mutants has joined man-made knockouts and other forms of trangenesis to correlate *specific* molecules to aspects of nervous system function, from which instant insight into their function could be obtained in many cases. From this process of phenotype-driven gene discovery, however, spontaneous mutants provided the ultimate sacrifice: they were both the incentive and a model for conceiving and automating large-scale, systematic efforts, such as chemical mutagenesis, whereby the mutation rate is increased some 1000-fold over the spontaneous rate (thus greatly increasing the chance that investigators will find mutants of interest to them). Thus, spontaneous mutants may have already made their major contribution on biomedical research. Nevertheless, the empirical lessons that spontaneous mutants have provided into the rates and consequence of various types of mutation, effects of strain background, search for candidate genes and mutations, and where to focus attention once genes are positionally identified should make for a very nice epitaph.

REFERENCES

1. Lander, E. S. and Botstein, D., Mapping Mendelian factors underlying quantitative traits using RFLP linkage maps, *Genetics,* 121, 185–199, 1989.
2. Ackerman, S. L., Kozak, L. P., Przyborski, S. A., Rund, L. A., Boyer, B. B., and Knowles, B. B., The mouse rostral cerebellar malformation gene encodes an UNC-5-like protein, *Nature,* 386, 838–842, 1997.
3. Sheldon, M., Rice, D. S., D'Arcangelo, G., Yoneshima, H., Nakajima, K., Mikoshiba, K., Howell, B. W., Cooper, J. A., Goldowitz, D., and Curran, T., Scrambler and yotari disrupt the disabled gene and produce a reeler-like phenotype in mice [see comments], *Nature,* 389, 730–733, 1997.
4. Zuo, J., De Jager, P. L., Takahashi, K. A., Jiang, W., Linden, D. J., and Heintz, N., Neurodegeneration in Lurcher mice caused by mutation in delta2 glutamate receptor gene [see comments], *Nature,* 388, 769–773, 1997.
5. Hamilton, B. A., Frankel, W. N., Kerrebrock, A. W., Hawkins, T. L., FitzHugh, W., Kusumi, K., Russell, L. B., Mueller, K. L., van Berkel, V., Birren, B. W., Kruglyak, L., and Lander, E. S., Disruption of the nuclear hormone receptor RORalpha in staggerer mice, *Nature,* 379, 736–739, 1996.
6. Thomas, K. R., Musci, T. S., Neumann, P. E., and Capecchi, M. R., Swaying is a mutant allele of the proto-oncogene *Wnt-1, Cell,* 67, 969–976, 1991.
7. Letts, V. A., Felix, R., Biddlecome, G. H., Arikkath, J., Mahaffey, C. L., Valenzuela, A., Bartlett II, F. S., Mori, Y., Campbell, K. P., and Frankel, W. N., The mouse stargazer gene encodes a neuronal Ca2+ channel γ subunit, *Nat. Genet.,* 19, 340–347, 1998.
8. Fletcher, C. F., Lutz, C. M., O'Sullivan, T. N., Shaughnessy, Jr., J. D., Hawkes, R., Frankel, W. N., Copeland, N. G., and Jenkins, N. A., Absence epilepsy in tottering mutant mice is associated with calcium channel defects, *Cell,* 87, 607–617, 1996.
9. Burgess, D. L., Jones, J. M., Meisler, M. H., and Noebels, J. L., Mutation of the Ca^{2+} channel β subunit gene *Cchb4* is associated with ataxia and seizures in the lethargic (*lh*) mouse, *Cell,* 88, 385–392, 1997.
10. Cox, G. A., Lutz, C. M., Yang, C.-L., Biemesderfer, D., Bronson, R. T., Fu, A., Aronson, P. S., Noebels, J. L., and Frankel, W. N., Sodium/hydrogen exchanger gene defect in slow-wave epilepsy mutant mice, *Cell,* 91, 139–148, 1997.
11. Gibson, F., Walsh, J., Mburu, P., Varela, A., Brown, K. A., Antonio, M., Beisel, K. W., Steel, K. P., and Brown, S. D., A type VII myosin encoded by the mouse deafness gene shaker-1, *Nature,* 374, 62–64, 1995.
12. Probst, F. J., Fridell, R. A., Raphael, Y., Saunders, T. L., Wang, A., Liang, Y., Morell, R. J., Touchman, J. W., Lyons, R. H., Noben-Trauth, K., Friedman, T. B., and Camper, S. A., Correction of deafness in shaker-2 mice by an unconventional myosin in a BAC transgene [see comments], *Science,* 280, 1444–1447, 1998.
13. Avraham, K. B., Hasson, T., Steel, K. P., Kingsley, D. M., Russell, L. B., Mooseker, M. S., Copeland, N. G., and Jenkins, N. A., The mouse Snell's waltzer deafness gene encodes an unconventional myosin required for structural integrity of inner ear hair cells, *Nat. Genet.,* 11, 369–375, 1995.
14. Bowes, C., Danciger, M., Kozak, C. A., and Farber, D. B., Isolation of a candidate cDNA for the gene causing retinal degeneration in the rd mouse, *Proc. Natl. Acad. Sci. USA,* 86, 9722–9726, 1989.
15. Burmeister, M., Novak, J., Liang, M.-Y., Basu, S., Ploder, L., Hawes, N. L., Vidgen, D., Hoover, F., Goldman, D., Kalnins, V. I., Roderick, T. H., Taylor, B. A., Hankin, M. H., and McInnes, R. R., Ocular retardation mouse caused by *Chx10* homeobox null allele: Impaired retinal progenitor proliferation and bipolar cell differentiation, *Nat. Genet.,* 12, 376–383, 1996.

16. Naggert, J. K., Fricker, L. D., Varlamov, O., Nishina, P. M., Rouille, Y., Steiner, D. F., Carroll, R. J., Paigen, B. J., and Leiter, E. H., Hyperproinsulinaemia in obese *fat/fat* mice associated with a carboxypeptidase E mutation which reduces enzyme activity, *Nat. Genet.,* 10, 135–142, 1995.

17. D'Arcangelo, G., Miao, G. G., Chen, S. C., Soares, H. D., Morgan, J. I., and Curran, T., A protein related to extracellular matrix proteins deleted in the mouse mutant reeler [see comments], *Nature,* 374, 719–723, 1995.

18. Zhang, Y., Proenca, R., Maffel, M., Barone, M., Leopold, L., and Friedman, J. M., Positional cloning of the mouse obese gene and its human homologue, *Nature,* 372, 425–432, 1994.

19. Chen, H., Charlat, O., Tartaglia, L. A., Woolf, E. A., Weng, X., Ellis, S. J., Lakey, N. D., Culpepper, J., Moore, K. J., Breitbart, R. E., Duyk, G. M., Tepper, R. I., and Morgenstern, J. P., Evidence that the diabetes gene encodes the leptin receptor: Identification of a mutation in the leptin receptor gene in db/db mice, *Cell,* 84, 491–495, 1996.

20. Noben-Trauth, K., Naggert, J. K., North, M. A., and Nishina, P. M., A candidate gene for the mouse mutation tubby, *Nature,* 380, 534–538, 1996.

21. Melvold, R. W., Wang, K., and Kohn, H. I., Histocompatibility gene mutation rates in the mouse: A 25 year review, *Immunogenetics,* 47, 44–54, 1997.

22. Russell, L. B. and Russell, W. L., Frequency and nature of specific-locus mutations induced in female mice by radiations and chemicals: A review, *Mutat. Res.,* 296, 107–127, 1992.

23. Pease, L. R., Schulze, D. H., Pfaffenbach, G. M., and Nathenson, S. G., Spontaneous H-2 mutants provide evidence that a copy mechanism analogous to gene conversion generates polymorphism in the major histocompatibility complex, *Proc. Natl. Acad. Sci. USA,* 80, 242–246, 1983.

24. Yun, T. J., Melvold, R. W., and Pease, L. R., A complex major histocompatibility complex D locus variant generated by an unusual recombination mechanism in mice, *Proc. Natl. Acad. Sci. USA,* 94, 1384–1389, 1997.

25. Fletcher, C. F., Copeland, N. G., and Jenkins, N. A., Genetic analysis of voltage-dependent calcium channels, *J. Bioenerg. Biomembr.,* 30, 387–398, 1998.

26. Mori, Y., Wakamori, M., Oda, S., Fletcher, C. F., Sekiguchi, N., Mori, E., Copeland, N. G., Jenkins, N. A., Matsushita, K., Matsuyama, Z., and Imoto, K., Reduced voltage sensitivity of activation of P/Q-type Ca2+ channels is associated with the ataxic mouse mutation rolling Nagoya (tg(rol)), *J. Neurosci.,* 20, 5654–5662, 2000.

26a. Letts, V. A., Lutz, C. M., and Frankel, W. N., unpublished results.

26b. Letts, V. A. and Frankel, W. N., unpublished results.

27. D'Arcangelo, G. and Curran, T., Reeler: New tales on an old mutant mouse, *Bioessays,* 20, 235–244, 1998.

28. Curran, T. and D'Arcangelo, G., Role of reelin in the control of brain development, *Brain Res. Rev.,* 26, 285–294, 1998.

29. Cox, G. A., Mahaffey, C. L., and Frankel, W. N., Identification of the mouse neuromuscular degeneration gene and mapping of a second site suppressor allele, *Neuron,* 21, 1327–1337, 1998.

30. King, D. P., Zhao, Y., Sangoram, A. M., Wilsbacher, L. D., Tanaka, M., Antoch, M. P., Steeves, T. D., Vitaterna, M. H., Kornhauser, J. M., Lowrey, P. L., Turek, F. W., and Takahashi, J. S., Positional cloning of the mouse circadian clock gene, *Cell,* 89, 641–653, 1997.

31. Sidman, R. L., Green, M. C., and Appel, S. H., *Catalog of the Neurological Mutants of the Mouse,* Harvard University Press, Cambridge, MA, 1965.

32. Noebels, J. L. Mutational analysis of inherited epilepsies. In *Basic Mechanisms of the Epilepsies: Molecular and Cellular Approaches,* Vol. 44, A. V. Delgado-Escueta, A. A. Ward, D. M. Woodbury, and R. J. Porter, Eds., Raven Press, New York, pp. 97–113, 1986.

33. Frankel, W. N. and Taylor, B. A. Mapping single locus mutations in mice: Towards gene identification of neurological traits. In *Handbook of Molecular Genetic Techniques for Brain Behavior Research,* W. E. Crusio and R. T. Gerlai, Eds., Elsevier, Amsterdam, pp. 62–81, 1999.

34. Noben-Trauth, K., Zheng, Q.-Y., Johnson, K. R., and Nishina, P. M., *mdfw*: A deafness susceptibility locus that interacts with deaf waddler (*dfw*), *Genomics,* 44, 266–272, 1997.

35. Johnson, K. R., Erway, L. C., Cook, S. A., Willott, J. F., and Zheng, Q. Y., A major gene affecting age-related hearing loss in C57BL/6J mice, *Hear Res.,* 114, 83–92, 1997.

36. Antoch, M. P., Song, E. J., Chang, A. M., Vitaterna, M. H., Zhao, Y., Wilsbacher, L. D., Sangoram, A. M., King, D. P., Pinto, L. H., and Takahashi, J. S., Functional identification of the mouse circadian clock gene by transgenic BAC rescue, *Cell,* 89, 655–667, 1997.

37. Hamilton, B. A., Smith, D. J., Mueller, K. L., Kerrebrock, A. W., Bronson, R. T., van Berkel, V., Daly, M. J., Kruglyak, L., Reeve, M. P., Nemhauser, J. L., Hawkins, T. L., Rubin, E. M., and Lander, E. S., The vibrator mutation causes neurodegeneration via reduced expression of PITP alpha: Positional complementation cloning and extragenic suppression, *Neuron,* 18, 711–722, 1997.

38. Wilson, S. M., Yip, R., Swing, D. A., O'Sullivan, T. N., Zhang, Y., Novak, E. K., Swank, R. T., Russell, L. B., Copeland, N. G., and Jenkins, N. A., A mutation in Rab27a causes the vesicle transport defects observed in ashen mice, *Proc. Natl. Acad. Sci. USA,* 97, 7933–7938, 2000.

39. Kusumi, K., Sun, E. S., Kerrebrock, A. W., Bronson, R. T., Chi, D.-C., Bulotsky, M. S., Spencer, J. B., Birren, B. W., Frankel, W. N., and Lander, E. S., The mouse pudgy mutation disrupts *Delta* homologue *Dll3* and initiation of early somite boundaries, *Nat. Genet.,* 19, 274–278, 1998.

40. Camper, S. A. and Saunders, T. L. Transgenic rescue of mutant phenotypes using large insert DNA fragments. In *Genetic Manipulation of Receptor Expression Function,* D. Accili, Ed., Wiley-Liss, Inc., New York, pp. 1–22, 2000.

41. Letts, V. A., Gervais, J. L. M., and Frankel, W. N., Remutation of the shaker-1 locus, *Mouse Genome,* 92, 116, 1994.

42. Cox, G. A., Mahaffey, C. L., Nystuen, A., Letts, V. A., and Frankel, W. N., The mouse fidgetin gene defines a new role for AAA family proteins in mammalian development, *Nat. Genet.,* 26, 198–202, 2000.

3 Conditional and Inducible Gene Targeting in the Nervous System

Mark Mayford and Eric Kandel

CONTENTS

3.1 INTRODUCTION

The genomes of both mice and humans consist of about 35,000 to 100,000 different genes, not accounting for possible splice variants. With this seemingly limited repertoire of molecular components, both mice and human are able to develop a functioning nervous system capable of the vast array of mental abilities with which we are all familiar. Our knowledge of the molecular mechanisms by which the mammalian brain develops and functions in the adult has progressed rapidly in the past decade in part due to technological advances in our availability to manipulate the mouse genetically. These methods are referred to as reverse genetic approaches.

In contrast to classical forward genetics where one starts with or screens for a phenotypic variant and then tracks down the genetic variation that produces it, in reverse genetics we begin with a gene of interest and generate animals, mice in this case, in which the gene is altered. The mouse is then examined for the phenotypic consequences, if any, of the genetic manipulation.

The reverse genetic approach is designed to ask the question: What is the role of a given gene in the animal? Starting with a known gene, one can generate a mouse either lacking the gene, expressing excessive levels of the gene product, or expressing a particular mutant variety of the gene. By examining the impact of these genetic manipulations on organismal function, one attempts to infer the role of the particular gene in the animal. While this may seem rather straightforward, it can, in fact, be quite difficult to obtain a meaningful interpretation of a genetically modified mouse. The primary difficulty is that most mutations are pleiotropic; they produce more than one phenotypic effect. For example, if you are interested in the function of a particular receptor in the adult brain that is also necessary for the development of the heart, the simple knockout will be lethal and no information will be gained on the neuronal function of that gene. Even if expression of the gene of interest is limited to the brain, it is usually present in diverse cells of different brain regions subserving different functions. For example, when analyzing complex behaviors such as learning and memory, it often is difficult to determine whether the phenotype is caused by a direct affect on cellular mechanisms of learning or due to indirect affects on sensory, motor, or motivational processes that can interfere with learning and memory. In addition, using standard knockout and transgenic techniques, it can also be difficult to determine whether the phenotype results from a direct requirement for the gene product in the adult animal or whether it results from an indirect affect of the gene on neural development.

These difficulties have led researchers to develop a second generation of techniques for genetic modification of the mouse. These new techniques have two advantages. One, they allow gene knockouts to be restricted so that deletion occurs only in specific brain areas. Two, they allow the temporal control of gene expression and gene deletion such that genetic manipulations can be made only in the adult brain. These two new techniques alleviate some of the difficulty in interpreting the results from genetically modified mice as they relate to complex behaviors. In this chapter we review the current status of the technology for alteration of genes in the brain of the mouse. We focus on the background and basic principles of transgenic mouse design and the interpretation of neurobiological and behavioral data. We will not discuss the basic techniques in mouse embryology and stem cell culture that are required for these genetic manipulations since these are often provided commercially and through CORE facilities and are reviewed extensively elsewhere.[1,2]

3.2 DETAILED METHODS

3.2.1 TRANSGENESIS

Transgenesis involves the introduction of a novel piece of DNA into the germ line cells of an animal. Once integrated, the transgene is transmitted to subsequent generations in a Mendelian fashion with 50% of offspring carrying the transgene and 50% lacking the transgene. The current methods for producing transgenic animals involves the microinjection of naked DNA into the pronucleus of a fertilized egg.[2] With this method, the experimenter has no control over the position in the genome at which the transgene integrates. In addition, the transgene usually integrates as multiple copies and both the levels and pattern of expression are sensitive to the integration locus.

The design of DNA constructs for the production of transgenic animals requires at least two components, a promoter element to drive transgene expression and the gene of interest, which is often a cDNA. When considering the selection of a promoter element one must consider when the transgene should be expressed, e.g., early or late in development, where in the brain and in what cell types the desired transgene should be expressed, and what levels of expression are required to obtain a functional effect. We know a great deal about the function of different anatomical structures in the brain, primarily through studies of lesions in neurological patients and experimental animals. Ideally, if one was interested in emotional responses mediated by the amygdala, for example, transgenic animals would be generated using a promoter that directed expression specifically to this structure. Unfortunately, the current repertoire of available brain-specific promoters is quite limited in terms of the cellular and anatomical restriction that can be obtained.

Once the appropriate promoter has been selected, generation of a transgenic construct involves straightforward molecular cloning techniques in which the gene to be expressed is placed downstream of the promoter element as shown in Figure 3.1A. When expressing a cDNA, it is necessary, at a minimum, to include a polyadenylation signal at the 3' end of the construct. In addition, the introduction of artificial introns, at both the 5' and 3' end of the transcript, recruit the splicing machinery, and can often enhance the levels of expression obtained.[3]

In theory one should be able to isolate promoter elements that recapitulate the expression pattern of any endogenous gene. Unfortunately, we are still lacking a complete understanding of the sequence elements that control the developmental onset, cell type specific patterns, and absolute level of expression of genes in the mouse. In general, one picks as a promoter a large (several Kbp) portion of genomic DNA upstream from the transcription start site of the gene of interest. This putative promoter element is then fused to a reporter gene and transgenic animals are produced by standard oocyte injection. Founder animals that arise from the injected oocytes, and carry the transgene, usually contain multiple copies of the transgene integrated at a single site in the genome. The multiple copies are integrated in series in a head-to-tail fashion. In some cases, the transgene will integrate in multiple copies and at multiple sites in the genome. This can usually be detected by careful southern blot analysis of the F_1 generation of mice arising from breeding of the founder animal. These are essentially two separate lines of transgenic mice and the multiple different integration loci should be bred apart from each other and maintained separately.

In many cases, two or three Kbp of genomic DNA upstream of the coding region will not carry sufficient regulatory elements to fully confer the desired patterns of transgene expression. Therefore, some investigators have resorted to using transgenic

Transgenic Animal

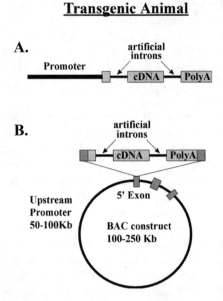

FIGURE 3.1 Transgenisis. (A) A standard transgenic construct includes a promoter element that normally consists of from 1 to 10 Kbp of genomic DNA upstream from and including the transcription start site. The cDNA to be expressed is flanked by a 5′ and 3′ intron and a polyadenylation signal. (B) To include larger promoter regions, a bacterial artificial chromosome (BAC) construct can be generated. The larger construct is more likely to recapitulate the endogenous pattern of gene expression.

constructs that are significantly larger than several Kbp, often generated using bacterial artificial chromosomes (BACs). BACs are constructs that can carry over 200 kbp of DNA. These constructs can be generated using homologous recombination in *Escherichia coli,* which allows precise manipulation of the BAC.[4] Generally, one obtains a BAC clone carrying the full genomic DNA of the gene whose pattern of expression one is trying to recapitulate in a transgenic animal. A cDNA for a reporter or effector gene is then inserted, using homologous recombination, into the first 5′ untranslated exon of the BAC construct as shown in Figure 3.1B. Transgenic animals are then generated using microinjection of the BAC, essentially as with a standard transgenic construct. The significantly larger fragments of promoter DNA that can be included in a BAC construct increase the likelihood that this type of construct will confer the expression pattern of the endogenous gene[5] and be less sensitive to the integration site of the transgene. In addition, BAC constructs can be useful for simple over expression of particular genes. For example, Takahashi used BAC transgenics to rescue the phenotype generated by the *clock* mutation, which was obtained through ENU mutagenesis.[6] The BAC transgene expressed sufficient clock protein to rescue the phenotype observed in the *clock* mutant mice. This use of BAC transgenics will likely become more common as the number of mutant animals arising from chemical mutagenesis programs increases.

3.2.2 GENE TARGETING (KNOCKOUTS)

While transgenic techniques are useful in investigating gene function, transgenics do not allow for the alteration or deletion of endogenous genes. In genetics, one often wants to ask in a first analysis: What is the phenotype of the null allele? What is the phenotype of an animal that completely lacks a particular gene product? This question requires the manipulation of endogenous genes rather than the introduction of exogenous transgenes. In the late 1980s, it became possible to generate targeted deletions or knockouts in specific genes using homologous recombination in mouse embryonic stem cells. By allowing the production of mice carrying null alleles in virtually any known gene, this technology has greatly expanded the use of mouse genetics in neurobiology.

The basic methodology for the generation of a knockout involves the alteration of the locus of interest in embryonic stem (ES) cells by homologous recombination.[1] The first step is the generation of a targeting construct that is to be inserted into the locus to be targeted. An example of such a construct is shown in Color Figure 3.2* which contains three key elements. First, the construct contains a large portion of genomic DNA derived from the locus to be targeted. While there are relatively few studies systematically investigating this variable, it is thought that longer stretches of genomic DNA favor homologous recombination.[7] In general one should include at least 6 Kbp of genomic DNA. Ideally, the genomic DNA is obtained from a library generated from the ES cell line used in the experiment, but at a minimum the DNA should be derived from a mouse strain that is isogenic with the ES cells used. This is critical since variation in DNA sequence between the targeting construct and the genomic locus will reduce significantly the efficiency of homologous recombination. One then has to decide on the nature of the mutation to be generated. Ideally, the entire gene coding sequence is deleted. However, because most genes are large and difficult to remove *in toto,* one tries more typically to delete an early coding exon so that no truncated protein products are synthesized in the knockout animal. In place of the first exon, one inserts a selectable marker such as a neomycin resistance gene carrying a weak promoter element of its own. This generates a simple targeting construct carrying positive selection for neomycin resistance and two long arms of chromosomal DNA homologous to the locus to be targeted.

An additional feature that is often included is a negative selection marker such as the Herpes Simplex thymidine kinase or the diphtheria toxin gene, which is placed distal to one of the arms of genomic DNA. When transfected into embryonic stem cells, the neomycin resistance marker is used to positively select transformants which picked up the targeting construct, and the negative selection marker is used to enrich for those cells that underwent a double recombination event so as to eliminate the marker as shown in the targeted locus of Color Figure 3.2. Those ES cell clones which carry the targeting vector integrated at the homologous site can be detected using southern blot analysis as shown in Color Figure 3.2B.* In some cases expression of the neomycin-resistance gene can itself produce phenotypic alterations.[8] Therefore, it is

---

* Color figures follow page 140.

now common practice to remove the neomycin-resistance gene using CRE recombinase as described in Section 3.2.3 on conditional knockouts.

Mouse ES cells are derived from early blastocyst stage embryos. They have the useful property that they can be grown as cell lines in culture while maintaining their pluripotency. Thus, ES cells that carry the targeted locus can be injected into wild-type blastocysts to generate an animal that is chimeric, with some cells arising from the wild-type blastocyst and some cells arising from the mutant ES cells. If the ES cells contribute to the production of germ line tissue, then breeding these animals will result in transmission of the mutant allele to 50% of the F_1 offspring that are derived from an ES cell gamete. This germ-line transmission is often detected using coat color indicators.

Once germ-line transmission is obtained, those animals will carry a single copy of the mutant locus. Interbreeding of two heterozygotes will generate "knockout" animals homozygous for the targeted locus. While the standard knockout approach has been extremely useful in understanding neural development and behavior, it is clear that it has a number of limitations which have led investigators to develop techniques for genetic manipulations that allow more anatomical control over the cell types in the brain which are altered as well as temporal control over the onset of the mutation.

3.2.3 CONDITIONAL KNOCKOUTS

The conventional knockout leads to the elimination of the gene of interest in every cell that expresses it, not only in the brain but in all organs of the body. The conditional knockout approach allows the gene of interest to be deleted specifically in the desired cell types. The most common strategy uses two different types of genetically modified mouse. In the first mouse, a cell type specific promoter is used to drive the expression of CRE recombinase. CRE recombinase is a sequence-specific recombinase that is able to recognize small 34 bp DNA elements called loxP sites and to initiate a recombination event between them.[9] If a particular segment of genomic DNA is flanked by loxP sites or "floxed," exposure to CRE recombinase will delete that segment of DNA. In the second mouse line, homologous recombination is used to flox the gene to be deleted. Typically, the LoxP sites are introduced into introns flanking an exon critical for the function of the gene. This is accomplished using a targeting construct containing three LoxP sites. Two sites flank the neomycin-resistance gene, which is inserted into an intronic region, while the third LoxP site is introduced into an upstream intron as shown in Color Figure 3.3A.* In this case, following the initial selection of recombinant ES cells, the neomycin-resistance gene remains in the intron. This can lead to phenotypic effects and therefore the neomycin-resistance gene is generally removed by transient transfection of the targeted ES cells with a plasmid expressing low levels of CRE recombinase. Three different types of recombination event can occur as shown in Color Figure 3.3A. In one case, the recombination occurs between the two distal LoxP sites (sites 1 and 3) and the entire neomycin-resistance gene plus the relevant exon

---

* Color figures follow page 140.

of the endogenous gene are deleted. These ES cells can be used to generate chimeric animals for production of a conventional knockout mouse. In the second case, recombination occurs between loxP sites 1 and 2 resulting in elimination of the relevant exon but not the neomycin-resistance gene. In the third case, recombination occurs between sites 2 and 3, which flank the neomycin-resistance gene. This recombination event leaves a single loxP site where the neomycin-resistance gene previously resided, resulting in an arrangement in which the only sequence alterations in the genomic DNA are the two loxP sites residing in introns. This should have a minimal effect on gene function in most cases, and mice generated carrying such a floxed locus should be essentially wild type. When mated to transgenic animals expressing the CRE recombinase from a cell type specific promoter, those cells that express CRE recombinase at sufficient levels will undergo a recombination event between the loxP sites (Color Figure 3.3B).* This results in the deletion of the relevant exon and generation of a functional knockout of the floxed gene. In the remaining tissues, the recombination event will not occur, and those cells will express a wild type version of the floxed gene.

This approach allows one to study the role of a particular gene in specific cell types or specific brain regions. It also allows investigation of the role of a gene that is lethal when deleted globally. While the conditional knockout approach provides anatomical specificity in the generation of mutant animals, it does not afford temporal control over the timing of the genetic modification. The timing is controlled by the pattern of expression of the CRE recombinase. For example, if the recombinase is active early in development, a knockout of the floxed gene will be generated in the CRE expressing cells and in all of the decedents of those cells. If the CRE recombinase is expressed from a promoter with delayed developmental expression, then the recombination event and onset of the gene deletion will be delayed to those later developmental time points. Thus, the technique affords some temporal control over the timing of gene deletion; however, this control is wholly governed by the expression pattern of the CRE recombinase. Ideally, one would like to have both anatomical and temporal control over genetic modification. This requires the use of an inducible system. One example of the use of an inducible system with transgenic animals is described in the next section.

3.2.4 INDUCIBLE TRANSGENICS

There are several inducible systems for regulating transgene expression that have been used successfully in cell culture.[10-12] We will discuss in detail the use of the tetracycline (TET) system, which has been applied successfully to study the brain in transgenic mice. Use of the TET system allows one to gain both anatomical and temporal control over the expression of a transgene. It is a binary system similar to that described for conditional knockouts using the CRE recombinase. The tTA molecule (commercially known as Tet-OFF) was developed by Herman Bujard and his colleagues and consists of the tetracycline repressor protein from *E. coli* fused to a VP16 transcriptional activation domain from the virus SV40.[10] The tetracycline

---

* Color figures follow page 140.

repressor binds to DNA sequences in the *E. coli* TET operator constitutively, but it will not bind when tetracycline is present. In eukaryotic cells, the VP16 domain confers transcriptional activation capacity on the protein. When a promoter is constructed with repeated iterations of the TET-operator sequences adjacent to a minimal transcription initiation site, the tTA will act as a tetracycline-suppressible transcription activator on these sequences. Thus, the TET operator minimal promoter element comprises a tetracycline-responsive element (TRE). When genes are fused downstream of the TRE, the expression can now be regulated in a tetracycline-dependent manner using the tTA transcription factor.

In mice, the use of the TET system to obtain anatomically restricted and temporally regulated transgene expression is outlined in Color Figure 3.4.* Two independent transgenic lines are produced using standard techniques.[13] In the first mouse, tTA is expressed from a cell type specific promoter. In general, this should not produce a phenotype since the TET repressor acts only on *E. coli* DNA sequences and should have minimal effects on the eukaryotic genome. However, at high levels in cell culture, the VP16 domain can have a generalized effect on cellular transcription perhaps by binding large amounts of the proteins involved in the basal transcription machinery.[14] In the second line of mouse, the TRE is linked to the transgene for which regulation is desired. In this second line, the TRE is inactive due to the absence of the tTA transcription factor. This mouse generally has a very low basal level of transgene expression. To obtain anatomically restricted and temporally regulated transgene expression, the two independent transgenes are introduced into the same mouse by mating. In double transgenic mice, those cells that express the tTA transcription factor will show activation of the TRE-linked transgene. Now, introduction of tetracycline analogs such as doxycycline into the animals' diet can be used to obtain suppression of the TRE-linked gene. However, in those cells not expressing tTA, the TRE-linked transgene is silent, both in the absence and the presence of tetracycline.

As shown in Color Figure 3.4B,* the TET system provides extremely tight regulation of transgene expression. In the absence of doxycycline, high-level expression of the TRE-linked transgene is obtained, while in the presence of doxycycline, the expression is suppressed to essentially background levels. One would often prefer to keep transgene expression suppressed during development and induced only in the adult animal. We have found that including doxycycline in the animals' food at 40 μg doxycycline per gram of mouse chow provides developmental suppression of TRE link transgene expression. Induction can be obtained within 1 to 2 days in the adult animal by switching to a nonmedicated food source.[14a]

An alternative to the standard tTA system for gene regulation, which allows gene induction in the presence of doxycycline, is also available.[11] The reverse tTA (rtTA) or TET-ON system is similar in all respects to the tTA system except that in this case addition of doxycycline induces TRE link gene expression. A series of mutations in the tetracycline repressor portion of the tTA molecule altered its DNA binding specificity such that it now bound to and activated transcription from TRE elements only in the presence of doxycycline. The rtTA system has been used successfully in the brain to obtain regulation of transgene expression.[15] One of the difficulties

---

* Color figures follow page 140.

that have been encountered with the rtTA system is that the inducibility of rtTA by doxycycline was reduced significantly by introduction of the mutations. It therefore requires significantly higher levels and prolonged treatments with doxycycline to obtain significant induction of TRE-linked transgenes using the rtTA system in the brain. However, a recently described rtTA mutant with an improved induction profile may alleviate some of these difficulties.[16]

3.2.5 INDUCIBLE KNOCKOUTS

It would be useful to obtain not only regulated transgene expression but also regulated gene knockout. In general, the approaches that have attempted to obtain inducible gene knockouts in the nervous system have employed techniques that make the activity of the CRE recombinase inducible. The combination of cell type specific CRE recombinase expression along with a floxed endogenous gene should in theory allow the inducible deletion of the floxed gene. Several groups have attempted to make the activity of the CRE recombinase protein itself responsive to exogenous ligands.

In one of the best examples, Tsujita et al. fused the ligand-binding domain of the human progesterone receptor to CRE recombinase.[17] The resulting chimeric protein showed recombinase activity only in the presence of synthetic progesterone antagonists. The chimeric progesterone CRE recombinase protein was expressed in transgenic animals using a promoter that limits expression to the cerebellar granule cells. When this transgene was introduced into mice carrying a floxed reporter locus, relatively low basal levels of CRE-mediated recombination were observed. However, upon injection of progesterone antagonists over several days, CRE-mediated recombination was observed at high efficiency in the cerebellar granule cells. While these experiments used a reporter gene to assess recombination efficiency, this technique should in theory be applicable to the inducible deletion of endogenous genes. While this approach looks promising, there remain several difficulties that are worth noting.

One difficulty is that many investigators have found that the receptor CRE fusion proteins show a significant amount of basal recombination. In many cases, significant gene deletion would be present even in the absence of an inducing stimulus.[18,19] A second problem that has been encountered using the CRE recombinase system in the brain, is that different floxed loci seem to have different sensitivities to recombinase activity. This leads to a situation where a given CRE transgenic mouse, for example one expressing CRE throughout the forebrain, may in one case give a targeted deletion throughout all forebrain neurons with one floxed mouse line, but when used with a second floxed line, give deletion only in a patchy or limited group of neurons within the forebrain. Moreover, the CRE-mediated gene knockout can accumulate over fairly long periods of time such that an animal at 2 months of age may have one pattern of gene deletion and then over the next several months that pattern can expand to include other brain areas. Thus, each new floxed knockout line of mouse must be tested with a specific CRE line to determine whether appropriate patterns of gene deletion can be obtained.

While there are some difficulties in using both the CRE and TET systems for regulating genetic manipulations, the binary nature of the systems offers a clear advantage. Each CRE and TET line need be generated and characterized only once

TABLE 3.1
Characterized CRE and TET Transgenic Lines Relevant to Neurobiology

Promoter	Expression	Effector Lines	Reference
CaMKIIα	Excitatory forebrain neurons	tTA[a], rtTA, CRE, LBD-CRE	13,15,20,21
Neuron specific enolase	Pan neuronal	tTA[a]	22
Nestin	Pan neuronal	CRE[a]	23
GluRe3	Cerebellar granule cells	LBD-CRE	17
Thy-1	Pan neuronal	LBD-CRE	21
Foxg1	Telencephalon	CRE	24
Emx-1	Dorsal telencephalon	CRE	25

Note: Transgenic mouse lines expressing either the CRE recombinase for use in conditional knockouts or the tTA transcription factor for use in TET regulated expression. LBD-CRE: steroid receptor-CRE fusion transgenics to allow inducible knockout.

[a] Mice available from The Jackson Labs, Bar Harbor, Maine.

Inducible Knockout

FIGURE 3.5 Inducible Gene Targeting Vector, a gene-targeting cassette that allows inducible knockout. The cassette is knocked into the locus of interest such that the expression of tTA is under the control of the endogenous genes promoter, and the coding region of the endogenous gene is placed under the control of a TRE (tetracycline responsive element). Arrows indicate transcription initiation sites.

and can then be used with multiple floxed or TRE lines. As more lines of CRE mice become available with specific anatomical restrictions, it should be possible to generate a single floxed mouse line and to then delete the floxed gene in a several different brain regions simply by mating to the appropriate CRE line of mouse. Table 3.1 shows the CRE and TET lines of transgenic mice relevant to neurobiology that have been described to date.

An alternative to the CRE recombinase system for obtaining inducible knockouts has recently shown some success.[26] The approach, as outlined in Figure 3.5, involves placing the endogenous gene to be targeted under the control of a tetracycline responsive promoter. A single targeting construct is used in which the tTA transcription factor is knocked into the five prime untranslated region of the targeted gene. This places tTA expression under the control of the target gene promoter. The targeting vector also contains a TRE that is placed upstream of the coding exons of the targeted gene. Successful homologous recombination in this case should lead to a situation in which tTA expression occurs with the same timing and pattern as the targeted gene while at the same time disrupting the targeted gene's transcription from its own promoter. The tTA expression in turn activates expression of the targeted gene via the TRE element. Homozygous mice in this case express the targeted gene normally, but in the presence of doxycycline the gene is suppressed. This approach offers the advantage that not only can the gene be inducibly suppressed, but also that this suppression is reversible. One concern in using this approach is that it is unlikely that the levels of expressions of the targeted gene, now placed under the control of a TRE, will be identical to the expression from the wild-type locus. Thus, homozygous mice are likely to either under or over express the targeted gene product and this may lead to a phenotypic alteration. In fact, in the one case where this technique was used to control expression of a K^+-channel, the genetically modified mice showed a 3-fold overexpression of the channel.[26] This produced a phenotypic alteration while complete suppression of the gene had no phenotypic effect.

3.2.6 POINT MUTATIONS

While the generation of the null mutation may be informative in terms of relating genes to biological function, one often wishes to produce more subtle genetic alterations. For many genes, a great deal is known about the function of individual protein domains. It may at times be useful to introduce a disruption that specifically affects one particular domain of a protein. Subtle manipulations such as point mutations or small deletions or substitutions can readily be introduced using standard knockout approaches combined with the CRE-lox system to remove the neomycin gene. In this case the desired point mutation is introduced into the appropriate exon using a standard targeting construct. The neomycin-resistance gene is placed in an adjacent intronic region flanked by loxP sites. Gene targeting transfections will generate an ES cell line carrying the mutant exon along with the neomycin-resistance gene. The neomycin-resistance gene is then removed using transient CRE transfection in the ES cells, leaving a locus in which only the point mutation and a single intronic loxP sequence remain.[27] Mice are then generated and the point mutation is made homozygous by mating.

3.2.7 GENETIC BACKGROUND

Single gene mutations do not act in isolation but in concert with the array of allelic variants at other loci in an individual animal. There are several quite separate issues relating to genetic background that must be considered in any experiment. The first is the overall background of control and mutant mice examined. In general knockout and transgenic mice are generated on a mixed background and this genetic variation must be similar in the control and experimental groups. One common mistake that has been made in the past has been to breed homozygote knockout mice to each other and to breed a second group of control wild-type mice. This alleviates the need to genotype the offspring, however it invalidates the direct comparison of the two groups. This is because if the mice started from a mixed genetic background, the separate inbreeding of the two groups will fix a vast number of different allelic variants in the two populations. With this breeding scheme, one is essentially generating two separate inbred lines of mice that never existed before and for which there is no appropriate control.

An approach for dealing with this issue was suggested at a recent Banbury Conference.[28] In general, for all behavioral experiments, the control group of mice should be derived from the same litters as the experimental group. In this way, even though the individual mice will carry different alleles, they will derive from the same parental alleles and should distribute randomly between control and experimental animals. In addition, the background should be clearly defined, for example F2 B6/129SvEv, and mice should not be compared across different levels of backcrossing. Finally, the mutant allele should be systematically backcrossed onto the two separate parental lines so that the mutant can be studied on an inbred background and a defined F_2 population can be generated in the future.

What does it mean if a mutant phenotype is sensitive to genetic background? It simply means that, as with all complex traits, the trait is not mediated solely by a single gene. While the identification of these modifier genes for a trait may be quite informative, perhaps more informative than your knockout, at present it is quite difficult to identify such modifiers and therefore one often attempts to minimize their effect. To do this one should work either on a pure inbred background if possible, or employ two inbred backgrounds that do not differ significantly in the trait under study to minimize the effect of modifiers. However, it is important to remember that modifiers will always be present and may be revealed by interaction with a knockout or transgene.

A final issue related to background is that the genomic region immediately surrounding a gene of interest will maintain its initial background character despite extensive backcrossing. If, for example, you generate a knockout of the NMDA receptor gene on a 129 ES cell line and then backcross this onto C57Bl6/J, even after many generations of backcrossing, the region of the chromosome that is closely linked to the knockout will derive from the 129 background and not Bl6. Thus, it is always possible that the phenotypes you observe on the Bl6 background derive not from your knockout but from the effect of 129 derived alleles in genes that are closely linked to the NMDA receptor. This is a difficult problem to deal with and is often discounted. One way of testing for this possibility is to generate or identify some marker in the 129 version of your gene that allows you to track and compare wild-type mice with the same linked loci.

3.3 SAMPLE DATA AND INTERPRETATIONS

3.3.1 GENERAL CONSIDERATIONS

The power of a knockout or transgenic mouse line is that it provides a source of animals with a single precise molecular alteration that can be studied across various levels of complexity from the molecular to the cellular to the behavioral. This can allow correlations to be made across these levels as discussed in the examples below.

The type of phenotyping tools employed in the analysis of a given mouse will of course depend on the nature of the genetic change produced and the hypothesis being tested; however, there are some basic aspects of characterization that are commonly employed in behavioral neuroscience.

The initial analysis of a new mouse line usually begins at the molecular level and simply asks the question: Did I cause the molecular change that I intended? In a knockout mouse one would ask whether the protein is completely absent or whether there is a truncated fragment expressed that could lead to altered or partial function. Genes do not work in isolation and it is likely that any genetic manipulation will result in some compensatory change in the expression of other related genes. This sort of compensatory action could have the effect of masking a potential phenotype, for example, through the upregulation of a gene with similar function. Thus, it is often useful to examine the expression of other isoforms of a deleted gene and of related genes in a known signaling pathway.

An extensive set of behavioral tests is available in the mouse and has been reviewed recently.[29] The tests will of course depend on the questions being asked but at a minimum should include some visual inspection of general health and grooming, weight gain, and some simple assessments of sensory and motor function. Beyond this initial assessment, there are tests available for learning and memory, anxiety, fear and aggression, social and sexual behavior, motivational state, consumatory behavior, drug and alcohol addiction, and psychiatric disorders. In addition, neurophysiological properties such as seizure activity and seizure threshold, basal synaptic transmission and forms of synaptic plasticity such as long-term potentiation (LTP) and long-term depression (LTD), can be measured both in brain slices and with chronically implanted electrodes in freely behaving animals. Much of the work with genetically modified mice, particularly using conditional modifications, has focused on learning and memory. Below we examine examples using conditional knockouts, inducible transgenics, and conditional cell type specific ablation.

3.3.2 CONDITIONAL KNOCKOUT OF THE NMDA RECEPTOR

Glutamate is the major excitatory neurotransmitter in the mammalian brain. A great deal of effort has focused on the role of the NMDA class of glutamate receptors in mediating long-lasting forms of synaptic plasticity such as long-term potentiation (LTP) and long-term depression (LTD). The NMDA receptor subunit NR1 is expressed throughout the nervous system and is required for the formation of functional receptors. Not surprisingly, a standard knockout of the NR1 gene leads to early neonatal lethality. Thus, using this approach, one is unable to obtain information about the role of functional NMDA receptors in the adult animal in synaptic plasticity and behavior. Tsien et al. used CRE-mediated conditional knockout to delete the NMDA R1 gene specifically in postnatal forebrain neurons.[30] They used the promoter for the alpha subunit of Ca2+/calmodulin-dependent protein kinase II (CaMKIIα) to express CRE recombinase specifically in excitatory neurons of the forebrain, that is the neocortex, hippocampus, striatum, and amygdala. They also generated a floxed locus of the NR1 gene using homologous recombination in ES cells. When they introduced the CRE-recombinase-expressing transgene into mice carrying the floxed NR1 locus, they generated animals that specifically deleted the NR1 gene in CA1 neurons of the hippocampus. The deletion did not occur until approximately 3 weeks postnatally. This highly specific localization of the gene deletion was surprising given the fact that the CaMKII promoter generally expresses more broadly in forebrain neurons. In fact, the CRE recombinase was shown to be expressed in neurons outside of the CA1 cells of the hippocampus, however, the deletion did not occur in some of those neurons in which the CRE recombinase was expressed. It is unclear why the deletion was more restricted than the pattern of expression of CRE recombinase. One possibility is that it reflects some threshold level of recombinase protein that is required to mediate a functional recombination event in a given cell. The CA1 neurons may have expressed the recombinase to a sufficiently high level to mediate recombination while the remaining forebrain neurons failed to reach this threshold level of expression. Alternately, it is possible that the NR1 locus is inaccessible to the CRE recombinase in forebrain neurons outside of the CA1 region of the hippocampus. It has become clear from a number of studies using conditional knockout with the CRE-loxP system that certain loci are accessible to gene deletion using a given CRE transgenic mouse line while other loci are inaccessible to gene deletion. This reflects a major difficulty in using the CRE-loxP system as it currently exists. Nevertheless, this represents an extremely powerful tool for understanding the role for specific molecules in neuronal function as exemplified by studies of the CA1 specific NMDA receptor deletion.

The NR1-CA1 knockout mice were studied at a number of different levels. First LTP and LTD were examined in the CA1 region of the hippocampus and found, as expected, to be absent in the knockout mice. However, when a different group of neurons, the dentate granule cells, were examined, they were found to have normal LTP. Thus, the deletion of the NR1 gene was specific to CA1 pyramidal cells and functionally eliminated LTP. The animals were then tested on a hippocampal-dependent form of learning and memory using the Morris water maze to test spatial memory. The water maze is a learning test in which animals are required to find a hidden escape platform submerged in a cloudy pool of water. The task requires the animals to use distal cues in the room to triangulate the location of the platform and is the best characterized test of hippocampal-dependent memory. While mice that carry the CRE transgene or the floxed NMDA receptor alone performed normally on this task, those mice carrying both genetic modifications, in which the NR1 subunit was functionally deleted in the CA1 neurons, were severely impaired in spatial memory. Thus, this highly specific genetic manipulation leads to selective deficits at both the level of neuronal physiology and whole animal behavior.

McHugh et al. examined the hippocampus in the NR1–CA1 knockout mice at the systems level by examining place cells.[31] In the hippocampus, the excitatory pyramidal cells code for the animals' spatial location. That is, a given cell will fire action potentials only when the animal is in a particular location in its environment. It is thought that the cumulative pattern of firing of place cells in the hippocampus represents a map of the animal's environment and encodes at any given moment the animal's position in that environment. In the NR1–CA1 knockout mice, the precision of this map was disrupted. Neurons that would normally coordinate their firing when the animal was in a particular location showed an uncoordinated firing pattern in the mutants.

Deletion of the NR1 gene specifically in the CA1 neurons of the hippocampus thus produces three distinct phenotypic alterations: impaired LTP, an impaired spatial map, and impaired spatial learning. One could imagine that the lack of LTP in the CA1 neurons leads to the alteration in place cell properties that in turn leads to an impairment in the encoding of spatial memories within the hippocampus. This is a reasonable interpretation and demonstrates the power of this approach, however a note of caution is warranted. While it is safe to say, at least in the genetic background used, that the NR1 knockout causes each of the three phenotypes, the relationship between the phenotypes is merely correlative. Thus, it cannot be said with certainty that the LTP deficit causes the place-cell deficit that in turn causes the learning deficit; they may arise as separate and independent consequences of the NR1 deletion.

3.3.3 TET REGULATION OF PROTEIN KINASE ACTIVITY

While NMDA receptor activation is the initial step in the production of long-lasting synaptic changes such as LTP that may be important for encoding long-term memories, it is the calcium signal carried through the NMDA receptor that likely triggers the biochemical transformations that mediate this information encoding. The tetracycline system for regulated transgene expression has been used to investigate the processing of this calcium signal. CaMKII is a protein kinase that is highly expressed in forebrain neurons and concentrated in postsynaptic dendritic spines. CaMKIIα has the capacity to respond to brief calcium signals and convert from a calcium-dependent protein kinase to a calcium-independent kinase through autophosphorylation. These biochemical attributes suggested a mechanism by which brief calcium signals could be converted into long-lasting changes in protein phosphorylation by CaMKII. Pharmacological and genetic evidence from knockout mice suggest that the CaMKIIα subunit may be critical for the production of LTP and the formation of long-term memories.

Mayford et al. used the TET system for transgene regulation to investigate the role of CaMKII activation in LTP and long-term memory.[13] The tTA transgene was expressed in forebrain neuron using the promoter for CaMKIIα itself. A mutant form of the kinase (CaMKII-Asp 286) was expressed from a TRE. Expression of the mutant CaMKII-Asp286 transgene resulted in an elevation in calcium-independent kinase activity and alteration in synaptic plasticity. Normally high-frequency stimulation produces maximal LTP in hippocampal slices, while intermediate-frequency stimulation produces moderate LTP, and low frequencies produce LTD. In the CaMKII-Asp286 mutant mice, high frequency LTP was comparable to wild-type animals, while LTP produced by stimulation in the lower frequency ranges (5–10 Hz) was impaired and LTD was enhanced in the transgenic animals. Thus, activation of the kinase produced a frequency-dependent impairment in LTP. The alteration of synaptic plasticity was in a frequency range of particular interest in regard to hippocampal function, since an endogenous oscillation known as the theta rhythm occurs in the hippocampus of animals during normal exploration and the theta rhythm is in the 5–10 Hz range. Thus, if LTP-like synaptic plasticity is induced in the hippocampus in freely behaving animals, it is likely induced by some pattern of activity in the theta frequency range.

In the transgenic animals, hippocampal spatial memory was severely impaired. As in the case with the NR1–CA1 knockout mice, a genetic manipulation that altered synaptic plasticity was also associated with an alteration in learning and memory capacity. In addition, the place-cell map of space was disrupted in the transgenic animals.[32] However, it is unclear whether these behavioral and cell physiological phenotypes arise from alterations in neuronal development or due to the acute alteration of CaMKII levels in the adult animal. To address this question, the expression of the CaMKII-Asp286 transgene was suppressed by feeding adult animals the tetracycline analog doxycycline. Following complete suppression of transgene expression such that CaMKII enzyme activity returned to wild-type levels, the behavioral and the electrophysiological phenotypes reverted to wild-type levels. Thus, the developmental activation of CaMKII did not produce an irreversible developmental alteration that impacted on synaptic plasticity or learning and memory in these transgenic animals. Moreover, although the relation between the behavioral and electrophysiological phenotypes is still only correlative, that correlation is stronger.

3.3.4 CALCINEURIN, MEMORY STORAGE, AND RECALL

The availability of inducible and reversible transgene expression is useful not only for investigating the developmental vs. acute adult effects of a genetic manipulation, but also for dissecting complex behavioral measures into their various components. For example, to execute a simple learning and memory task, an animal has to perceive the relevant stimuli and encode the appropriate information during learning, but it also has to access that stored information during a recall test. Moreover, the nature of the information may be altered during a little-understood process known as consolidation, where memories are transformed from relatively labile short-term forms to more permanent long-term forms. Which of these processes is altered by a given genetic manipulation?

To begin to address these questions, Mansuy et al. used the rtTA system to overexpress calcineurin in forebrain neurons in a manner similar to that just described for CaMKII.[15] Calcineurin is a Ca2+ activated protein phosphatase that is concentrated at synapses. Calcineurin acts to oppose the induction of stable long-lasting LTP and transgenic mice that overexpress calcineurin show impairments in hippocampal LTP and a defect in spatial learning in the water maze. If expression of the calcineurin transgene is suppressed during learning, the animals acquire spatial memory normally, but when the transgene is now activated, the retrieval of the memory is impaired. The actual memory was still present since when the transgene was again suppressed, the memory reappeared. Thus, the use of regulated genetic modification allows one to explore the various phases of memory acquisition, storage, consolidation, and retrieval.

3.3.5 TARGETED NEURONAL ABLATION

The central nervous system contains a vast array of different cell types that vary in morphology, connectivity, and neurotransmitter phenotypes. One of the difficulties in understanding brain function is elucidating the role of small cell populations within a larger neuronal circuit. An example of a genetic approach to understanding the role of a minor cell type within a larger network is provided by the recent work of Nakanishi and his colleagues.[33]

The striatum is a major component of the forebrain involved in control of motor activity, learning and memory, mood, and motivation. The primary cell type found in the striatum is the GABAergic medium spiny neuron, which makes up approximately 95% of the cells in the striatum. However, approximately 1–2% of striatal neurons are cholinergic and it has been difficult to address of the role of this small subpopulation of interneurons. Nakanishi used immunotoxin-mediated cell targeting to address the role of these neurons in striatal function. In this case a transgenic mouse is created in which the human interleukin-2α-receptor is expressed from a promoter that drives expression specifically into cholinergic striatal neurons as well as to neurons in several other brain regions. Those neurons that express the transgene are now sensitive to an immunotoxin consisting of a monoclonal antibody to the IL2 receptor fused to a bacterial toxin. Following stereotactic injection of the immunotoxin specifically into the striatum, more than 80% of cholinergic neurons were eliminated in the transgenic mice. Those neurons in the striatum not expressing the transgene were unaffected. Thus, a very minor population of neurons can be specifically ablated within a given brain structure through a combination of transgenic expression and stereotactic delivery of exogenous toxins. The studies showed that the local cholinergic interneurons were critical for controlling striatal activity and their ablation resulted in significant abnormalities in motor function. The combination of transgenic manipulation to target specific cell populations combined with the application of exogenous effector molecules by stereotactic injection is likely to be quite useful in elucidating local circuit function in control of animal behavior. One could imagine the use of similar techniques to locally delete or over express specific transgenes in selected cell populations using CRE- or TET-regulated approaches.

3.4 DISCUSSION

In the last quarter century, genetic studies in invertebrates have been particularly successful in elucidating the molecular mechanisms involved in everything from development and cell-fate determination to circadian rhythmicity. The mouse is the only mammalian species that is routinely accessible to genetic manipulation and as such will serve as the model for genetic studies in higher organisms. Can the power of the genetic approach be applied to the mouse to understand complex questions of mammalian brain function?

The examples given here and throughout this volume are cause for optimism. It is now possible to specifically delete single genes in restricted cell types within the brain at specific points during development. In addition, the expression of gene products can be controlled in specific cell types in a temporally regulated manner. The capacity for directed or reverse genetic manipulation of the mouse approaches or in some areas surpasses that of classical genetic model organisms. The ability for such selective manipulations of the mouse genome combined with analysis of the resulting mutant animals at multiple levels of complexity from the single neuron to the behaving animal are beginning to yield insights into a broad array of brain functions from neuronal development to complex behaviors such as learning and memory.

In the future, progress in several areas can be anticipated to have a major impact on the field. Much of what we know about functional domains in the mammalian brain from both human and experimental animal studies is anatomical and based on lesion or functional imaging studies. We currently lack promoter elements that would allow genetic manipulation of highly specific anatomical structures or specific cell types within these structures. The characterization of new highly specific promoter elements to produce for example, CRE or tTA effector lines of mice should allow for a more detailed dissection of complex behaviors. The goal would be to attain a situation similar to that in *Drosophila* with the GAL4 system where once a particular gene of interest is modified, for example floxed, that gene could then be knocked out in a series of different brain structures simply by mating to highly specific CRE-expressing transgenic lines. An alternative approach would be to deliver the CRE molecule to specific brain regions using stereotactic application of viral vectors expressing the recombinase. Inducing molecules such as doxycycline could also be delivered locally to induce genetic modification using the TET or CRE systems.

A second important area for development in mouse genetics in fact has relatively little to do with genetics as a whole but involves adapting to the mouse phenotyping techniques currently applied to other organisms. The rat has been the model organism for the majority of studies of mammalian brain function from the behavioral to the electrophysiological level. It will be critical to adapt, as much as possible, the techniques applied to the rat for analysis of genetically modified mice. Thus, developing methods for multi-unit recording in freely behaving mice as well as adapting the myriad of complex behavioral tasks will be critical.

Finally, the sequencing of the mouse and human genomes will add greatly to our repertoire of candidate genes accessible to genetic manipulation. Currently, we know between 10 and 20% of the total number of genes in the mouse genome based on recognizable cDNA sequence. The genome sequencing projects will both open up the remaining 80 to 90% of mammalian genes to study and facilitate the rapidity with which genetic manipulations are achievable. The use of phenotype-driven, ENU-based mutagenic screening combined with targeted manipulation of candidate genes should vastly increase our rate of understanding of the genetic mechanisms which control complex behavior.

REFERENCES

1. Joyner, A., *Gene Targeting: A Practical Approach*, Oxford University Press, Oxford, 1992.
2. Hogan, B., Beddington, R., Costantini, F., and Lacy, E., *Manipulating the Mouse Embryo*, CSHL Press, Cold Spring Harbor, NY, 1994.
3. Palmiter, R., Sandgren, E., Avarbock, M., Allen, D., and Brinster, R., Heterologous introns can enhance expression of transgenes in mice, *Proc. Natl. Acad. Sci. USA*, 88, 478, 1991.

4. Yang, X. W., Model, P., and Heintz, N., Homologous recombination based modification in *Escherichia coli* and germline transmission in transgenic mice of a bacterial artificial chromosome, *Nat. Biotechnol.*, 15, 859, 1997.

5. Yang, X. W., Wynder, C., Doughty, M. L., and Heintz, N., BAC-mediated gene-dosage analysis reveals a role for Ziprol (Ru49/Zfp38) in progenitor cell proliferation in cerebellum and skin, *Nat. Genet.*, 22, 327, 1999.

6. Antoch, M. P., Song, E. J., Chang, A. M., Vitaterna, M. H., Zhao, Y., Wilsbacher, L. D., Sangoram, A. M., King, D. P., Pinto, L. H., and Takahashi, J. S., Functional identification of the mouse circadian Clock gene by transgenic BAC rescue, *Cell*, 89, 655, 1997.

7. Hasty, P., Rivera-Perez, J., and Bradley, A., The length of homology required for gene targeting in embryonic stem cells, *Mol. Cell Biol.*, 11, 5586, 1991.

8. Moran, J. L., Levorse, J. M., and Vogt, T. F., Limbs move beyond the radical fringe, *Nature*, 399, 742, 1999.

9. Sauer, B. and Henderson, N., Site-specific DNA recombination in mammalian cells by the Cre recombinase of bacteriophage P1, *Proc. Natl. Acad. Sci. USA*, 85, 5166, 1988.

10. Gossen, M. and Bujard, H., Tight control of gene expression in mammalian cells by tetracycline-responsive promoters, *Proc. Natl. Acad. Sci. USA*, 89, 5547, 1992.

11. Gossen, M., Freundlieb, S., Bender, G., Muller, G., Hillen, W., and Bujard, H., Transcriptional activation by tetracyclines in mammalian cells, *Science*, 268, 1766, 1995.

12. No, D., Yao, T. P., and Evans, R. M., Ecdysone-inducible gene expression in mammalian cells and transgenic mice, *Proc. Natl. Acad. Sci. USA*, 93, 3346, 1996.

13. Mayford, M., Bach, M. E., Huang, Y. Y., Wang, L., Hawkins, R. D., and Kandel, E. R., Control of memory formation through regulated expression of a CaMKII transgene, *Science*, 274, 1678, 1996.

14. Gilbert, D. M., Heery, D. M., Losson, R., Chambon, P., and Lemoine, Y., Estradiol-inducible squelching and cell growth arrest by a chimeric VP16-estrogen receptor expressed in *Saccharomyces cerevisiae*: Suppression by an allele of PDR1, *Mol. Cell Biol.*, 13, 462, 1993.

14a. Bejar, R., Unpublished results.

15. Mansuy, I. M., Winder, D. G., Moallem, T. M., Osman, M., Mayford, M., Hawkins, R. D., and Kandel, E. R., Inducible and reversible gene expression with the rtTA system for the study of memory, *Neuron*, 21, 257, 1998.

16. Urlinger, S., Baron, U., Thellmann, M., Hasan, M. T., Bujard, H., and Hillen, W., Exploring the sequence space for tetracycline-dependent transcriptional activators: Novel mutations yield expanded range and sensitivity, *Proc. Natl. Acad. Sci. USA*, 97, 7963, 2000.

17. Tsujita, M., Mori, H., Watanabe, M., Suzuki, M., Miyazaki, J., and Mishina, M., Cerebellar granule cell-specific and inducible expression of Cre recombinase in the mouse, *J. Neurosci*, 19, 10318, 1999.

18. Kellendonk, C., Tronche, F., Monaghan, A. P., Angrand, P. O., Stewart, F., and Schutz, G., Regulation of Cre recombinase activity by the synthetic steroid RU 486, *Nucleic Acids Res.*, 24, 1404, 1996.

19. Kellendonk, C., Tronche, F., Casanova, E., Anlag, K., Christian, O., and Schutz, G., Inducible site-specific recombination in the brain, *J. Mol. Biol.*, 285, 175, 1999.

20. Tsien, J. Z., Chen, D. F., Gerber, D., Tom, C., Mercer, E. H., Anderson, D. J., Mayford, M., Kandel, E. R., and Tonegawa, S., Subregion- and cell type-restricted gene knockout in mouse brain, *Cell*, 87, 1317, 1996.

21. Kellendonk, C., Tronche, F., Casanova, E., Anlag, K., Opherk, C., and Schutz, G., Inducible site-specific recombination in the brain, *J. Mol. Biol.*, 285, 175, 1999.
22. Chen, J., Kelz, M. B., Zeng, G., Sakai, N., Steffen, C., Shockett, P. E., Picciotto, M. R., Duman, R. S., and Nestler, E. J., Transgenic animals with inducible, targeted gene expression in brain, *Mol. Pharmacol.*, 54, 495, 1998.
23. Tronche, F., Kellendonk, C., Kretz, O., Gass, P., Anlag, K., Orban, P. C., Bock, R., Klein, R., and Schutz, G., Disruption of the glucocorticoid receptor gene in the nervous system results in reduced anxiety, *Nat. Genet.*, 23, 99, 1999.
24. Hebert, J. M. and McConnell, S. K., Targeting of cre to the Foxg1 (BF-1) locus mediates loxP recombination in the telencephalon and other developing head structures, *Dev. Biol.*, 222, 296, 2000.
25. Iwasato, T., Datwani, A., Wolf, A. M., Nishiyama, H., Taguchi, Y., Tonegawa, S., Knopfel, T., Erzurumlu, R. S., and Itohara, S., Cortex-restricted disruption of NMDAR1 impairs neuronal patterns in the barrel cortex, *Nature*, 406, 726, 2000.
26. Bond, C. T., Sprengel, R., Bissonnette, J. M., Kaufmann, W. A., Pribnow, D., Neelands, T., Storck, T., Baetscher, M., Jerecic, J., Maylie, J., Knaus, H. G., Seeburg, P. H., and Adelman, J. P., Respiration and parturition affected by conditional overexpression of the Ca2+-activated K+ channel subunit, SK3, *Science*, 289, 1942, 2000.
27. Giese, K. P., Fedorov, N. B., Filipkowski, R. K., and Silva, A. J., Autophosphorylation at thr286 of the alpha calcium-calmodulin kinase II in LTP and learning, *Science*, 279, 870, 1998.
28. Banbury, C., Mutant mice and neuroscience: Recommendations concerning genetic background, Banbury Conference on Genetic Background in Mice, *Neuron*, 19, 755, 1997.
29. Crawley, J. N., *What's Wrong with My Mouse?*, Wiley-Liss, New York, 2000.
30. Tsien, J. Z., Huerta, P. T., and Tonegawa, S., The essential role of hippocampal CA1 NMDA receptor-dependent synaptic plasticity in spatial memory, *Cell*, 87, 1327, 1996.
31. McHugh, T. J., Blum, K. I., Tsien, J. Z., Tonegawa, S., and Wilson, M. A., Impaired hippocampal representation of space in CA1-specific NMDAR1 knockout mice, *Cell*, 87, 1339, 1996.
32. Rotenberg, A., Mayford, M., Hawkins, R. D., Kandel, E. R., and Muller, R. U., Mice expressing activated CaMKII lack low frequency LTP and do not form stable place cells in the CA1 region of the hippocampus, *Cell*, 87, 1351, 1996.
33. Kaneko, S., Hikida, T., Watanabe, D., Ichinose, H., Nagatsu, T., Kreitman, R. J., Pastan, I., and Nakanishi, S., Synaptic integration mediated by striatal cholinergic interneurons in basal ganglia function, *Science*, 289, 633, 2000.

Section 2

4 Functional Identification of Neural Genes

Lawrence H. Pinto and Joseph S. Takahashi

CONTENTS

4.1 GENETIC DISSECTION AND THE REQUIREMENTS FOR ITS APPLICATION

Genetic dissection is the use of a series of mutations that affect a biological process in order to identify the functional elements of the process;[1] this permits elucidation of their function and interactions.[2] Genetic dissection has been used routinely to study biochemical pathways. A series of single gene mutations, affecting enzymes within a particular pathway, can each perturb the product of the pathway, e.g., by reducing the function of the essential enzymes. Each affected enzyme can then be identified biochemically by taking advantage of knowledge of the substrates and products in the pathway.

Some complex neurally mediated processes such as behaviors are more difficult to study than biochemical pathways. They require communication between cell types within the nervous system and the appropriate interaction of multiple pathways and the molecular components of these pathways such as neurotransmitters, ion channels, and transcription factors. Moreover, the identification of all of the genes which underlie a complex neuronal process often cannot be assessed biochemically. This is because certain neuronal signals that control behaviors spread rapidly throughout the nervous system, making biochemical identification difficult. One powerful way to study genes that control such processes is to clone them and study their gene products.

Forward genetics is the use of mutant phenotypes to identify essential genes. This approach has facilitated rapid progress in diverse areas.[3] The demonstration that the *Shaker* gene encoded a potassium channel[4] and the subsequent cloning of this gene[5,6] facilitated the identification of several other families of ion channels and led to a detailed knowledge of their structure–function relationship.[7] Likewise, the field of circadian biology was advanced from descriptive to mechanistic analyses with the identification of the *per* gene.[2] The forward genetic approach has also been of great utility in the elucidation of developmental processes. The identification of mutants with aberrant pattern (body plan) formation in the fly[8] and zebrafish[9] has permitted an exponential increase in our understanding of the processes involved. Forward genetics differs from reverse genetics, in which a gene is modified (e.g., with a "knockout," see Keverne[9a]) in that knowledge of the gene sequence in not needed for the forward genetic approach. The emphasis of this review will lie in the discovery of those genes which are essential for a particular neuronal phenotype of interest, rather than on the generation of mutants and their classification into general phenotypes. This latter activity has been reviewed recently.[10-15]

In order to apply forward genetics to the study of a complex neuronal phenotype, two principal requirements need to be satisfied. First, the population of animals used for the study should be genetically homogeneous, or isogenic, thereby permitting an induced mutation in the DNA sequence to be identified readily. Isogenicity is also helpful in reducing the variability of the phenotype under study, thus allowing animals with an abnormal phenotype to be identified reliably. Second, a reliable quantitative assay needs to be established for the neuronal phenotype to be studied. Ideally, this assay should be neither labor intensive nor time consuming, although this is not always possible. It is still possible, but more difficult, to use this approach if the assay is lethal or renders the animal sterile.

The mouse satisfies both of the above requirements for many neuronal phenotypes. First, inbred strains are available for the mouse. Inbred strains are produced by 20 or more generations of crosses either between siblings or between offspring and the older parent. Thus, each member of a strain is considered to be homozygous at each locus across the genome. This also reduces variation among mice under investigation. However, it should be kept in mind that the reproducibility from animal-to-animal within a given inbred strain varies among the strains and that subtle differences in lab environment can alter the performance of a given strain.[16,17] For example, in our experience the circadian period of animals of the C57BL/6J strain is much more consistent than that of either the 129 strain commonly used for gene targeting or the BALB/cJ strain. The second requirement, for a reliable assay, is more readily satisfied for some neuronal phenotypes than for others. The mouse's size and consistent pattern of activity make it ideal for studying circadian behavior. Learning and memory may be studied in mice using several tests, including context-dependent fear conditioning.[18,19] Mice are also amenable to the study of auditory-driven[20] and olfactory-driven[21] behavior, and they can be tested for addiction to alcohol and drugs of abuse.[22,23] However, the study of visually driven behaviors needs to be approached with care. Several mouse strains (e.g., C3H/HeJ) carry the mutant allele at the *rd* locus, resulting in the progressive degeneration of photoreceptors starting about PND (postnatal day) 10. The eyes of mice track moving visual targets using the optokinetic nystagmus reflex,[24] but no rapid assays for visual perception have as yet been reported.[25] In addition, several common mutations that cause ataxia actually result from degeneration of the inner ear as the primary phenotype.[26] Thus, mice satisfy the requirements for many neuronal phenotypes, but caution must be exercised in design of the screening assay and in choosing the appropriate inbred strain for each particular study. The resources for mutagenizing mice,[27,28] their phenotypic characterization,[13,14,16,25,29-31] and the availability of mapping tools[32] and cryopreservation of their gametes[33,34] have been the topic of a recent volume.[34a]

4.2 ADVANTAGES AND DISADVANTAGES OF THE FORWARD GENETIC APPROACH

There are several advantages to the forward genetic approach:

1. It requires no prior knowledge of the mechanism or components of the neuronal phenotype to be studied and it makes no assumptions about the genes that are involved.
2. It is often possible to isolate a number of mutant alleles of one gene that alter the function of the gene product in different ways. From these various alterations it is often possible to understand the function of the gene product better than by studying only the wild-type gene product.
3. Once a single gene that is essential for a neuronal phenotype has been found, it is possible to use this gene to facilitate the search for other genes that are essential (see Modifier Screens below).
4. In some instances, point mutations may be more informative than targeted null mutations.
5. Gain-of-function alleles can be identified.
6. Finally, it is possible to achieve certain economies in the process of screening for mutants. For example, one set of mutagenized animals can be screened for several neuronal phenotypes.

There are three primary disadvantages to the forward genetic approach, two theoretical and one practical. Theoretically, should the neuronal phenotype under study require the activity of a protein that is essential for life, then mutation of the gene encoding the essential protein could have a negative effect on the general state of health of the animal. Therefore, it may be difficult to tell if the altered performance of the animal in the phenotype under study is due to a specific effect in the phenotypic mechanism or is due to general deterioration. An example of this is seen in the *Wheels* mutation, which was identified by screening of mutagenized animals. Heterozygous animals demonstrate a lengthened circadian period.[35] However, the *Wheels* mutation, which is homozygous lethal, also demonstrates abnormal development of the inner ear, leading to the conclusion that the circadian defects are secondary to the inner ear defects.[36] In brief, even for a behavior as basic as the circadian clock, a mutation that alters the behavior may not be fully informative about the mechanism for the behavior. However, the presence of morphological abnormalities in a mutant animal does not necessarily preclude the possibility that a disrupted gene product participates in a specific behavior. This has been demonstrated for the role of FYN tyrosine kinase in hippocampal LTP. In *fyn*-deficient mice, hippocampal LTP is impaired, but the analysis of this finding is complicated by the presence of a number of neurological defects such as uncoordinated hippocampal architecture and reduced neurite outgrowth.[37,38] To address this complication, FYN was expressed using the calcium/calmodulin-dependent protein kinase promoter to achieve high levels of FYN expression in the forebrain but not the rest of the brain. The result was restoration of hippocampal LTP in the presence of persistent morphological abnormalities of the hippocampus.[39] One way to distinguish a primary from a secondary defect is to characterize several phenotypes. If only one is altered, this finding strengthens the argument that the single defect is primary. A second theoretical shortcoming of the approach is that one function may be performed by two gene products, either of which is sufficient to sustain the function. Thus, removal of either gene product will not alter the function subserved by both. In other words, a gene must be essential for the process under study in order for it to be detected by this approach. The number of vital genes in a region can be estimated by studying the frequency of lethal mutations linked to markers in the region.[40] The principal practical shortcoming is that, having identified a mutant phenotype, the mutated gene still needs to be cloned and its gene product identified.

4.3   INDUCTION OF MUTATIONS

In order to serve as the basis for a forward genetic approach to test for phenotypes in postnatal animals, a mutagen needs to satisfy several requirements. These include: high mutation rate in premeiotic cells, allow survival and fertility of the mutagenized animal, and make only small perturbations in the genome (to facilitate detection of the altered nucleotide sequence). The first large-scale induced mutation experiments in mice were done using X-radiation with the intention of determining the susceptibility to ionizing radiation. Unfortunately, this method of mutagenesis not only produces mutants at a low rate but also induces large chromosomal perturbations such as deletions and inversions (see Table 4.1).[41-43] Other mutagens such as chlorambucil

TABLE 4.1
Properties of Mutagens in Mice

Agent	Best Dose	Best Target	Mutation Rate per Locus ($\times 10^{-5}$)	Predominant Mutation
X-rays	6 Gy	Spermatogonia	13	Small deletions
	5 Gy + % Gy	Spermatogonia	50	Small deletions
	3 Gy	Postmeiotic	33	Deletions, translocations
	4 Gy	Oocytes	19	Deletions, translocations
ENU	250 mg/kg	Spermatogonia	66	Intragenic point mutations
	4 × 100 mg/kg	Spermatogonia	150	Intragenic point mutations
Procarbazine	600 mg/kg	Spermatogonia	5	Small, intragenic
		Postmeiotic	22	Deletions, translocations
Chlorambucil	10 mg/kg	Postmeiotic	127	Deletions, translocations
Transgene			~0.1	
Gene trap			~1	
None			0.5–1.0	

Note: References for the various mutation rates are given in the text.

behave in a similar manner.[43,44] A wide range of mutagens is available (see Table 4.1). However, ENU (N-ethyl-N-nitrosourea) most closely satisfies the above require- ments. It induces mutations at a high rate[27,45-48] and acts premeiotically. This results in the production of many gametes per mutagenized animal. One important advan- tage of ENU is that it induces point mutations, most often by transversion.[28,49-51] As a result, both gain- and loss-of-function mutants can be produced. Lethal mutations, which are typically null or loss-of-function alleles, are less frequent with ENU because the mutations are more subtle (i.e., intragenic point mutations). The fre- quency of occurrence of lethal mutations has been studied for the T/t-H-2 region of chromosome 17 of the mouse[40] and found to be 11 lethal mutations in 280 gametes studied. The induction of point mutations offers the advantages that the mutation will most likely have an effect on the transcription unit in which it occurs and the base substitution can be identified by sequence comparison. This contrasts with deletion mutations which can exert an effect far from the deleted sequence.

ENU mutagenesis is successful in adult males but not females,[52] probably due to the differing mechanisms for gametogenesis in male and female mammals. Sperm are produced by division and differentiation of spermatogonial stem cells throughout the life of the animal, whereas ova are produced by maturation of oocytes which undergo meiosis I in the perinatal period. Thus, administration of ENU to adult males will affect spermatogonial stem cells, but administration to females will affect oocytes, in which DNA repair mechanisms are very active.

Sperm are not effectively mutagenized by ENU, and thus it is necessary to eliminate from the mutagenized animal those sperm which are mature or are in the process of maturation at the time ENU is administered. This can be achieved by allowing 1–2 month passage of time after administration of the mutagen and then testing for (temporary) sterility.

Embryonic stem cells can be effectively mutagenized by ENU[53] and the related compound ethylmethanesulphonate (EMS).[54] This offers the advantage of being able to screen for phenotypes of interest directly in the embryonic stem cell. Mutant stem cells could then be incorporated into chimeras which are then tested for germ line transmission of the mutation. However, this approach will have limited usefulness for studying complex phenotypes involving multiple neurons, since its advantage lies in the ability to test a single embryonic stem cell.

When administering ENU, it is necessary to measure the actual concentration of the compound spectrophotometrically,[40] as purity varies from lot to lot. It is also possible to take advantage of the increase in induced mutation rate that results from repetition of doses of ENU.[55] The investigator also needs to be observant of the strain that is being injected with ENU, since some strains of mice are more tolerant of the compound than others,[27] and a dose established for one strain may result in permanent sterility when administered to another strain.

4.4　DETECTION OF MUTATIONS

4.4.1　Breeding Schemes

Mutations induced by ENU must be detected in the progeny of the mutagenized male. ENU will affect only the paternally derived chromosomes. Consequently, only semidominant and dominant mutations can be found by screening first-generation (G1) progeny. This section will consider various breeding schemes, their utility, and their efficiency in identifying ENU-induced mutations.

Semidominant mutations are those for which the phenotype of the heterozygote is intermediate between wild-type and homozygous mutants. An example of semi-dominance can occur with oligomeric proteins. If one monomer of the oligomer is abnormal, it is easy to visualize how the entire oligomer might have reduced function. Two examples which illustrate this are the homotetrameric K channel[56] and the heteromeric CLOCK-BMAL transcription factor complex.[57] Semidominant muta-tions can be detected within G1 progeny. However, for some phenotypes, the very nature of a semidominant mutation requires that the screening test be quantitative. It is noteworthy that animals hemizygous for a mutation within the Y-chromosome also can be identified within G1 progeny.

Recessive mutations will not be detected in the heterozygous mutant progeny of a single generation cross. Instead, the third generation progeny (G3) must be tested. Two different breeding schemes for ENU mutagenesis and screening for dominant, semi-dominant, or recessive mutations are shown in Figure 4.1. These are a backcross scheme and an intercross scheme. In the case of a screen for dominant mutations, the first generation, G1, mouse is examined, and each G1 mouse represents one gamete. In the backcross scheme, to screen for recessive mutations, each male G1 mouse is used to found a three-generation pedigree. The G2 females are backcrossed to the G1 male in order to isolate G3 progeny that are homozygous for the mutagenized gamete (Figure 4.1, left). In the intercross breeding scheme, two G1 mice are used to found the pedigree, and G2 progeny are intercrossed with siblings (Figure 4.1, right) so that two mutagenized gametes are represented in the same pedigree.

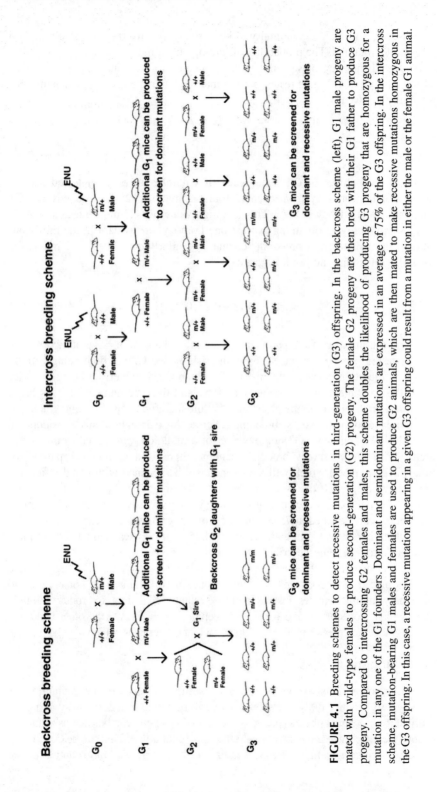

FIGURE 4.1 Breeding schemes to detect recessive mutations in third-generation (G3) offspring. In the backcross scheme (left), G1 male progeny are mated with wild-type females to produce second-generation (G2) progeny. The female G2 progeny are then bred with their G1 father to produce G3 progeny. Compared to intercrossing G2 females and males, this scheme doubles the likelihood of producing G3 progeny that are homozygous for a mutation in any one of the G1 founders. Dominant and semidominant mutations are expressed in an average of 75% of the G3 offspring. In the intercross scheme, mutation-bearing G1 males and females are used to produce G2 animals, which are then mated to make recessive mutations homozygous in the G3 offspring. In this case, a recessive mutation appearing in a given G3 offspring could result from a mutation in either the male or the female G1 animal.

However, not every G3 animal will be homozygous for the recessive mutation of the founder animal. The probability of finding one homozygote in this generation, termed the "Efficiency of scanning," "E" is related to the number of G2 matings and the number of G3 mice tested per G2 mating.[40] For the backcross scheme, this probability is given by

$$E = 1 - \left[0.5 + 0.5\left(0.75^n\right)\right]^k \tag{4.1}$$

where k = the number of G2 females and n = number of progeny tested per G2 female. For five progeny from each of four G2 matings (20 G3 progeny per G1 pedigree), the efficiency is 85%. This is a high value per G3 mouse tested.

The probability of obtaining at least one homozygous mutant in the third generation with an intercross breeding scheme is similarly calculated as 1 minus the probability of finding no homozygotes:

$$E = 1 - \left[0.75 + 0.25\left(0.75^n\right)\right]^k \tag{4.2}$$

where k = the number of G2 pairs (rather than females) and n = the number of G3 progeny tested per pair. In this scheme, the risk of two G2 not both being carriers of a given mutation is greater than that of a given G2 daughter not being a carrier (hence the coefficient of 0.75 rather than 0.5), and thus the number of pairs has a greater role in determining the probability. While the efficiency of scanning is lower in comparison to the backcross breeding scheme, this calculation only considers the probability of obtaining a G3 homozygote for a mutation transmitted from one G1 parent. With an intercross breeding scheme there is a second G1 parent (or mutagenized gamete) which will simultaneously be scanned with equal efficiency in the same pedigree. If one were to consider the case of 36 G3 progeny (6 G3 from each of 6 G2 pairs), the efficiency of scanning is only 75%. However, in this scheme, two genomes are scanned at 75% efficiency, or 1.5 genomes are scanned per 36 G3 mice, which is relatively comparable to the 0.85 genomes scanned per 20 G3 mice in the backcross scheme described above.

There are some practical advantages to the intercross scheme. The reproductive demands placed on the G1 males is less. Rotation of G1 males is not required to impregnate multiple females. Instead, pairs are established and left to produce multiple litters, which procedurally and for record keeping is simpler. On the other hand, the intercross scheme depends on the fertility of mutant G1 females, which may be limited.

4.4.2 MODIFIER SCREENS

A modifier screen is based on the principle that the presence of one mutation in a pathway will enhance the sensitivity to a second mutation in that pathway. If a gene involved in a neuronal phenotype, m, is identified, this gene can be used as a starting point to identify other genes involved in the same phenotype. This can be done using a modifier screen that induces a new mutation, "*". Animals homozygous for the known mutation that affects the phenotype under study (m/m) are crossed with mutagenized animals (*/+) to produce a fraction of offspring that are compound heterozygotes (+/*; +/m). If the newly mutated gene (*) and the known gene (m) interact, these compound heterozygotes will have a phenotype which differs from that of both the homozygous mutant (m/m) and the wild type (+/+). This can be illustrated with a simple example. A neuronal phenotype requires ca. 50% throughput of a pathway, that includes the product of gene m, and that heterozygotes demonstrate ca. 50% activity of the product of gene m. Animals heterozygous for this recessive mutation (m/+) will have a phenotype that cannot be readily distinguished from wild type. The newly induced mutation (*) reduces the throughput of the steps for which it is responsible by an additional 50%. The combination of the defects of genes m and * will reduce overall throughput of the pathway to 25% of normal, which is less than the amount required for a normal phenotype. Consequently, the double heterozygotes (+/*; +/m) can be detected in the screening test. This approach facilitated the elucidation of the series of gene products needed to form the specialized seventh photoreceptor of the fly eye, starting with one mutation, *sevenless*, that served as the "sensitizing" mutation[58-60] or starting point. One allele of the *tim* gene, a component of the circadian oscillator, was also identified as a result of its interaction with the another circadian gene *per*.[61] Modifier screens have the additional advantage of being able to reveal alleles of the original gene.

4.4.3 BREEDING FOR MODIFIER SCREENS

To detect modifying mutations, the first-generation progeny of the mutagenized male (+/+) and a female that is homozygous for the sensitizing mutation (m/m) are examined. In order to detect a modifying gene, sufficient progeny must be examined to ensure the generation of one compound heterozygote (+/*; +/m). For a 0.9 probability of detecting one such animal, it is necessary to screen eight progeny. Although this appears to be tedious, an advantage results from this breeding scheme, based on the principle that a mutagenized parent carrying mutation m will give birth to more than one progeny bearing the mutation if enough progeny are examined. Thus, if one putative mutant is identified in the screen, then it ought to be possible to find additional affected progeny from the same breeding pair. This finding will provide evidence that the first individual was correctly identified as a mutant.

In most cases, modifiers alter quantitative variables, and false positives may occur if the experimental conditions are not constant. For phenotypes that are characterized quantitatively, it is imperative that the assay employed in a modifier screen be quantitative and that the conditions of rearing and testing be constant. However, other phenotypes, for example, the *vibrator* phenotype[62] allow modifiers to be scored with visual observation. Extra care must be exercised in the choice of the strain of mouse to be mutagenized in modifier screens. It is possible that the strain in which the sensitizing mutation occurs will differ from the strain chosen for mutagenesis. In this case the sensitizing mutation will be detected in a hybrid animal, and mapping the mutation on a hybrid background might prove difficult (see below). For this reason, it is advisable to perform a control experiment, without mutagen, to be sure that the choice of strains will permit mapping.

4.5   SELECTION OF AN APPROPRIATE
MUTAGENESIS SCHEME

The efficiency of the strategy to generate and detect mutations must be carefully considered. It is important to consider what fraction of the genes intrinsic to a process will be found by screening a given number of mutagenized mice. The goal of many genetic studies is to achieve "saturation," the theoretical definition of which is the point at which each gene in the genome has been mutated at least once in a fashion that produces an identifiable phenotype. However, the induced mutation rate varies widely from locus to locus,[55] and thus it is possible that many mutations will occur at one locus before a single mutation appears at another locus. Thus, another perspective is called for in this analysis.

4.5.1   THE NUMBER OF ANIMALS NEEDED TO BE SCREENED
IN A GENOME-WIDE SCAN

A simplified description of the process of mutagenesis can be given by the Poisson distribution in which there are many target genes (N), each of which has a small probability of being mutagenized (p). The number of mutagenized genes in the genomes of a population of animals, $M_a(n)$, will thus be given by the Poisson distribution

$$M_a(n) = \frac{a^n e^{-a}}{n!} \qquad\qquad (4.3)$$

where $a = Np$ and n is the number of mutations for which the distribution is being evaluated.

Let us estimate the number of animals required to be screened in order to find one mutant. The simplest case is that of a one-generation screen for semidominant mutations. In this case the probability of finding one or more mutations in a single gene when a number of G1 animals equal to $(1/p)$ is screened equals $(1 -$ the probability of finding no mutants), or $(1 - M_1(0))$, or about 0.73 $(1 - 1/e)$. Thus, if M genes are thought to be involved in a particular phenotype, only $1/pM$ animals would need to be screened for a 0.73 probability of finding the first mutant.

How many mutagenized animals need to be screened in order to identify nearly every gene that is essential in a given process? Because the mutation rate varies from gene to gene, it is not possible to induce a single mutation in every gene without inducing many mutations in the most mutable genes. Thus, it is more useful to consider how many mice need to be screened in order to find mutations in 95% of the genes that have a probability p of being mutagenized. In order to achieve mutations with an identifiable phenotype in 95% of the genes $(1 - M_3[0]) \sim 0.95$, it is necessary to screen $(3/p)$ animals (Equation 4.2). Thus, to find 95% of the genes involved in a process which has a high induced mutation rate of $p \sim 2 \times 10^{-3}$,[48] it is necessary to screen only about 600 gametes. However, by screening this number of gametes there is only a small chance of finding a gene with a low mutation rate of $p \sim 10^{-4}$.

Is it more efficient to screen for semidominant or for recessive mutations? For semidominant mutations, the induced mutation rate is lower than for recessive mutations.[49,63] Assuming an induced mutation rate of p ~ 1.5×10^{-4} per locus per gamete for semidominant mutations, it would be necessary to screen 15,000 mice to identify 95% of the genes involved. The basis for this calculation is as follows: Favor[63] and Favor et al.,[49] found rates of about 1.5×10^{-3} for all dominant mutations that cause cataracts; the rate per locus will be much less, for example, if 10 genes are involved, the rate will be 1.5×10^{-4}. A calculation of the number of animals required to detect a recessive mutation can yield a surprising result. The induced-mutation rate for some recessive mutations at some loci is rather high, and for these mutations p ~ 2×10^{-3}.[48] It requires the screening of about 20 G3 mice in each pedigree in a backcross scheme (see Figure 4.1) to detect a recessive mutation in this scheme (Equation 4.1); thus, to identify a single mutation in such a highly mutable gene would require screening 20/p or about 10,000 mice. This gives rise to the ironic conclusion that detecting a recessive mutation in a readily mutable gene that is essential to a given process may require less effort than finding the dominant mutations that can affect the same process, even though additional breeding is required to render the recessive mutation homozygous. Moreover, finding a recessive mutation in one of several highly mutable genes might require fewer mice than finding a single gene with a dominant mutation. Two other complications that must be kept in mind when considering mutagenesis schemes are that there may exist genes for which it is not practically possible to generate dominant mutations, and that certain recessive mutations may be lethal.

4.5.2 WHAT IS THE MOST EFFICIENT MUTAGENESIS STRATEGY?

The answer to this question depends on the goal of the study and the starting information available to the investigator. For example, if it is known that genes relevant to a phenotype of interest are found in a particular region of the genome, then directing mutagenesis to that region may be sensible.[64,65] This can be done by breeding mutagenized males to mice that are hemizygous for a deletion in the appropriate region or by taking advantage of a deletion in the region of interest.[65] The advantage of this method is that the use of overlapping deletions can aid in the localization of genes, but the disadvantage is that mice hemizygous for the deletion must be identified with molecular techniques before screening.[66-68] However, if nothing is known about the genes controlling the phenotype under study, then genome-wide mutagenesis will reveal these genes with less effort than targeted mutagenesis to a series of regions (Table 4.2).

Gene trap strategies offer the advantage of leaving a unique DNA marker that can facilitate the cloning of the affected gene[69] and the detection of its expression pattern. However, this strategy requires manipulation of embryonic stem cells and a three-generation breeding scheme to detect recessive mutations, and the presently available technology does not offer a high mutation rate in mice.[70-72] In spite of this, many gene trap mutants are available[69,71,73,74] because the number of genes involved in complex processes such as development is large, and the rate at which mutants are found is proportional to the number of genes involved (Equation 4,2). However,

TABLE 4.2
Efficiency of Various Breeding Schemes in Mutagenesis

Screen	No. of Gametes for 3-fold Poisson Coverage[a]	No. of Mice per Gamete	Total No. of Mice	Genes Scanned per Gamete	Genes Scanned per Mouse
Dominant, G_1[c]	2,000	1	2,000	75	75
	20,000[a]	1	20,000	7.5	7.5
Specific Locus, G_1[a]	2,000	1	2,000	75	75
Genome-Wide Recessive, G_3[a]	2,000	20	40,000	75	3.75
Recessive over Deletion, G_1[a] (10 cM deletions with 160 deletion lines)	$2,000 \times 160 = 320,000$	2[b]	640,000	0.63	0.31
Gene Trap, G_3					
Assuming 10^{-4} forward mutation frequency	30,000	20	600,000	10	0.5
Assuming 10^{-5} forward mutation frequency	300,000	20	6,000,000	1	0.05

[a] Assuming forward mutation frequency of 1.5×10^{-3} per locus per gamete with ENU; this is unrealistically low for dominant mutations but reasonable for recessive mutations at some loci (see text). Genes scanned per gamete calculated on the assumption of 50,000 genes per genome and a backcross breeding scheme.

[b] Only half of the animals produced will be hemizygous for the deletion.

[c] Assuming a dominant mutation frequency of 1.5×10^{-4} per locus per gamete (see text).

this approach is less likely to be useful for studying processes for which the number of genes is small. Thus, although the efficiency of screening (the number of genes screened per animal tested) is low for the gene trap strategy (Table 4.2), this approach has great utility for the study of many processes.

4.6 DISTINGUISHING A MUTANT FROM A FALSE-POSITIVE ANIMAL

False positives present a great practical problem for genetic screening projects. The first step, after identifying a putative mutant, is to establish whether its phenotype is heritable. A breeding program which will render the putative mutant allele in the heterozygous (semidominant screen) or homozygous (recessive screen) condition will achieve this. Testing of sufficient progeny will permit demonstration of the expected ratio of mutant:wild-type phenotypes in the population. This generally requires the testing of 20–30 animals. However, if the rate of false positives is high, the number of mice needing to be retested may become excessive. Fortunately, both the backcross and intercross breeding schemes (Figure 4.1) present an opportunity to confirm heritability of a phenotype by simply screening additional animals from the cross that produced the original putative mutant. In the backcross scheme, each G3 animal is the progeny of a G2 dam and the mutant G1 sire. If one putative mutant arises from such a cross, then additional mutants should occur when its siblings are tested. In fact, each sibling will have a 25% chance to be homozygous for the same mutation. The intercross scheme will also produce many G3 offspring from the same G2 sire and G2 dam that produced the original putative mutant. If the abnormal phenotype of a putative mutant results from a mutation and not from an environmental variable, then each of its siblings will have a 25% chance to inherit the same mutant alleles. It is not possible to apply a similar approach to a semidominant G1 screen, since each G1 animal results from a different gamete in the mutagenized G0 animal. Thus, two commonly used breeding schemes provide a ready guard against the consequences of incorrectly identifying an animal as a mutant.

When should an animal with an abnormal complex phenotype such as a behavior be considered a putative mutant? The answer to this question depends on several considerations. First, if the abnormal phenotype is qualitatively different from wild type (e.g., moves its eyes opposite to, rather than in the same direction as, a visual target), then the animal should be considered a putative mutant. If the animal simply does not perform the behavior at all, it is possible that the health of the animal has been compromised or that the animal has a problem with an essential physiological system that prevents it from performing the behavior. If a quantitative index of the animal's performance differs from that of wild type, then the question can only be resolved by determining where the animal's performance lies with respect to the statistical distribution of that index for wild type. Thus, it is imperative to possess thorough quantitative information about wild-type animals prior to beginning a screening process. If the putative mutant's performance does not differ by more than three standard deviation units from the mean for wild type, it is likely that establishing heritability and mapping the mutant gene will prove difficult.

4.7 GENETIC MAPPING

This procedure is necessary to demonstrate conclusively that the trait under study is indeed heritable and is essential to the next steps in identifying the gene by positional cloning. The range of mapping tests for the mouse has been reviewed recently.[32] Simple sequence polymorphisms (SSLP) and single nucleotide polymorphisms (SNP)[75,76] provide a rapid, sensitive approach to genetic mapping in the mouse. Over 6,500 SSLP have been identified between mouse strains most frequently used in genetic mapping.[77-79] These SSLP permit localization of new markers to an interval of 0.1 cM on average.[79] Recently, over 2,800 SNPs have been identified among inbred strains of the mouse[80] at over 1,700 sequence tagged sites. One breeding strategy that is used to take advantage of these polymorphisms is the cross-backcross system.[81] The initial mating is a cross of the mutant animal on the founder strain with an animal from another strain for which there exists a set of SSLPs or SNPs that are polymorphic with the founder strain (the counterstrain). This cross will produce F1 animals that are heterozygous for both the mutation and the polymorphisms. To map semidominant mutations, the F1 animals are then backcrossed to a wild-type animal of the founder strain to produce second-generation offspring that will bear one chromosome of each pair that results from recombination in the F1 parent. These progeny are then tested for the mutant phenotype, and a set of polymorphic SSLPs or SNPs located uniformly over the genome are characterized. This characterization is usually displayed as the haplotype of each animal, and these haplotypes will fall into groups. The mutant gene can be assigned to a region of the genome by noting the association of a haplotype carrying polymorphisms characteristic of the founder strain in a certain region with the mutant phenotype. This then allows assignment of the mutant gene to one certain region of the genome. For mapping recessive mutations, either a backcross to a homozygous mutant or an intercross is used.

There is a practical problem that occurs in the mapping of mutations. It is possible that the genetic background of the counterstrain may modify the phenotype that results from the mutant gene in such a way as to make difficult the establishment of the genotype of an individual animal from its phenotype. Each of the N2 progeny of the F1 animals in a mapping cross will possess a different combination of genes from the founder strain and counterstrain, and thus, if different alleles of genes capable of modifying the mutant gene exist in these strains, the resulting offspring will exhibit more variable phenotypes. A well-known example of this occurs in crosses between BTBR and C57BL/6J strains: both strains have normal insulin resistance but male F1 hybrids between these strains display severe insulin resistance.[82] This would make it difficult to use these strains together in a mapping scheme for this particular phenotype. However, difficulties involving interstrain interactions can usually be solved by the appropriate breeding of animals of questionable genotype (i.e., "test crosses"). For example, an animal in a mapping cross may be thought to be heterozygous for a semidominant mutation, but certain aspects of the phenotype may render this assignment ambiguous. If so, then a test cross of the questionable animal with a wild-type animal of the founder strain should produce progeny, half of which bear the mutant phenotype. It should be noted that there is a potential benefit that can arise from the ambiguity of mapping crosses. It is very possible that modifying alleles can be identified and even mapped from these crosses.[62] Finally, it should be noted that mapping with SNPs, although usually done with two strains, could in theory be done with one strain and a substrain differing only at single nucleotides spaced throughout the genome. This would eliminate troublesome interstrain interactions.

It is possible to generate genetic markers in organisms for which no molecular genetic markers presently exist. This can be done by applying the approach of genetically directed representational difference analysis.[83,84] In principle, this approach generates a marker that can be used to identify regions of synteny in organisms for which genetic maps and genomic sequence are available. Knowledge of the genes encoded in the syntenic regions of the other, better studied, species will allow the candidate gene approach to be applied to the discovery of the affected gene in the original species. This approach has been applied successfully to identify the *tau* mutation in the hamster, an organism for which few genetic tools were available.[84]

4.8 MOLECULAR IDENTIFICATION OF MUTANTS

Two approaches are currently used to identify mutant genes. The first is the candidate gene approach. If the mutant gene maps to a region of the genome in which a gene that is likely to participate in the process under study is found (e.g., a certain neurotransmitter receptor), then it is possible to test the hypothesis that the mutation affects this gene. The sequences of the genomic DNA of the wild-type gene and the mutant gene can be compared if the gene has already been cloned. If not, the sequences of their cDNA can be compared. If these results demonstrate a difference in the encoding gene, the conclusion that the mutation exerts its effect through this gene can be strengthened by manipulation of the gene with gene targeting.

The recent availability of a draft sequence of the human genome and the planned availability of a draft sequence of the mouse genome shortly after the publication of this volume will offer expanded opportunities to apply the candidate gene approach. Once the gene in question is mapped with sufficient resolution to allow it to be mapped to within a region encoding several genes, those genes encoding predicted protein products that may be of interest to the process in question can be considered as candidate genes. However, in many cases the mutation will not map to a genomic region that contains candidate genes, and it will be necessary to employ positional cloning to identify the mutant gene.

4.9 POSITIONAL CLONING

The gene of interest may map to a region of the genome containing genes which, when mutated, might be expected to yield the mutant phenotype. However, if no such candidate genes are identified, it will be necessary to employ the approach of positional cloning to find the mutant gene. The strategy for cloning genes by position starting from information contained in their map positions has been reviewed.[3,85-87] The genomic resources available for the mouse are substantial. Among them are ample DNA polymorphisms, five YAC libraries covering the genome ca. 20-fold, and BAC libraries covering the genome ca. 40-fold, all of which are in the public domain (Table 4.3). The following is a brief review of one set of steps leading to positional cloning that we have found useful.

4.9.1 HIGH-RESOLUTION MAPPING

The first step in positional cloning is to map the gene with high enough resolution to allow the identification of cloned DNA fragments that span the region containing the gene, e.g., YAC and BAC clone. It is necessary to use about use about 1000 meioses in order to obtain sufficient map resolution to identify a YAC that spans the mutant gene.[88] However, the greater the number of meioses used, the smaller the genomic interval that will have to be examined at the molecular level (see below).

4.9.2 FUNCTIONAL RESCUE OF THE MUTATION

The advantage of functional rescue of a mutation lies in the demonstration that unit of DNA used for rescue contains the genetic information that was altered by the

TABLE 4.3
The Mouse as a Genetic Organism

- Genome size: 3,000 Mb, 1600 cM
- Genetic maps: >20,000 loci mapped, >6500 SSLP, >11,000 STS, > 2,800 SNP
- Physical maps: 5 YAC libraries (~20X, 240–820 kb aver.)
                  2 P1 libraries (6.5X, 70–80 kb aver.)
                  5 BAC/PAC libraries (~40X, 120–195 kb aver.)
                  YAC contig map
                  Radiation hybrid map
                  Wash U/HHMI Mouse EST Project
- Informatics: MGD, The Jackson Laboratory (www.informatics.jax.org)
                  MIT Mouse Genome Center (www-genome.wi.mit.edu)
- Synteny and rapid transfer to human genome
- YAC/BAC germ-line expression and mutant rescue in mice

mutation. Thus, it is best to achieve functional rescue using the smallest possible sequence of DNA. If the gene of interest is very large, this functional unit must be a YAC.[89] However, for most genes, bacterial artificial chromosome (BAC) clones are large enough to contain the transcription unit. In fact, mutations in three important behavioral genes were rescued functionally as a prelude to their cloning: the *per* gene of *Drosophila*,[90,91] the *Clock* gene of the mouse,[92] and the vibrator gene of the mouse.[62] Functional rescue in mice can be achieved by constructing transgenic animals with a mutant genotype into which the YAC or BAC[92] has been incorporated. Because even high-resolution mapping lacks the precision necessary to fix the position of the mutant gene within the region represented by a BAC clone, trial and error will be necessary to identify a BAC mapping near the mutated gene which is capable of functional rescue. An important control is to be sure that incorporation of other BAC that map nearby does not provide functional rescue. These experiments make possible the demonstration that the rescuing BAC does indeed encode the gene of interest.

4.9.3 CLONING THE GENE

Having confirmed that the BAC (or YAC) contains the gene of interest, all effort can be focused on sequences contained in the BAC in order to obtain the sequence of the gene and deduce the function of its gene product from the sequence. The details of this are beyond the scope of the present review, but it is useful to outline the steps that are often used.[93] First, shotgun sequencing of the BAC will identify sequences that have homology to known genes. These sequences can be identified by searching several databases, including that for expressed sequence tags. However, even if the sequence of the entire BAC were known, it is likely that several transcription units will occur within the BAC, and sequence information alone will not allow the gene of interest to be identified. Since ENU induces point mutations, it would not be reasonable to attempt to find a single point mutation among the 100–200 kb contained in the BAC. A more feasible approach is to analyze the mRNA that is expressed in the tissue which is likely to be affected by the mutation. This can be done by sequencing the cDNA corresponding to mRNA that are both expressed in this tissue and encoded by sequences contained in the BAC. Two practical ways for doing this are to screen cDNA libraries using the BAC as a probe and to selectively enrich the cDNA clones by hybridization to the BAC. It must be demonstrated that cDNA found in this way map to the physical interval of the genome that includes the BAC. These cDNA can be tested by Northern or *in situ* hybridization to tissue known to be involved in the phenotype under study. However, a wide pattern of expression does not preclude the possibility that the gene is involved in the phenotype under study, for the same protein can participate in different processes in different tissues. Thus, the combined approach of genomic sequencing, hybridization to the rescuing BAC, and enrichment of cDNA will identify the genes expressed in the critical region of the genome and expressed in the tissue of interest.

To find the nucleotide substitution in the mutant gene, it may be sufficient to sequence the cDNA of the mutant animal and compare it with that of the wild-type animal or to sequence genomic DNA from the mutant animal and compare this sequence with that from the wild-type animal. However, the finding of a single-base substitution is not sufficient to demonstrate that this substitution is the cause of the mutant phenotype. One proof would be the identification of a second allele of the mutant gene and the demonstration of an alteration in the sequence of the same cDNA. Alternatively, the predicted gene products of wild-type and mutant must be compared to determine if the function of the altered gene product can be understood in terms of the known biology of the process under study. For example, if the base substitution that is found causes a truncation of the predicted gene product and that truncation would be expected to result in observed phenotype, then the investigator is on solid ground to state that the gene has been found and that the mutation has been identified. However, this latter evidence would need to be supported with genomic rescue in order to constitute proof.

4.10 EXAMPLE OF THE USE OF GENETIC DISSECTION TO ANALYZE A COMPLEX NEURONAL PHENOTYPE

Circadian behavior in mammals is driven by a primary oscillator located in the suprachiasmatic nucleus, the SCN.[94,95] This oscillator does not depend on the generation of action potentials,[96] and the electrical activity of dissociated SCN neurons in culture has a period that is intrinsic to each cell, independent of its neighbors.[97-99] These observations have led to the hypothesis that the primary circadian oscillator is a property of individual cells. The finding of circadian variations in mRNA of a circadian gene that is described below, *period* (*per*),[100] and the transcriptional regulation of this gene's expression by its own gene product[101,102] led to the conclusion that this intracellular oscillator results from an intracellular feedback loop. Support for the existence of an intracellular oscillator comes from the observation that application of protein synthesis inhibitors to mammals can alter the circadian period and phase, and that there is a critical period of protein synthesis early in the subjective day.[103] This finding further suggests that the intracellular feedback loop involves protein synthesis, and subsequent mutagenesis studies have identified many of these proteins.

The first mutation known to affect the circadian clock in mammals, *Tau*, occurred spontaneously in the golden hamster.[104] Unfortunately, this gene has still not been cloned because of inadequate genetic resources for the hamster such as genetic markers and inbred strains with well-characterized DNA polymorphisms. In contrast, a study using ENU mutagenesis in the mouse allowed a circadian gene, *Clock*, to be identified.[105] When mutated, this gene alters the circadian period, or in the homozygote, abolishes circadian rhythmicity after about 2 weeks in constant darkness. The gene responsible was cloned in a recent study.[93] Transgenic technology in the mouse allowed functional rescue by a single BAC.[92] This latter study confirmed that the BAC contained the sequence of wild-type allele of the gene and permitted identification of the *Clock* gene. The gene encodes a transcription factor with a basic helix-loop-helix DNA binding domain, a PAS domain, and a Q-rich activation domain. The partner of the CLOCK protein, BMAL1, has been identified using the yeast two-hybrid system.[57] It has been demonstrated that the CLOCK-BMAL1 heterodimer is capable of activating transcription of two other clock genes, *per* and *tim* in *Drosophila*[106] and the *mPer1* gene in mammals.[57,107] The PER and TIM proteins also form a heterodimer[108] and interactions between the two proteins permit translocation of the heterodimer to the nucleus.[109] Once in the nucleus, the PER-TIM heterodimer inhibits the activation of transcription of *per* and *tim* genes by the CLOCK-BMAL1 heterodimer[106] in *Drosophila* and inhibits the transcription of the *mPer1* gene in mammals.[107] Thus, these four proteins provide a framework for a transcriptional autoregulatory feedback loop oscillator. This system requires that the half-life of the PER-TIM protein complex to be short, so that transcriptional regulation can alter the levels of the complex. In *Drosophila* the rapid turnover of cytoplasmic PER protein monomers results from the association of the monomers with a kinase[110] encoded by the *double-time* (*dbt*) gene. The first circadian gene to be found, *per*, was identified by forward genetics in *Drosophila*.[2] Subsequently, the *tim*[111,112] and *dbt*[113] genes were identified in *Drosophila* by forward genetics. The human and mouse orthologs of *per* have been identified and cloned in recent studies.[114-117] Thus, almost all of the proteins that are currently thought to play a role in the primary intracellular circadian oscillator were first identified by forward genetics.

4.11 SYNOPSIS OF STEPS TO IDENTIFY A NEURONAL GENE

The following steps outline one approach to cloning a neuronal gene. The actual steps that will be followed will depend on the problem under study and the nature of the mutation that is isolated. This example is intended for a species for which genetic mapping tools are available but no candidate gene has been identified.

1. Choose a strain of mouse which demonstrates the phenotype robustly.
2. Develop a rapid screening test for the phenotype and establish the failure rate of the test in the strain of interest. Attempt to alter the behavior with nongenetic intervention to show that the test is sensitive.

3. Choose a counterstrain with SSLP or SNP polymorphisms and apply mapping crosses to determine whether the phenotype is modified significantly by the genetic background of the counterstrain.
4. Inject ENU into mice of the target strain and perform breeding to produce progeny that are appropriate for the semidominant or recessive mutations being sought.
5. Screen progeny. Use test crosses to establish heritability.
6. Map the gene to within about 5 cM.
7. Decide if the phenotype is of great enough interest to merit cloning the gene and if the phenotype is robust enough to allow the next steps.
8. Perform high-resolution mapping (1,000–3,000 informative meioses).
9. It is assumed at this point that no candidate genes have been identified from the mapping.
10. Construct a YAC and BAC contig spanning the critical interval containing the mutation.
11. Perform functional rescue of the mutation.
12. Clone gene with combination of sequencing, hybridizing to the rescuing BAC, cDNA selection, and Northern blotting of tissues expected to express the gene. The predicted gene product must have a predicted structure consistent with the function that was altered by the mutation.
13. Identify the mutation by comparing the sequence of genomic DNA from wild-type and mutant mice. The predicted mutant gene product must be consistent with the observed defect in the mutant animal.
14. Confirm with a second allele if available.

4.12 SUMMARY

Identification of the molecular components of complex biological processes can be aided greatly by induced random mutagenesis, screening for phenotypes that affect these processes, and identification of the genes that are affected, an approach known as genetic dissection. The strategy of genetic dissection is to identify the molecular components of a complex neuronal phenotype by using a three-step process. First, random mutations are induced within a genome and animals demonstrating defects in the phenotype of interest are identified by screening the mutagenized individuals. Second, the affected gene is genetically mapped or localized in the genome. Last, the gene is identified at the molecular level using candidate gene and positional cloning approaches.

ACKNOWLEDGMENTS

We thank Drs. William Dove, Alexandra Shedlovsky, and Martha Hotz Vitaterna for critical advice. This work was supported by the NSF, NIH, and an Unrestricted Grant in Neuroscience from the Bristol-Myers Squibb Foundation. J. S. T. is an Investigator in the Howard Hughes Medical Institute.

REFERENCES

1. Hotta, Y. and Benzer, S., Abnormal electroretinograms in visual mutants of drosophila, *Nature*, 222, 354, 1969.
2. Konopka, R. J. and Benzer, S., Clock mutants of *Drosophila melanogaster*, *Proc. Natl. Acad. Sci. USA*, 68, 2112, 1971.
3. Takahashi, J. S., Pinto, L. H., and Vitaterna, M. H., Forward and reverse genetic approaches to behavior in the mouse, *Science*, 264, 1724, 1994.
4. Wu, C.-F., Ganetzky, B., Haugland, F., and Liu, A.-X., Potassium currents in *drosophila:* Different components affected by mutations of two genes, *Science*, 220, 1076, 1983.
5. Papazian, D. M., Schwarz, T. L., Tempel, B. L., Jan, Y. N., and Jan, L. Y., Cloning of genomic and complementary DNA from shaker, a putative potassium channel gene from drosophila, *Science*, 237, 749, 1987.
6. Tempel, B. L., Papazian, D. M., Schwarz, T. L., Jan, Y. N., and Jan, L. Y., Sequence of a probable potassium channel component encoded at shaker locus of drosophila, *Science*, 237, 770, 1987.
7. Doyle, D. A., Cabral, J. M., Pfuetzner, R. A., Kuo, A., Gulbis, J. M., Cohen, S. L., Chait, B. T., and MacKinnon, R., The structure of the potassium channel: Molecular basis of k+ conduction and selectivity, *Science*, 280, 69, 1998.
8. Nusslein-Volhard, C. and Wieschaus, E., Mutations affecting segment number and polarity in drosophila, *Nature*, 287, 795, 1980.
9. Brand, M., Heisenberg, C. P., Warga, R. M., Pelegri, F., Karlstrom, R. O., Beuchle, D., Picker, A., Jiang, Y. J., Furutani-Seiki, M., van Eeden, F. J., Granato, M., Haffter, P., Hammerschmidt, M., Kane, D. A., Kelsh, R. N., Mullins, M. C., Odenthal, J., and Nusslein-Volhard, C., Mutations affecting development of the midline and general body shape during zebrafish embryogenesis, *Development*, 123, 129, 1996.
9a. Keverne, E. B., An evaluation of what the mouse knockout experiments are telling us about mammalian behavior, *BioEssays*, 19, 1091, 1997.
10. Hrabé de Angelis, M., Flaswinkel, H., Fuchs, H., Rathkolb, B., Soewarto, D., Marschall, S., Heffner, S., Pargent, W., Wuensch, K., Jung, M., Reis, A., Richter, T., Allesandrini, F., Jakob, T., Fuchs, E., Kolb, H., Kremmer, E., Schaeble, K., Rolinski, B., Roscher, A., Peters, C., Meitinger, T., Strom, T., Steckler, T., Holsboer, F., Gekeler, F., Schindewolf, C., Jung, T., Avraham, K., Behrendt, H., Ring, J., Zimmer, A., Schughart, K., Pfeffer, K., Wolf, E., and Balling, R., Genome-wide, large-scale production of mutant mice by enu mutagenesis, *Nat. Genet.*, 25, 444, 2000.
11. Nolan, P. M., Peters, J., Strivens, M., Rogers, D., Hagan, J., Spurr, N., Gray, I. C., Vizor, L., Brooker, D., Whitehill, E., Washbourne, R., Hough, T., Greenaway, S., Hewitt, M., Liu, X., McCormack, S., Pickford, K., Selley, R., Wells, C., Tymowska-Lalanne, Z., Roby, P., Glenister, P., Thornton, C., Thaung, C., Stevenson, J.-A., Arkell, R., Mburu, P., Hardisty, R., Kiernan, A., Erven, A., Steel, K. P., Voegeling, S., Guenet, J.-L., Nickols, C., Sadri, R., Naase, M., Isaacs, A., Davies, K., Browne, M., Fisher, E., Martin, J., Rastan, S., Brown, S. D. M., and Hunter, J., A systematic, genome-wide, phenotype driven mutagenesis programme for gene function studies in the mouse, *Nat. Genet.*, 25, 440, 2000.
12. Nolan, P. M., Peters, J., Vizor, L., Strivens, M., Washbourne, R., Hough, T., Wells, C., Glenister, P., Thornton, C., Fisher, E., Rogers, D., Hagan, J., Reavill, C., Gray, I., Wood, J., Spurr, N., Browne, M., Rastan, S., Hunter, J., and Brown, S. D. M., Implementation of a large-scale enu mutagenesis program: Towards increasing the mouse mutant resource, *Mammal. Genome*, 11, 500, 2000.

13. Pretsch, W., Enzyme-activity mutants in mus musculus. I. Phenotypic description and genetic characterization of ethylnitrosourea-induced mutations, *Mammal. Genome*, 11, 537, 2000.

14. Rathkolb, B., Decker, T., Fuchs, E., Soewarto, D., Fella, C., Heffner, S., Pargent, W., Wanke, R., Balling, R., Hrabé de Angelis, M., Kolb, H. J., and Wolf, E., The clinical-chemical screen in the Munich enu mouse mutagenesis project: Screening for clinically relevant phenotypes, *Mammal. Genome*, 11, 543, 2000.

15. Soewarto, D., Fella, C., Teubner, A., Rathkolb, B., Pargent, W., Heffner, S., Marschall, S., Wolf, E., Balling, R., and Hrabé de Angelis, M., The large-scale Munich enu-mouse-mutagenesis screen, *Mammal. Genome*, 11, 507, 2000.

16. Tarantino, L. M., Gould, T. J., Druhan, J. P., and Bucan, M., Behavior and mutagenesis screens: The importance of baseline analysis of inbred strains, *Mammal. Genome*, 11, 555, 2000.

17. Crabbe, J. C., Wahlsten, D., and Dudek, B. C., Genetics of mouse behavior: Interactions with laboratory environment, *Science*, 284, 1670, 1999.

18. Owen, E. H., Logue, S. F., Rasmussen, D. L., and Wehner, J. M., Assessment of learning by the Morris water task and fear conditioning in inbred mouse strains and F1 hybrids: Implications of genetic background for single gene mutations and quantitative trait loci analyses, *Neuroscience*, 80, 1087, 1997.

19. Bourtchuladze, R., Frenguelli, B., Blendy, J., Cioffi, D., Schutz, G., and Silva, A. J., Deficient long-term memory in mice with a targeted mutation of the camp-responsive element-binding protein, *Cell*, 79, 59, 1994.

20. Gibson, F., Walsh, J., Mburu, P., Varela, A., Brown, K. A., Antonio, M., Beisel, K. W., Steel, K. P., and Brown, S. D., A type vii myosin encoded by the mouse deafness gene shaker-1, *Nature*, 374, 62, 1995.

21. Zhao, H., Ivic, L., Otaki, J. M., Hashimoto, M., Mikoshiba, K., and Firestein, S., Functional expression of a mammalian odorant receptor, *Science*, 279, 237, 1998.

22. Grisel, J. E., Belknap, J. K., O'Toole, L. A., Helms, M. L., Wenger, C. D., and Crabbe, J. C., Quantitative trait loci affecting methamphetamine responses in bxd recombinant inbred mouse strains, *J. Neurosci.*, 17, 745, 1997.

23. Picciotto, M. R., Zoli, M., Rimondini, R., Lena, C., Marubio, L. M., Pich, E. M., Fuxe, K., and Changeux, J. P., Acetylcholine receptors containing the beta2 subunit are involved in the reinforcing properties of nicotine, *Nature*, 391, 173, 1998.

24. Mangini, N. J., Vanable, J. W., Jr., Williams, M. A., and Pinto, L. H., The optokinetic nystagmus and ocular pigmentation of hypopigmented mouse mutants, *J. Compar. Neurol.*, 241, 191, 1985.

25. Pinto, L. H. and Enroth-Cugell, C., Tests of the mouse visual system, *Mammal. Genome*, 11, 531, 2000.

26. Steel, K. P. and Brown, S. D., Genes and deafness, *Trends Genet.*, 10, 428, 1994.

27. Justice, M. J., Carpenter, D. A., Favor, J., Neuhauser-Klaus, A., Angelis, M. H. d., Soewarto, D., Moser, A., Cordes, S., Miller, D., Chapman, V., Weber, J. S., Rinchik, E. M., Hunsicker, P. R., Russell, W. L., and Bode, V. C., Effects of enu dosage on mouse strains, *Mammal. Genome*, 11, 484, 2000.

28. Noveroske, J. K., Weber, J. S., and Justice, M. J., The mutagenic action of n-ethyl-n-nitrosourea in the mouse, *Mammal. Genome*, 11, 478, 2000.

29. Flaswinkel, H., Allesandrini, F., Rathkolb, B., Decker, T., Kremmer, E., Servatius, A., Jakob, T., Soewarto, D., Marschall, S., Fella, C., Behrendt, H., Ring, J., Wolf, E., Balling, R., Hrabé de Angelis, M., and Pfeffer, K., Identification of immunological relevant phenotypes in enu mutagenized mice, *Mammal. Genome*, 11, 526, 2000.

30. Rolinski, B., Arnecke, R., Dame, T., Kreischer, J., Olgemoller, B., Wolf, E., Balling, R., Hrabé de Angelis, M., and Roscher, A. A., The biochemical metabolite screen in the Munich enu mouse mutagenesis project: Determination of amino acids and acyl-carnitines by tandem mass spectrometry, *Mammal. Genome*, 11, 547, 2000.

31. Schindewolf, C., Lobenwein, K., Trinczek, K., Gomolka, M., Soewarto, D., Fella, C., Pargent, W., Singh, N., Jung, T., and Hrabé de Angelis, M., Comet assay as a tool to screen for mouse models with inherited radiation sensitivity, *Mammal. Genome*, 11, 552, 2000.

32. Wells, C. and Brown, S. D. M., Genomics meets genetics: Towards a mutant map of the mouse, *Mammal. Genome*, 11, 472, 2000.

33. Glenister, P. H. and Thornton, C. E., Cryoconservation-archiving for the future, *Mammal. Genome*, 11, 565, 2000.

34. Nakagata, N., Cryopreservation of mouse spermatozoa, *Mammal. Genome*, 11, 572, 2000.

34a. *Chemical Mutagenesis in Mice, Mammal. Genome, 11, 7, 2000.*

35. Pickard, G. E., Sollars, P. J., Rinchik, E. M., Nolan, P. M., and Bucan, M., Mutagenesis and behavioral screening for altered circadian activity identifies the mouse mutant, wheels, *Brain Res.*, 705, 255, 1995.

36. Nolan, P. M., Kapfhamer, D., and Bucan, M., Random mutagenesis screen for dominant behavioral mutations in mice, *Methods*, 13, 379, 1997.

37. Beggs, H. E., Soriano, P., and Maness, P. F., Ncam-dependent neurite outgrowth is inhibited in neurons from fyn-minus mice, *J. Cell Biol.*, 127, 825, 1994.

38. Grant, S. G., O'Dell, T. J., Karl, K. A., Stein, P. L., Soraino, P., and Kandel, E. R., Impaired long-term potentiation, spatial learning, and hippocampal development in fyn mutant mice, *Science*, 258, 760, 1992.

39. Kojima, N., Wang, J., Mansuy, I. M., Grant, S. G. N., Mayford, M., and Kandel, E. R., Rescuing impairment of long-term potentiation in fyn-deficient mice by introducing fyn transgene, *Proc. Natl. Acad. Sci. USA*, 94, 4761, 1997.

40. Shedlovsky, A., Guenet, J. L., Johnson, L. L., and Dove, W. F., Induction of recessive lethal mutations in the t/t-h-2 region of the mouse genome by a point mutagen, *Genet. Res.*, 47, 135, 1986.

41. Russell, W. L., *X-ray-induced mutations mice*, 16, Cold Spring Harbor Symposia on Quantitative Biology XVI, Cold Spring Harbor, New York, 1951, 327–335.

42. Russell, W. L., Russell, L. B., and Cupp, M. B., Dependence of mutation frequency on radiation dose rate in female mice, *Proc. Natl. Acad. Sci. USA*, 45, 18, 1959.

43. Russell, L. B., Hunsicker, P. R., Cacheiro, N. L. A., Bangham, J. W., Russell, W. L., and Shelby, M. D., Chlorambucil effectively induces deletion mutations in mouse germ cells, *Proc. Natl. Acad. Sci. USA,* 86, 3704, 1989.

44. Rinchik, E. M., Bangham, J. W., Hunsicker, P. R., Cacheiro, N. L., Kwon, B. S., Jackson, I. J., and Russell, L. B., Genetic and molecular analysis of chlorambucil-induced germ-line mutations in the mouse, *Proc. Natl. Acad. Sci. USA*, 87, 1416, 1990.

45. Russell, W. L., Hunsicker, P. R., Carpenter, D. A., Cornett, C. V., and Guinn, G. M., Effect of dose fractionation on the ethylnitrosourea induction of specific-locus mutations in mouse spermatogonia, *Proc. Natl. Acad. Sci. USA*, 79, 3592, 1982.

46. Russell, W. L., Hunsicker, P. R., Raymer, G. D., Steele, M. H., Stelzner, K. F., and Thompson, H. M., Dose–response curve for ethylnitrosourea-induced specific-locus mutations in mouse spermatogonia, *Proc. Natl. Acad. Sci. USA*, 79, 3589, 1982.

47. Russell, W. L., Kelly, E. M., Hunsicker, P. R., Bangham, J. W., Maddux, S. C., and Phipps, E. L., Specific-locus test shows ethylnitrosourea to be the most potent mutagen in the mouse, *Proc. Natl. Acad. Sci. USA*, 76, 5818, 1979.

48. Shedlovsky, A., McDonald, J. D., Symula, D., and Dove, W. F., Mouse models of human phenylketonuria, *Genetics*, 134, 1205, 1993.

49. Favor, J., Neuhauser-Klaus, A., and Ehling, U. H., The induction of forward and reverse specific-locus mutations and dominant cataract mutations in spermatogonia of treated strain dba/2 mice by ethylnitrosourea, *Mut. Res.*, 249, 293, 1991.

50. Favor, J., Neuhauser-Klaus, A., Ehling, U. H., Wulff, A., and van Zeeland, A. A., The effect of the interval between dose applications on the observed specific-locus mutation rate in the mouse following fractionated treatments of spermatogonia with ethylnitrosourea, *Mut. Res.*, 374, 193, 1997.

51. Pearce, S. R., Peters, J., Ball, S., Morgan, M. J., Walker, J. I., and Faik, P., Sequence characterization of enu-induced mutants of glucose phosphate isomerase in mouse, *Mammal. Genome*, 6, 858, 1995.

52. Ehling, U. H. and Neuhauser-Klaus, A., Induction of specific-locus mutations in female mice by 1-ethyl-1-nitrosourea and procarbazine, *Mut. Res.*, 202, 139, 1988.

53. Chen, Y., Yee, D., Dains, K., Chatterjee, A., Cavalcoli, J., Schneider, E., Om, J., Woychik, R. P., and Magnuson, T., Genotype-based screen for enu-induced mutations in mouse embryonic stem cells, *Nat. Genet.*, 24, 314, 2000.

54. Munroe, R. J., Bergstrom, R. A., Zheng, Q. Y., Libby, B., Smith, R., John, S. W., Schimenti, K. J., Browning, V. L., and Schimenti, J. C., Mouse mutants from chemically mutagenized embryonic stem cells in process citation, *Nat. Genet.*, 24, 318, 2000.

55. Hitotsumachi, S., Carpenter, D. A., and Russell, W. L., Dose-repetition increases the mutagenic effectiveness of n-ethyl-n-nitrosourea in mouse spermatogonia, *Proc. Natl. Acad. Sci. USA*, 82, 6619, 1985.

56. MacKinnon, R., Aldrich, R. W., and Lee, A. W., Functional stoichiometry of shaker potassium channel inactivation, *Science*, 262, 757, 1993.

57. Gekakis, N., Staknis, D., Nguyen, H. B., Davis, F. C., Wilsbacher, L. D., King, D. P., Takahashi, J. S., and Weitz, C. J., Role of the clock protein in the mammalian circadian mechanism, *Science*, 280, 1564, 1998.

58. Fortini, M. E., Simon, M. A., and Rubin, G. M., Signaling by the sevenless protein tyrosine kinase is mimicked by ras1 activation, *Nature*, 355, 559, 1992.

59. Simon, M. A., Bowtell, D. D., Dodson, G. S., Laverty, T. R., and Rubin, G. M., Ras1 and a putative guanine nucleotide exchange factor perform crucial steps in signaling by the sevenless protein tyrosine kinase, *Cell*, 67, 701, 1991.

60. Simon, M. A., Dodson, G. S., and Rubin, G. M., An sh3-sh2-sh3 protein is required for p21ras1 activation and binds to sevenless and sos proteins *in vitro*, *Cell*, 73, 169, 1993.

61. Rutila, J. E., Zeng, H., Le, M., Curtin, K. D., Hall, J. C., and Rosbash, M., The timsl mutant of the drosophila rhythm gene timeless manifests allele-specific interactions with period gene mutants, *Neuron*, 17, 921, 1996.

62. Hamilton, B. A., Smith, D. J., Mueller, K. L., Kerrebrock, A. W., Bronson, R. T., van Berkel, B., Daly, M. J., Kruglyak, L., Reeve, M. P., Nemhauser, J. L., Hawkins, T. L., Rubin, E. M., and Lander, E. S., The *vibrator* mutation causes neurodegeneration via reduced expression of pitpα: Postitional complementation cloning and extragenic suppression, *Neuron*, 18, 711, 1997.

63. Favor, J., The frequency of dominant cataract and recessive specific-locus mutations in mice derived from 80 or 160 mg ethylnitrosourea per kg body weight treated spermatogonia, *Mut. Res.*, 162, 69, 1986.

64. Schimenti, J. and Bucan, M., Functional genomics in the mouse: Phenotype-based mutagenesis screens, *Genome Res.*, 8, 698, 1998.

65. Rinchik, E. M. and Carpenter, D. A., N-ethyl-n-nitrosourea mutagenesis of a 6- to 11-cm subregion of the fah–hbb interval of mouse chromosome 7: Completed testing of 4557 gametes and deletion mapping and complementation analysis of 31 mutations, *Genetics*, 152, 373, 1999.

66. Brown, S. D. and Peters, J., Combining mutagenesis and genomics in the mouse — closing the phenotype gap, *Trends Genet.*, 12, 433, 1996.

67. Justice, M. J., Zheng, B., Woychik, R. P., and Bradley, A., Using targeted large deletions and high-efficiency n-ethyl-n-nitrosourea mutagenesis for functional analyses of the mammalian genome, *Methods*, 13, 423, 1997.

68. Rinchik, E. M., Carpenter, D. A., and Selby, P. B., A strategy for fine-structure functional analysis of a 6- to 11-centimorgan region of mouse chromosome 7 by high-efficiency mutagenesis, *Proc. Natl. Acad. Sci. USA*, 87, 896, 1990.

69. Evans, M. J., Carlton, M. B., and Russ, A. P., Gene trapping and functional genomics, *Trends Genet.*, 13, 370, 1997.

70. Friedrich, G. and Soriano, P., Insertional mutagenesis by retroviruses and promoter traps in embryonic stem cells, *Meth. Enzymol.*, 225, 681, 1993.

71. Hill, D. P. and Wurst, W., Screening for novel pattern formation genes using gene trap approaches, *Meth. Enzymol.*, 225, 664, 1993.

72. Wurst, W., Rossant, J., Prideaux, V., Kownacka, M., Joyner, A., Hill, D. P., Guillemot, F., Gasca, S., Cado, D., Auerbach, A., et al., A large-scale gene-trap screen for insertional mutations in developmentally regulated genes in mice, *Genetics*, 139, 889, 1995.

73. Zambrowicz, B. P., Friedrich, G. A., Buxton, E. C., Lilleberg, S. L., Person, C., and Sands, A. T., Disruption and sequence identification of 2,000 genes in mouse embryonic stem cells, *Nature*, 392, 608, 1998.

74. Hicks, G. G., Shi, E. G., Li, X. M., Li, C. H., Pawlak, M., and Ruley, H. E., Functional genomics in mice by tagged sequence mutagenesis, *Nat. Genet.*, 16, 338, 1997.

75. Wang, D. G., Fan, J. B., Siao, C. J., Berno, A., Young, P., Sapolsky, R., Ghandour, G., Perkins, N., Winchester, E., Spencer, J., Kruglyak, L., Stein, L., Hsie, L., Topaloglou, T., Hubbell, E., Robinson, E., Mittmann, M., Morris, M. S., Shen, N., Kilburn, D., Rioux, J., Nusbaum, C., Rozen, S., Hudson, T. J., Lander, E. S., et al., Large-scale identification, mapping, and genotyping of single-nucleotide polymorphisms in the human genome, *Science*, 280, 1077, 1998.

76. Zhao, L. P., Aragaki, C., Hsu, L., and Quiaoit, F., Mapping of complex traits by single-nucleotide polymorphisms, *Am. J. Hum. Genet.*, 63, 225, 1998.

77. Dietrich, W., Katz, H., Lincoln, S. E., Shin, H. S., Friedman, J., Dracopoli, N. C., and Lander, E. S., A genetic map of the mouse suitable for typing intraspecific crosses, *Genetics*, 131, 423, 1992.

78. Dietrich, W. F., Miller, J. C., Steen, R. G., Merchant, M., Damron, D., Nahf, R., Gross, A., Joyce, D. C., Wessel, M., and Dredge, R. D., A genetic map of the mouse with 4,006 simple sequence length polymorphisms, *Nat. Genet.*, 7, 220, 1994.

79. Dietrich, W. F., Miller, J., Steen, R., Merchant, M. A., Damron-Boles, D., Husain, Z., Dredge, R., Daly, M. J., Ingalls, K. A., O'Connor, T. J., Evans, C. A., DeAngelis, M. M., Levinson, D. M., Kruglyak, L., Goodman, N., Copeland, N. G., Jenkins, N. A., Hawkins, T. L., Stein, L., Page, D. C., and Lander, E. S., A comprehensive genetic map of the mouse genome, *Nature*, 380, 149, 1996.

80. Lindblad-Toh, K., Winchester, E., Daly, M. J., Wang, D. G., Hirschhorn, J. N., Laviolette, J. P., Ardlie, K., Reich, D. E., Robinson, E., Sklar, P., Shah, N., Thomas, D., Fan, J. B., Gingeras, T., Warrington, J., Patil, N., Hudson, T. J., and Lander, E. S., Large-scale discovery and genotyping of single-nucleotide polymorphisms in the mouse, *Nat. Genet.*, 24, 381, 2000.

81. Green, E. L., *Genetics in Probability and Animal Breeding Experiments*, Oxford University Press Inc., New York, 1981, 271.

82. Ranheim, T., Dumke, C., Schueler, K. L., Cartee, G. D., and Attie, A. D., Interaction between btbr and c57bl/6j genomes produces an insulin resistance syndrome in (btbr × c57bl/6j) F1 mice, *Arterioscl. Thromb. Vasc. Biol.*, 17, 3286, 1997.

83. Lisitsyn, N. A., Segre, J. A., Kusumi, K., Lisitsyn, N. M., Nadeau, J. H., Frankel, W. N., Wigler, M. H., and Lander, E. S., Direct isolation of polymorphic markers linked to a trait by genetically directed representational difference analysis, *Nat. Genet.*, 6, 57, 1994.

84. Lowrey, P. L., Shimomura, K., Antoch, M. P., Yamazaki, S., Zemenides, P. D., Ralph, M. R., Menaker, M., and Takahashi, J. S., Positional syntenic cloning and functional characterization of the mammalian circadian mutation tau, *Science*, 288, 483, 2000.

85. Stubbs, L., Long-range walking techniques in positional cloning strategies, *Mammal. Genome*, 3, 127, 1992.

86. Silver, L. M., *Mouse Genetics: Concepts and Applications*, Oxford University Press, New York, 1995, 362.

87. Copeland, N. G., Jenkins, N. A., Gilbert, D. J., Eppig, J. T., Maltais, L. J., Miller, J. C., Dietrich, W. F., Weaver, A., Lincoln, S. E., Steen, R. G., et al., A genetic linkage map of the mouse: Current applications and future prospects, *Science*, 262, 57, 1993.

88. Guenet, J.-L., The mouse genome, *Genomes, Molecular Biology and Drug Discovery,* Browne, M. J., and Thurlby, P. L., Eds., Academic Press, San Diego, 1996, 27–51.

89. Huxley, C., Exploring gene function: Use of yeast artificial chromosome transgenesis, *Methods*, 14, 199, 1998.

90. Bargiello, T. A., Jackson, F. R., and Young, M. W., Restoration of circadian behavioural rhythms by gene transfer in drosophila, *Nature*, 312, 752, 1984.

91. Zehring, W. A., Wheeler, D. A., Reddy, P., Konopka, R. J., Kyriacou, C. P., Rosbash, M., and Hall, J. C., P-element transformation with period locus DNA restores rhythmicity to mutant, arrhythmic *Drosophila melanogaster,* Cell, 39, 369, 1984.

92. Antoch, M. P., Song, E. J., Chang, A. M., Vitaterna, M. H., Zhao, Y., Wilsbacher, L. D., Sangoram, A. M., King, D. P., Pinto, L. H., and Takahashi, J. S., Functional identification of the mouse circadian clock gene by transgenic bac rescue, *Cell*, 89, 655, 1997.

93. King, D. P., Zhao, Y., Sangoram, A. M., Wilsbacher, L. D., Tanaka, M., Antoch, M. P., Steeves, T. D., Vitaterna, M. H., Kornhauser, J. M., Lowrey, P. L., Turek, F. W., and Takahashi, J. S., Positional cloning of the mouse circadian clock gene, *Cell*, 89, 641, 1997.

94. Hastings, M. H., Central clocking, *Trends Neurosci.*, 20, 459, 1997.

95. Moore, R. Y., Organization of the mammalian circadian system, Ciba Foundation Symposium, 88–99, 1995.

96. Schwartz, W. J., Gross, R. A., and Morton, M. T., The suprachiasmatic nuclei contain a tetrodotoxin-resistant circadian pacemaker, *Proc. Natl. Acad. Sci. USA*, 84, 1694, 1987.

97. Welsh, D. K., Logothetis, D. E., Meister, M., and Reppert, S. M., Individual neurons dissociated from rat suprachiasmatic nucleus express independently phased circadian firing rhythms, *Neuron*, 14, 697, 1995.

98. Liu, C., Weaver, D. R., Strogatz, S. H., and Reppert, S. M., Cellular construction of a circadian clock: Period determination in the suprachiasmatic nuclei, *Cell*, 91, 855, 1997.

99. Herzog, E. D., Takahashi, J. S., and Block, G. D., Clock controls circadian period in isolated suprachiasmatic nucleus neurons, *Nat. Neurosci.*, 1, 708, 1998.

100. Hardin, P. E., Hall, J. C., and Rosbash, M., Feedback of the drosophila period gene product on circadian cycling of its messenger RNA levels, *Nature*, 343, 536, 1990.

101. Hardin, P. E., Hall, J. C., and Rosbash, M., Circadian oscillations in period gene mrna levels are transcriptionally regulated, *Proc. Natl. Acad. Sci. USA*, 89, 11711, 1992.

102. So, W. V. and Rosbash, M., Post-transcriptional regulation contributes to drosophila clock gene mrna., *EMBO J.*, 16, 7146, 1997.

103. Takahashi, J. S. and Turek, F. W., Anisomycin, an inhibitor of protein synthesis, perturbs the phase of a mammalian circadian pacemaker, *Brain Res.*, 405, 199, 1987.

104. Ralph, M. R. and Menaker, M., A mutation of the circadian system in golden hamsters, *Science*, 241, 1225, 1988.

105. Vitaterna, M. H., King, D. P., Chang, A. M., Kornhauser, J. M., Lowrey, P. L., McDonald, J. D., Dove, W. F., Pinto, L. H., Turek, F. W., and Takahashi, J. S., Mutagenesis and mapping of a mouse gene, clock, essential for circadian behavior, *Science*, 264, 719, 1994.

106. Darlington, T. K., Wager-Smith, K., Ceriani, M. F., Staknis, D., Gekakis, N., Steeves, T. D. L., Weitz, C. J., Takahashi, J. S., and Kay, S. A., Closing the circadian loop: Clock-induced transcription of its own inhibitors per and tim, *Science*, 280, 1599, 1998.

107. Sangoram, A. M., Saez, L., Antoch, M. P., Gekakis, N., Stanknis, D., Whiteley, A., Fruechte, E. M., Vitaterna, M. H., Shimomura, K., King, D. P., Young, M. W., Weitz, C. J., and Takahashi, J. S., Mammalian circadian autoregulatory loop: A timeless ortholog and mper1 interact and negatively regulate clock-bmal1-induced transcription, *Neuron*, 21, 1101, 1998.

108. Gekakis, N., Saez, L., Delahaye-Brown, A. M., Myers, M. P., Sehgal, A., Young, M. W., and Weitz, C. J., Isolation of timeless by per protein interaction: Defective interaction between timeless protein and long-period mutant perl, *Science*, 270, 811, 1995.

109. Saez, L. and Young, M. W., Regulation of nuclear entry of the drosophila clock proteins period and timeless, *Neuron*, 17, 911, 1996.

110. Kloss, B., Price, J. L., Saez, L., Blau, J., Rothenfluh, A., Wesley, C. S., and Young, M. W., The drosophila clock gene double-time encodes a protein closely related to human casein kinase i-epsilon, *Cell*, 94, 97, 1998.

111. Sehgal, A., Price, J. L., Man, B., and Young, M. W., Loss of circadian behavioral rhythms and per RNA oscillations in the drosophila mutant timeless, *Science*, 263, 1603, 1994.

112. Myers, M. P., Wager-Smith, K., Wesley, C. S., Young, M. W., and Sehgal, A., Positional cloning and sequence analysis of the drosophila clock gene, timeless, *Science*, 270, 805, 1995.

113. Price, J. L., Blau, J., Rothenfluh, A., Abodeely, M., Kloss, B., and Young, M. W., Double-time is a novel drosophila clock gene that regulates period protein accumulation, *Cell*, 94, 83, 1998.

114. Albrecht, U., Sun, Z. S., Eichele, G., and Lee, C. C., A differential response of two putative mammalian circadian regulators, mper1 and mper2, to light, *Cell*, 91, 1055, 1997.

115. Shearman, L. P., Zylka, M. J., Weaver, D. R., Kolakowski, L. F., Jr., and Reppert, S. M., Two period homologs: Circadian expression and photic regulation in the suprachiasmatic nuclei, *Neuron*, 19, 1261, 1997.

116. Sun, Z. S., Albrecht, U., Zhuchenko, O., Bailey, J., Eichele, G., and Lee, C. C., Rigui, a putative mammalian ortholog of the drosophila period gene, *Cell*, 90, 1003, 1997.

117. Tei, H., Okamura, H., Shigeyoshi, Y., Fukuhara, C., Ozawa, R., Hirose, M., and Sakaki, Y., Circadian oscillation of a mammalian homologue of the drosophila period gene, *Nature*, 389, 512, 1997.

5 Studying Brain Development and Wiring Using a Modified Gene Trap Approach

Kevin J. Mitchell, Lisa V. Goodrich,
Philip A. Leighton, Xiaowei Lu, Kathy Pinson,
Paul Scherz, Olivia G. Kelly, Joel Zupicich,
Paul Wakenight, Peri Tate, Judy Mak,
Edivinia Pangilinan, Helen Rayburn,
Danielle Rottkamp, Joe Zhong, William C. Skarnes,
and Marc Tessier-Lavigne

CONTENTS

5.1 RATIONALE

The exquisitely complex circuitry of the mammalian brain emerges from a series of distinct developmental processes. First, cell identities are established along the newly formed neural tube through an interplay of cell-intrinsic genetic programs and molecular interactions between cells.[1] This results in a program of gene expression in each cell that, by supplying a repertoire of cell surface receptors and signal transduction proteins, directs its subsequent migration, the guidance of its axon to appropriate regions, and its choice of synaptic targets. The cell's integration into functional circuits and its survival in turn rely on expression of the appropriate neurotransmitters, receptors, and ion channels. While electrical activity and competition between neurons influence the ultimate circuitry at a fine level, a remarkably complex wiring diagram can form in the absence of these processes, based solely on the developmental program encoded in the genome.[2] A major goal in developmental neuroscience is to identify the genes involved in this program, especially those encoding surface proteins likely to have direct roles in cell–cell interactions underlying development of the final circuitry of the brain.

We review here the application of a modified gene trapping method that permits a systematic phenotypic screen to identify such genes in mice.[3,3a] This approach is based in concept on large-scale genetic screens that have been used with great success, for example, in fruit flies to dissect the genetic programs underlying patterning of the body axis[4] or wiring of the simple ventral nerve cord.[5] However, instead of chemical mutagenesis, the screen is based on an insertional mutagenesis method, gene trapping, that has several practical advantages in terms of efficiency and the identification of phenotypes in mutant mice. In this section, we first introduce the idea of gene trapping, then discuss how we have applied it to identify ligands and receptors that regulate neural development. We also describe the rationale for modifying the vector in a way that facilitates the identification of brain wiring mechanisms.

5.1.1   GENE TRAPPING AS AN EFFICIENT METHOD OF MUTAGENESIS

In gene trapping approaches, large numbers of insertional mutations can be isolated in embryonic stem (ES) cells using a DNA construct that integrates at random in the genome.[6] When the gene trap vector inserts into the intron of a gene, a fusion transcript is created by splicing of the upstream regions of the "trapped" gene to vector sequences, usually coding for β-geo, a fusion between the reporter protein β-galactosidase and the selectable marker neomycin phosphotransferase.[7] Because the insertion provides a molecular tag, sequence is easily obtained for the trapped gene in each cell line by 5′ RACE and direct sequencing.[8] With the imminent completion of the human and mouse genomic sequences, this tag can readily be used to identify the trapped gene. This obviates the time-consuming task of positional cloning that would be required with chemical mutagenesis and at the same time provides a means to prescreen insertions prior to the generation of mutant mice.

5.1.2   MODIFIED GENE TRAP VECTORS CAN HELP FOCUS
ON PARTICULAR CLASSES OF GENES

Gene trap screens can be further refined by prescreening for specific classes of genes based on the subcellular localization of reporter-gene fusion products.[9,10] To focus on genes directly involved in cell–cell interactions, we use the "secretory trap" approach to enrich for insertions in genes encoding transmembrane or secreted proteins.[9] The secretory trap vector is plasmid based (rather than retroviral) and contains a β-geo reporter that has a type II membrane-spanning domain at its N-terminus and that is flanked by a splice acceptor and polyadenylation signal[19] (Color Figure 5.1).* Insertion into an intron generates a fusion transcript between the β-geo gene and upstream exons of the trapped gene. Due to the presence of the transmembrane domain in the β-geo protein, β-galactosidase activity is predicted to be retained only in those fusions that incorporate an N-terminal signal sequence of a target gene, i.e., fusions with genes whose products are targeted to the secretory pathway.

In contrast to some gene trap vectors,[7,11] the secretory trap vector that we employ lacks its own translation initiation signal. Therefore, productive fusions that yield drug-resistant colonies should only be made if the upstream exons of the trapped gene contain protein-coding sequence. In an earlier study,[12] we recognized the importance of selecting only this class of insertions for phenotypic analysis. Insertions in 5′ untranslated regions or regions which spliced inefficiently, rarely induce recessive lethal phenotypes. In contrast, secretory trap insertions that disrupt coding regions effectively abolish wild-type transcripts and reliably produce phenotypic nulls.[3a]

Because the gene trap method establishes stable lines of mice for each mutation, it is possible to screen for defects at all stages of development. This provides a complementary approach to screens using chemical mutagens (e.g., ENU), which can analyze larger numbers of mutations but which, for recessive mutations, are

---

* Color figures follow page 140.

normally limited to analysis at one stage of development.[13,14] ENU screens usually rely on backcrossing females to a founder male to homozygose potential mutations. The females must then be sacrificed to examine embryos for defects at a specific stage. In contrast, the molecular tag generated by the gene trap insertions makes it possible to identify essential genes simply by the absence of homozygotes at weaning, even in cases that have no obvious morphological defects. Litters at various embryonic stages can then be genotyped to determine the precise lethal phase and nature of the defect. The generation of stable lines also presents an opportunity to concurrently screen mutant lines for defects in several tissues or processes. In our case, these include all aspects of brain development from formation and patterning of the neural tube to cell migration, axon guidance, cell survival, and the emergence of functional circuits.

5.1.3 Adding an Axonal Marker to the Gene Trap Vector Facilitates Identification of Wiring Mechanisms

We are especially interested in using this method to identify new genes involved in axon guidance. However, in screening for defects in something as complex as the wiring of the mammalian brain another problem arises. Many axon guidance mutants display subtle wiring defects affecting small numbers of axons, leaving the majority of the nervous system unaffected (see e.g., References 15 and 16). Without a way of focusing on axons most likely to be affected, it would be necessary to scan through large portions of the nervous system, using stains, tracers, or molecular markers to examine each axonal population. Finding defects would be challenging and in many cases impossible, since many axon tracts lack specific markers to visualize them against the background of a largely normal pattern of projections. Compounding this problem is another one: there is still relatively little known about the normal wiring pattern of the brain and its development, making it often difficult to detect or assess wiring defects.

To tackle the problem of identifying subtle wiring defects against a background of normal axonal projections, we have added a second marker, human placental alkaline phosphatase (PLAP), to the secretory trap vector (Color Figure 5.1a). PLAP is a GPI-linked cell surface protein which, when expressed transgenically in neurons, labels axons in their entirety.[17,18] In the modified vector, β-*geo* is followed by an internal ribosome entry site (IRES)[19,20] and the *PLAP* gene. When a gene is trapped, a bicistronic messenger RNA is produced under the control of the target gene promoter and enhancer elements. This mRNA directs the production of two proteins: the β-geo fusion to the endogenous protein and the PLAP protein, which is translated independently using the IRES (Color Figure 5.1a). In mice derived from mutant ES cells, neurons expressing the trapped gene are labeled in their cell bodies by β-geo (which is retained within an intracellular compartment[9]), exhibiting a blue reaction product when processed for X-gal histochemistry (Color Figure 5.1b). In the same neurons, axons are labeled on the cell surface with PLAP, exhibiting a purple reaction product when processed for alkaline phosphatase histochemistry (Color Figure 5.1c). A direct anatomical screen for differences in the projection patterns of PLAP-labeled axons in homozygotes and heterozygotes can thus identify cell-autonomous axon guidance defects. Importantly, the PLAP histochemical marker allows one to quickly identify and focus on axons that normally express the trapped gene.

A large-scale screen was designed to take advantage of the features of the PLAP vector. The screen takes part in three stages (Color Figure 5.1d): the first is to trap secretory genes in ES cells, the second is to select lines by sequence, and the third is to generate mice and analyze them for phenotypes. The first level of phenotypic analysis is for gross defects: intercross offspring are genotyped and assessed for viability and any overt phenotypes such as ataxia or size differences. Subsequently, crosses are performed to obtain wild-type, heterozygous and homozygous progeny at three ages: E11.5, E15.5, and birth (P0), ages chosen to encompass a large number of axon guidance events during mouse development. Embryos are initially examined for morphological defects and neonates are also examined for physiological defects in breathing, suckling, or movement. Alternating coronal/transverse sections through the entire brain and portions of the spinal cord are processed by X-gal staining (to reveal cell bodies expressing the trapped gene) or PLAP staining (to reveal their axons). Axon tracts stained in the heterozygote are identified by comparison to known anatomy and compared to the staining pattern in the homozygotes to reveal any defects associated with the mutations. In addition to highlighting wiring defects, the β-gal and PLAP markers can also reveal differences in brain patterning, cell migration, or cell survival.

5.2 DETAILED METHODS

5.2.1 GENERATION OF THE PLAP SECRETORY TRAP VECTOR

A detailed map of the PLAP vector is presented in Figure 5.2. This vector consists of the secretory trap vector pGT1TM, a modification of pGT1.8TM,[9] with IRES-PLAP added. To fuse the translational initiator of the EMCV IRES[19] to PLAP,[21] the IRES was digested at the *Nco*I site encompassing the endogenous viral initiator, the overhang filled in, and ligated to PLAP digested with *Sph*I and incubated with T4 DNA polymerase to remove the 3′ overhang. Translation beginning at this initiator ATG will produce fully wild-type PLAP. Using linkers, *Nhe*I sites were placed on both ends of the IRES-PLAP cassette, which was then inserted into the *Xba*I site of pGT1TM. Two Flp recognition target (FRT) sequences from yeast were included, one just upstream of the IRES, with an in-frame ER retention signal (KDEL), and another just downstream of the CD4 transmembrane domain of pGT1tm. These sites are designed to allow site-directed recombination to excise the β-*geo* gene in ES cells and to recreate a single FRT site that can be used to insert other genetic elements under the control of the endogenous gene's promoter.[22,23] Importantly, the transmembrane ER retention sequences should ensure that the gene is still disrupted and that the fusion protein remains sequestered intracellularly. (Note that this recombination system has not been tested in the context of this vector.) Finally, the *Hin*dIII site of pGT1tm was changed to *Sal*I for linearization of the PLAP vector. Vectors in all three reading frames were generated and were named pGT0TMpfs, pGT1TMpfs, and pGT2TMpfs. Some genes may only be accessible with one or two of these vectors, depending on the organization of the intron–exon boundaries.

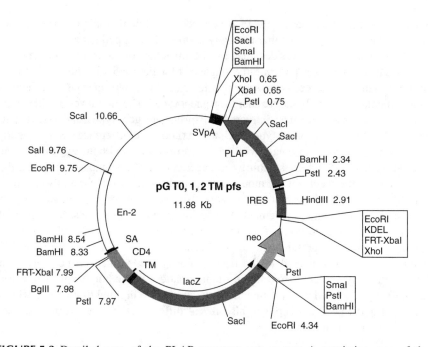

FIGURE 5.2 Detailed map of the PLAP secretory trap vector: A restriction map of the plasmid-based PLAP secretory trap vectors is shown. (Three vectors were constructed in three reading frames: pGT0,1,2 TMpfs.) Genetic elements are identified inside the circle, and enzyme sites used in construction of the vector or present in polylinkers are denoted outside the circle, along with the positions of FRT sites (FRT-XbaI) and an ER-retention signal (KDEL). The unique *Sal*I site is used to linearize the vector for electroporation. En-2, intron from the mouse engrailed locus; SA, splice acceptor; CD4, TM, transmembrane and flanking sequences from the CD4 gene; lacZ, β-galactosidase sequences fused to: neo, neomycin phosphotransferase sequences; IRES, internal ribosome entry site; PLAP, human placental alkaline phosphatase; SVpA, polyadenylation signal. See Methods for details of the construction of this vector.

5.2.2 ISOLATION OF GENE TRAP INSERTIONS AND GENERATION OF MUTANT MICE

Detailed protocols for the culture of the ES cells and for the entire electroporation procedure have recently been published for the original secretory trap vector.[12] Where there are deviations from these methods for the PLAP vector, the changes are noted below.

Briefly, feeder-independent E14Tg2A.4 or CGR8.8 embryonic stem cells are thawed and expanded over several passages to provide about 10^8 cells. The gene trap vector is linearized and electroporated into the ES cells. The electroporated cells are plated and grown in the presence of neomycin to select drug-resistant colonies. These colonies are picked to individual wells, allowed to grow to confluence, and passaged into duplicate wells. One set is stained for β-gal activity and screened visually to identify cells displaying a punctate subcellular pattern of X-gal staining, characteristic of genuine "secretory" insertions. Positive clones are expanded over several passages and again duplicated. RNA is prepared from one set while the other is frozen. A 5'RACE-PCR direct sequencing protocol is used to obtain a sequence tag from the upstream exons of the trapped gene in each cell line. This tag is used to identify the gene where possible or to extend the coding sequence using expressed sequence tags or genomic sequences. Lines selected on this basis for phenotypic analysis (i.e., genes encoding novel transmembrane or secreted proteins and known genes likely to play a role in the processes of interest) are thawed, expanded, and injected into blastocysts to generate chimeric mice. These mice are subsequently bred to generate stable lines that can be maintained as heterozygotes by molecular genotyping and intercrossed to generate homozygous mutant animals.

5.2.3 Preparation of DNA for Electroporation

150 μg of PLAP vector DNA (pGT0,1, or 2TMpfs) is linearized with *Sal*I overnight, precipitated with ethanol, washed with 70% ethanol, allowed to air-dry in a sterile laminar flow hood, resuspended in 0.1 ml of sterile PBS, vortexed, and allowed to resuspend over 4 h or more.

5.2.4 Screening X-Gal-Stained Cells for the "Secretory" Pattern of β-Gal Activity

Each electroporation typically yields around 300 drug-resistant colonies. These colonies are picked and clones are stained in 24 well dishes for X-gal activity as described.[12] They are then visually inspected on a microscope for the presence and subcellular distribution of X-gal staining. Typically, 20 to 30% of the cell lines will show some amount of X-gal staining, although this number can vary widely from electroporation to electroporation. Approximately 10% of the total number of colonies usually exhibit the "secretory" pattern of X-gal staining: a large number of discrete dots of various size in the perinuclear region and throughout the cytoplasm (Color Figure 5.3).* This should not be confused with some other common patterns that can arise including a diffuse spray throughout the cytoplasm or a single small dot per cell. These patterns normally do not represent productive insertions. The ES cell cultures in each well typically show some degree of differentiation with clumps of true embryonic stem cells growing on top of large spread-out differentiated cells (Color Figure 5.3). X-gal staining can be present in all cells of all types in the dish or in as few as 1%; both distributions can represent insertions in genuine secretory genes.

5.2.5 Generation of Sequence Tags

Preparation of RNA from ES cell lines, 5' RACE and direct sequencing are performed as described[8,12] using the primers below. For productive clones, this protocol typically generates sequence tags of 50–600 nucleotides (using a Cy5-labeled sequencing primer and running reactions on an ALF Express automated sequencer).

---

* Color figures follow page 140.

1st strand synthesis:	oligo 166:	5′ CGCCAGGGTTTCCCAGTCACGAC 3′
2nd strand synthesis:	oligo 56:	5′ GGTTGTGAGCTCTTCTAGATGGTTTT-TTTTTTTTTTTTT 3′
1st PCR:	oligo 59:	5′ GGTTGTGAGCTCTTCTAGATGG 3′
	oligo 92:	5′ CCAGAACCAGCAAACTGAAGGG 3′
2nd PCR:	oligo 67:	5′ AGTAGACTTCTGCACAGACACC 3′
	oligo 105:	5′ GGTTGTGAGCTCTTCTAGATGG 3′ (5′ biotinylated)
Sequencing:	oligo MD1:	5′ AAGAAGGAGCCTTCTCTGCC 3′

5.2.6 HISTOLOGY

1. Collect embryos/brains, fix at 4°C in 4% paraformaldehyde in 1 X PBS for 6 h to overnight. (Longer fixation makes sectioning easier but can drastically reduce β-gal activity; 6 h is optimal).
2. Wash three times with 1X PBS (1 h each). (Can store for weeks in PBS at 4°C).
3. Carefully dry tissue surface using Kimwipes and embed in 5% low-melt agarose in 1 X PBS (chilled on ice to speed up polymerization).
4. Section with vibrating microtome (e.g., Leica VT1000S). Alternate 100 μm sections through the entire brain can be collected on two sets of Superfrost Plus slides (Fisher Scientific) and air dried for 1 h and then stored in PBS. (Sections for PLAP staining can be post-fixed for 1 h in 4% formaldehyde or 4% PFA to reduce background [optional].)

5.2.7 X-GAL STAINING SECTIONS

1. Wash sections on slides two times in 0.1 M phosphate buffer (10′ each).
2. Prepare the required amount of X-gal staining buffer immediately before use by diluting a 50 mg/ml X-gal stock solution (dissolved in dimethyl-formamide) to a final concentration of 1 mg/ml in 0.1 M phosphate buffer containing 2 mM magnesium chloride, 5 mM potassium ferrocyanide (Sigma), and 5 mM potassium ferricyanide (Sigma). Filter the X-gal staining buffer to prevent crystal formation.
3. Transfer sections to staining solution and allow reaction to proceed at 37°C until desired intensity is achieved.
4. Rinse sections twice with phosphate buffer and post-fix for 1 h with 0.1 M phosphate buffer containing 5 mM EGTA, 2 mM magnesium chloride, and 0.2% glutaraldehyde (Sigma).
5. If no counterstain is desired, rinse off fix in water and mount in aqueous solution: Aqua PolyMount from PolySciences, Inc.
6. Alternatively, sections can be counterstained with Nuclear Fast Red (Vector Laboratories) and coverslipped with Cytoseal 60 (Stephens).

5.2.8 PLAP Staining Sections

1. Heat sections on slides at 65°C for 45 min. in PBS.
2. Allow to cool completely.
3. Transfer to AP buffer (100 mM Tris pH9.5, 100 mM NaCl, 50 mM MgCl$_2$) containing 0.1 mg/ml 5-bromo-4-chloro-3-indolyl phosphate and 1 mg/ml nitroblue tetrazolium).
4. Allow reaction to proceed at 37°C until desired intensity is achieved (15 min. to overnight; 4°C is best if leaving overnight).
5. Stop reaction with PBS/50 mM EDTA, pH 5.1 (PBS/EDTA).
6. Clear sections by passing through a methanol series and into benzyl benzoate:benzyl alcohol (BB:BA), 2:1. Rehydrate by passing back through the methanol series into PBS/EDTA. (5′ each 25%, 50%, 75%, 100% MeOH, then BB:BA for 1 h to overnight, then back through MeOH series, 5′ each and into PBS/EDTA.)
7. Coverslip with Aqua PolyMount from PolySciences Inc.

Notes:

1. Longer fixation in paraformaldehyde or formaldehyde further reduces background caused by endogenous phosphatases.
2. Heating at 65°C should only be done in PBS, since heating in AP buffer will destroy the tissue. Care must be taken to allow the samples to cool down completely before addition of the AP buffer.
3. Homozygotes will stain much darker than heterozygotes. For best comparison, heterozygotes should be allowed to stain two to three times longer.
4. Both the X-gal and PLAP staining reactions can be performed efficiently in small volumes in slide-mailers (SPI Supplies, West Chester, PA) or larger staining dishes.

5.3 POSSIBLE METHODOLOGIC VARIATIONS

It is possible in theory to prescreen insertions for neuronal expression by either differentiating cell lines into neurons in culture or by examining β-gal and PLAP expression in chimeric embryos. We have chosen not to pursue either of these methods for several reasons. First, a lack of expression in the particular type of neurons generated *in vitro* cannot be used to exclude expression in all neurons *in vivo*. Second, examining expression in chimeric embryos necessarily involves sacrificing the animals; positive clones must then be reinjected to establish lines of mice. Since the majority of lines examined (nearly 90%) show some expression in the nervous system, this duplication of effort far outweighs the slight degree of selection. We find it more efficient in the long term to simply establish heterozygous lines and screen them for expression and phenotype at the same time.

TABLE 5.1
Distribution of PLAP Secretory Trap Insertions

Genuine Secretory Genes:	185	Aberrant Events:	107
Known:	**155**	No ORF	4
Transmembrane	66	Unspliced	48
Multi-TM	7	5′ UTR	32
Secreted	33	Non-secretory gene	23
ER/Golgi/Lysosome	49		
Novel:	**30**		
Predicted:　TM	16		
Multi-TM	8		
Secreted	1		
Unclassified	5		

Note: Number of sequences.

5.4 EXAMPLES OF THE METHOD APPLIED TO ACTUAL DATA

This section summarizes, amplifies, and in some cases simply reproduces both results and discussion from two papers.[3,3a]

5.4.1 SPECTRUM OF GENES HIT

The ability of the PLAP vector to trap desired genes was tested by electroporating it into ES cells, then amplifying and sequencing the fusion transcripts from 466 X-gal positive clones (obtained over multiple experiments). By selecting cell lines that showed the distinctive subcellular distribution of β-gal activity (perinuclear with multiple cytoplasmic dots [Color Figure 5.3]), we achieved a significant enrichment for productive insertions in secreted and membrane proteins. Of 385 clones showing this secretory staining pattern, 292 yielded productive sequence (Table 5.1). Of these, 155 (53%) correspond to known transmembrane or secretory genes in the public database, and another 30 (10%) represent novel genes, 25 of which are predicted to encode transmembrane or secreted proteins. Due to repeat hits, mainly in a small number of "hotspot" loci, these 185 known and novel sequences represent 115 different genes.

The remaining 107 sequences (28%) that gave a "secretory" staining pattern were false positives. These can arise due to inefficient splicing to the vector or insertion in the 5′UTR of genes. In these cases, translation from internal ATG start sites or cryptic start sites (CTG) within the transmembrane-encoding sequence of the vector may lead to expression of an active β-gal protein. These false positives can be rapidly identified by sequencing and are not pursued for phenotypic analysis. Additional clones that gave a "nonsecretory" pattern of X-gal staining for the most part either gave no sequence (32/81) or represented aberrant splicing or translation events (43/81).

A selected list of the known genes trapped with the PLAP vector is presented in Table 5.2 (for a complete and current list, see www.genetrap.org). They include single-pass transmembrane receptors and adhesion molecules, multiple-membrane-spanning proteins such as ion channels, secreted and extracellular matrix proteins, and proteins localized to the membranes of intracellular compartments. This latter class includes examples that have been linked to human disorders leading to mental retardation or neurodegeneration (e.g., alpha-mannosidase[24] and N-acetyl glucosamine 6-sulfatase[25]) and also some members with surprisingly specific roles in development (e.g., EXT-1, see below).

Importantly, for the study of brain wiring, 13 of these gene products are known or predicted axon-guidance molecules, including two Eph receptors, several cell adhesion molecules, and a plexin family member. In addition, several of the novel genes contain modules that have been found in axon-guidance molecules, such as immunoglobulin domains, fibronectin type-III domains, plexin repeats, leucine-rich repeats, and cadherin domains.[3a]

5.4.2 INSERTION HOTSPOTS AND TARGET POOL SIZE

Certain loci appear to be hotspots for insertion of the gene trap vector, with extreme examples including *Lamc1* (*laminin γ1*), *Ptprk* (*protein tyrosine phosphatase, receptor K*), *Hspg2* (*perlecan*) and *Cdh1* (*E-cadherin*). The reason for this is not known, but it does not appear to reflect particularly high expression levels in ES cells. Insertions in these loci can represent up to a third of the total known genes but can be rapidly screened out by sequence. Because there are also some weaker hotspots, it is difficult to use the standard measure of the number of repeat hits as a means of estimating the target pool size, since they may not represent a true trend toward saturation. A better measure in this case is simply the proportion of unique hits per electroporation, which has remained constant at approximately one third (for known genes). The fact that the rate of insertion in genes that have not been hit before is not decreasing indicates that the screen is not approaching saturation, at least for now.

5.4.3 BUILDING A MOLECULAR WIRING DIAGRAM OF THE BRAIN

We have observed a remarkable variety and specificity of axonal staining patterns in the lines derived from insertions in known and novel genes (Color Figure 5.4).* For example, while *Ptprk* and *neogenin* are both expressed in cortical axons that cross in the corpus callosum and in the anterior commissure, only *neogenin* labels the subcortical commissures in the hippocampus and septum (Color Figure 5.4a, c). In contrast, commissural axons are not labeled in *Adam23* or *Crim1* animals at P0, whereas the striatum shows strong expression for *Adam23* but not for *Ptprk* or *neogenin* (Color Figure 5.4 a–d). Color Figure 5.4 also shows examples of expression patterns for two genes encoding novel transmembrane proteins. *LST16* encodes a

---

* Color figures follow page 140.

TABLE 5.2
Examples of Known Genes Trapped
with the PLAP Secretory Trap Vector

Transmembrane

Adam19	a disintegrin and metalloprotease domain 19
Adam23	a disintegrin and metalloprotease domain 23
Atrn	attractin
Cd98	CD98 antigen
Crim1	cysteine-rich motor neuron 1
Cdh1	cadherin 1 (E-cadherin)
Cdh3	cadherin 3 (P-cadherin)
Emb	embigin
Eng	endoglin
EphA2	Eph receptor A2
EphA4	Eph receptor A4
Fat1	fat 1 cadherin
Gp330	glycoprotein 330 (LRP2)
Icam	intercellular adhesion molecule
Itga5	integrin alpha 5
Itga6	integrin alpha 6
Jag1	jagged 1
Jcam	junction cell adhesion molecule
Kit	kit oncogene
Ldlr	low density lipoprotein receptor
Lifr	leukemia inhibitory factor receptor
Lisch7	liver-specific bHLH-Zip transcription factor
Lrp6	LDL receptor-related protein 6
Mfge8	milk fat globule-EGF factor 8 protein
Neo1	neogenin
Notch 1	Notch gene homolog 1, (Drosophila)
Notch 2	Notch gene homolog 2, (Drosophila)
Notch 3	Notch gene homolog 3, (Drosophila)
Plxn1	plexin-A1
Ptprf	protein tyrosine phosphatase, receptor (LAR)
Ptprk	protein tyrosine phosphatase, receptor K
Ptprs	protein tyrosine phosphatase, receptor S
Pvs	poliovirus sensitivity
Sdfr1	stromal cell derived factor receptor 1
Selel	selectin, endothelial cell, ligand
Sema6A	semaphorin 6A
Tmeff1	TM protein EGF-like, follistatin-like 1

Multi-transmembrane

Gabbr1	GABA B receptor, 1a
Kcnk5	potassium channel, subfamily K, member 5
LGR4	(human G protein-coupled receptor, GPR48)
N33	(human putative tumor suppressor)
TM9SF3	transmembrane protein p76

TABLE 5.2 (continued)
Examples of Known Genes Trapped
with the PLAP Secretory Trap Vector

Secreted

Agrn	agrin
Cyr61	cysteine rich protein 61
Fbln1	fibulin 1
Fbn	fibronectin
Gpc3	glypican 3
Gpc4	glypican 4
Hsgp2	perlecan (heparan sulfate proteoglycan 2)
Lama1	laminin, alpha 1
Lama5	laminin, alpha 5
Lamb1-1	laminin beta 1
Lamc1	laminin, gamma 1
Leftb	left-right determination factor B
Ntn1	netrin 1
Pros1	protein S (alpha)
Spint2	serine protease inhibitor, Kunitz type 2 (Hai-2)
Tfpi	tissue factor pathway inhibitor

ER/Golgi/Lysosomal

Asph	aspartate beta-hydroxylase
ATP6K	ATPase, H+ transporting lysosomal protein
Beta3gnt	beta-1,3-N-acetylglucosaminyltransferase 1
Canx	calnexin
Cd36l2	CD36 antigen-like 2
Ctsc	cathepsin C
Erj3	DNAJ1/2 related chaperone
Ero1l	ERO-1 like (*S. cerevisiae*)
Ext1	exostoses (multiple) 1
Glc6S	glucosamine (N-acetyl)-6-sulfatase
Grp58	glucose regulated protein, 58kD/Erp60
Hs6st	heparan sulfate 6-O-sulfotransferase 1
Man2a1	mannosidase 2, alpha 1
P4hb	prolyl 4-hydroxylase, beta polypeptide
Pcsk3	proprotein convertase subtilisin type 3 (furin)
Plod2	procollagen lysine, oxoglutarate dioxygenase 2
Ppib	peptidylprolyl isomerase B
Ptdss2	phosphatidylserine synthase 2
Rpn2	ribophorin 2
S1p	site-1 protease
Stch	stress 70 protein chaperone
Tor1b	torsin family 1, member B

predicted transmembrane protein with a combination of leucine-rich and immuno-globulin repeats (Color Figure 5.4e), which labels the habenula, piriform cortex, and barrel cortex (Color Figure 5.4f, g). *KST37* encodes a predicted nidogen/plexin homology transmembrane protein (Color Figure 5.4h) which is expressed in various peripherally projecting axons (Color Figure 5.1c), as well as the fimbria and stripes of Purkinje cells in the cerebellum (Color Figure 5.4j, k). Overall, the majority of lines examined (88%) exhibit staining in the nervous system and many help mark and define specific axon tracts.

5.4.4 PHENOTYPIC ANALYSIS

Prior to screening for axon guidance defects, heterozygotes are intercrossed and neonatal litters examined for obvious morphological or behavioral defects. Because early embryonic defects often lead to resorption of the embryo, such phenotypes are not apparent at birth, except as a decrease in litter size. However, the molecular tag generated by the gene trap insertions makes it possible to identify essential genes simply by the absence of homozygotes at weaning. Earlier litters can then be genotyped to determine the stage at which homozygotes die and analyzed for morphological defects. Of 100 insertions of either the original secretory trap vector or the modified PLAP secretory trap vector transmitted to the germline of mice, about one third caused lethal phenotypes in homozygous offspring with examples that perturb all stages and many aspects of embryonic and postnatal development.[3a]

These include a number with defects in nervous system development and/or function, affecting, for example: neural tube closure (*KST27*, Color Figure 5.5a),* CNS midline patterning (*Ext1*, Color Figure 5.5b), mid- and hindbrain patterning (*Lrp6*,[26] Color Figure 5.5d), axon guidance (*netrin-1*,[27] *neuropilin-2*,[16] *Sema6A* [see below], *EphA4* [see below]), neuromuscular junction formation (*agrin*[28]), forebrain cholinergic innervation (*Ptprf* (*LAR*)[29]), locomotion (*EphA4*, see below), and motor control (*Adam23*, Color Figure 5.5e).

For several of these lines, comparison with other mutant phenotypes suggests a linkage in the same biochemical processes. For example, the spectrum of abnormalities observed in mutants of the LDL-receptor-related protein *Lrp6*, including the loss of specific structures in the mid- and hindbrain (Color Figure 5.5d), suggested that Lrp6 may function as a coreceptor in Wnt protein signaling,[26] a hypothesis that was subsequently confirmed biochemically.[30] Similarly, the CNS and limb defects in mutants of the glycosyltransferase *Ext1* (Color Figure 5.5b) are reminiscent of defects in *Sonic*[31] and *Indian*[32] *Hedgehog* mutants, consistent with the proposed function of *tout-velu*, the Drosophila homologue of *Ext1* in promoting diffusion of the Hedgehog morphogen.[33,34] The phenotype caused by an insertion in the disintegrin-metalloproteinase family member *Adam23,* where homozygous pups display tremor and ataxia and die by 2 weeks of age, is very similar to the phenotype reported for the knockout of the highly related gene, *Adam22*,[35] suggesting that these two genes may function together in a nonredundant fashion. Adam23 has been shown to promote cell adhesion through binding of integrin $\alpha v \beta 3$[36] and is highly expressed in the cerebellum

---

* Color figures follow page 140.

(Color Figure 5.5f) and basal ganglia (Color Figure 5.4d). Both these areas are potential loci for movement disorders.

5.4.5 AXON-GUIDANCE PHENOTYPES IN **PLAP** GENE TRAP MUTANTS

In addition to identifying mutants with morphological or neurological defects, a major goal of the screen is to discover novel axon-guidance molecules by studying wiring patterns in mutant animals. Both neonatal lethal and fully viable lines are screened since axon-guidance molecules are not necessarily expected to cause lethality or even overt gross phenotypes. Axon-guidance defects have been observed in two mutant lines, *Sema6A* and *EphA4*. Preliminary analysis of these two mutants confirms (1) that we are able to identify novel axon-guidance phenotypes, and (2) that characterization of PLAP staining can also be used to extend the analysis of genes that have already been knocked out.

5.4.6 *SEMAPHORIN6A* MUTANTS HAVE DEFECTS IN THALAMOCORTICAL PROJECTIONS

Sema6A belongs to a subfamily of semaphorins[37] characterized by an extracellular semaphorin domain, a transmembrane domain and a long cytoplasmic tail.[38,39] Members of this class, including *Sema6A*, can repel some axons *in vitro,*[39,40] consistent with a traditional role as guidance signals. However, the length of the cytoplasmic tail, which includes an Evl-binding site in Sema6A,[41] and a Src-binding site in Sema6B,[42] suggests that these semaphorins may also function as receptors.

We isolated an insertion in the *Sema6A* gene that completely abolishes wild-type *Sema6A* transcripts. Staining for β-gal activity duplicates the published *in situ* hybridization pattern for Sema6A[38] with particularly strong expression in the dorsal thalamus (Color Figure 5.6a–b).* PLAP staining labels a number of specific axonal populations in this line including thalamocortical axons (Color Figure 5.6c, d). In homozygotes, these axons show a dramatic phenotype, failing to turn up through the internal capsule and instead projecting down toward the amygdala (Color Figure 5.6g–i). This defect is fully penetrant but is specific to caudal thalamocortical axons; rostral projections appear normal in every (n = 12) animal (Color Figure 5.6j–m). Misrouting of thalamocortical axons was confirmed by injection of the lipophilic axonal tracer, DiI, into the dorsal thalamus (Color Figure 5.6f,i). Thus, the PLAP marker provides a powerful tool to detect axon guidance defects *in vivo*.

5.4.7 REVISITING DEFECTS IN *EPHA4* MUTANTS USING THE **PLAP** MARKER

An additional test case was provided by an insertion in the *EphA4* gene. Null mutations generated by gene targeting of *EphA4*[43,44] exhibit a hopping-kangaroo gait and display defects in the corticospinal tract (CST) and anterior commissure (AC).

---

* Color figures follow page 140.

Analysis of the gene trap *EphA4* allele using the PLAP marker (Color Figure 5.7)* showed the same phenotypes previously described in the targeted alleles using antibodies and dye tracers, confirming the mutagenicity of the insertion and also demonstrating the usefulness of the PLAP axonal marker. In addition, the β-gal and PLAP expression patterns prompted a reinterpretation of the cell-autonomy of the observed phenotypes. The fact that EphA4 is expressed in the CST but not the aAC, as revealed by the PLAP marker, suggests that it functions cell-autonomously in directing the development of the CST but not the aAC. Indeed, Kullander et al.[44a] have obtained direct evidence in support of such roles through the analysis of *EphA4* kinase-deficient and null mice.

5.5 DATA INTERPRETATION

5.5.1 BACKGROUND STAINING

While most endogenous alkaline phosphatase activity can be eliminated by heat inactivation, some background staining is still seen in the nervous system, especially at early embryonic stages. This is usually restricted to specific populations of radial glia in the spinal cord and brain but may also include some neurons. To ensure that observed staining patterns reflect expression of the trapped gene, the alkaline phosphatase stain should be compared with the X-gal stain, which has much lower background (essentially none). Wild-type littermates can also be stained in parallel to identify regions of background staining at specific stages.

5.5.2 SECRETORY TRAP INSERTIONS GENERATE NULL ALLELES

In order to correctly interpret mutant phenotypes, it is crucial to know whether the alleles induced are true nulls. Several lines of evidence suggest that gene trap insertions that fall within the coding sequence of the gene typically generate phenotypic and molecular null alleles.[3a,16,26,28,45] However, residual levels of transcript have been detected in some cases[27,29] and it is advisable to check in each case (most quantitatively by RNase protection analysis) whether wild-type transcripts and protein are completely abolished before attempting to interpret phenotypes, or the lack of a phenotype, in any given line.

A potential confounding factor, in theory, is that the fusion protein that is generated between upstream exons of the trapped gene and the β-*geo* gene could conceivably have residual or dominant activities. However, we have found that these fusions are sequestered in intracellular compartments, generating the characteristic punctate subcellular staining pattern.[9] Fusions with secreted or transmembrane proteins are therefore unlikely to have any effect at the cell surface. It is theoretically possible however, that fusions involving Golgi or endoplasmic reticulum proteins will have residual activity in some cases. This may explain why the defects we observe in our *Ext1* gene trap allele are less severe than a recently reported targeted knockout of *Ext1* where no homozygous embryos were observed beyond E8.5.[46]

---

* Color figures follow page 140.

5.5.3 GENETIC BACKGROUND

In our breeding protocol, we generally backcross twice to C57/Bl6 to dilute potential effects of random mutations that may have occurred in the ES cell clone. However, these lines are certainly not fully inbred and some variability between animals may still arise. In addition we have observed differences in lethality and other phenotypes upon outbreeding to different strains (e.g., in the cases of *neuropilin-2, fat 1 cadherin,* and *glypican 3*; data not shown).

5.5.4 CELL AUTONOMY

Expression of the PLAP marker in misrouted axons can provide strong evidence that a guidance defect is cell autonomous. Both the β-gal and PLAP markers are extremely sensitive, as shown by the example of *EphA4*, where expression was readily detected in CST neurons in contrast to previous findings by antibody or *in situ* hybridization.[43] Thus, EphA4 may indeed act as a receptor in the CST, a model further supported by analysis of *EphA4* kinase-deficient mice.[44a]

On the other hand, expression of PLAP does not necessarily prove a cell-autonomous function for the gene in question, especially in cases where it is also expressed in the environment of the growing axons. For example, *Sema6A* is expressed both in thalamocortical axons and in the amygdala at the time when these axons are growing (E12.5-E14.5, Color Figure 5.6a). Thus it is impossible to distinguish fully between autonomous (receptor) and non-autonomous (ligand) activities of this semaphorin without further experimentation.

5.6 DISCUSSION

We have developed and applied a modified gene trap method to perform a phenotypic screen in mice for defects in various aspects of brain development, especially axon guidance. The most important properties of this method are the following:

1. Insertions effectively block production of wild-type transcript and generate phenotypic null or severe hypomorphic alleles.
2. Prescreening insertions using the selective properties of the secretory trap vector and on the basis of sequence allows one to focus phenotypic analysis on genes most likely to be directly involved in the processes of interest.
3. The generation of stable lines that can be genotyped molecularly permits screening for defects in various processes at all stages of development.
4. The addition of the axonal marker PLAP in the secretory trap vector provides a means to label subpopulations of axons *in vivo* and to screen for genes required for their normal projections.

These properties confer significant advantages over screens using chemical mutagens such as ENU. Indeed, without the presence of a cell-autonomous axonal marker, it would be practically impossible to screen through random mutants for wiring defects. On the other hand, the gene trapping method does have several limitations.

First, it is not possible to screen through the very high numbers of mutations that can be produced in ENU screens, though this problem is offset by the ability to focus on specific classes of genes. Second, the processes mediating insertion of the gene trap vector into the genome are not well understood. Integration is clearly not completely random and many genes may be inaccessible with this method due to their chromatin configuration or lack of expression in ES cells. In addition, the vector we use relies on splicing and thus cannot trap genes that lack introns. Thus, the gene trap screen is not designed to saturate the genome. However, the pool of accessible genes can be estimated at several thousand (encoding transmembrane and secreted proteins) and includes many genes expressed in highly restricted patterns in the nervous system.

The use of the PLAP marker in conjunction with the gene trap vector solves, at least for cell-autonomous mechanisms, the most vexing problem facing a genetic screen for brain wiring mechanisms: how to examine only those axons most likely to be affected by the mutation among the many normally projecting axons. As shown for *Sema6A* and *EphA4*, one can readily identify defects in the trajectories of axons expressing the trapped gene simply by comparing the wiring of PLAP-stained axons in heterozygous and homozygous embryos. Since the method does not rely on advance knowledge of the type of molecules that will be involved, axon-guidance mechanisms in mammals can be identified without bias for the kinds of proteins, their expression levels, or the tissues in which they are produced.

In addition to their usefulness as mutagens, each insertion has added value in several ways. First, the insertions identify expressed sequences, including many encoding novel transmembrane or secreted genes. Second, each insertion incorporates site-specific recombination elements designed to allow insertion of any desired sequence into the locus, under the control of the endogenous gene's promoter and enhancer sequences (see Methods). Third, the incorporation of reporter genes makes these mouse lines useful even in the absence of a mutant phenotype. In particular, insertions of the PLAP secretory trap vector can provide markers for previously untraceable or even unknown axonal tracts. In addition, this approach will accelerate the enumeration of the various membrane proteins made by particular axonal tracts and help build a molecular map of axonal projections. As the complement of surface proteins made by axons is defined, it may be possible to uncover a code of surface receptors that defines particular axonal trajectories, much as recent studies have defined a transcriptional code (the LIM code) involved in specifying motor neuron trajectories.[47,48] Finally, the ongoing generation of a bank of mouse lines in which particular axonal populations are labeled with PLAP (which can be accessed through www.genetrap.org) will provide a resource for the scientific community to help elucidate the normal wiring diagram of the mammalian brain.

REFERENCES

1. Edlund, T. and Jessell, T. M., Progression from extrinsic to intrinsic signaling in cell fate specification: A view from the nervous system, *Cell,* 96, 211–224, 1999.
2. Goodman, C. S. and Shatz, C. J., Developmental mechanisms that generate precise patterns of neuronal connectivity, *Cell,* 72 (Suppl), 77–98, 1993.

3. Leighton, P. A., Mitchell, K. J., Goodrich, L. V., Lu, X., Pinson, K., Scherz, P., Skarnes, W. C., and Tessier-Lavigne, M., Defining brain wiring patterns and mechanisms through gene trapping in mice, *Nature,* 410, 174–179, 2001.

3a. Mitchell, K. J., Pinson, K., Kelly, O. G., Brennan, J., Zupicich, J., Scherz, P., Leighton, P. A., Goodrich, L. V., Lu, X., Avery, B. J., Tate, P., Dill, K., Pangilinan, E., Wakenight, P., Tessier-Lavigne, M., and Skarnes, W. C., Functional analysis of secreted and trans-membrane proteins critical to mouse development, *Nature Genetics,* 28, 241–249, 2001.

4. Nüsslein-Volhard, C. and Wieschaus, E., Mutations affecting segment number and polarity in Drosophila, *Nature,* 287, 795–801, 1980.

5. Seeger, M., Tear, G., Ferres-Marco, D., and Goodman, C. S., Mutations affecting growth cone guidance in Drosophila: Genes necessary for guidance toward or away from the midline, *Neuron,* 10, 409–426, 1993.

6. Brennan, J. and Skarnes, W. C., Gene trapping in mouse embryonic stem cells, *Meth. Mol. Biol.,* 97, 123–138, 1999.

7. Friedrich, G. and Soriano, P., Promoter traps in embryonic stem cells: A genetic screen to identify and mutate developmental genes in mice, *Genes Dev.,* 5, 1513–1523, 1991.

8. Townley, D. J., Avery, B. J., Rosen, B., and Skarnes, W. C., Rapid sequence analysis of gene trap integrations to generate a resource of insertional mutations in mice, *Genome Res.,* 7, 293–298, 1997.

9. Skarnes, W. C., Moss, J. E., Hurtley, S. M., and Beddington, R. S., Capturing genes encoding membrane and secreted proteins important for mouse development, *Proc. Natl. Acad. Sci. USA,* 92, 6592–6596, 1995.

10. Tate, P., Lee, M., Tweedie, S., Skarnes, W. C., and Bickmore, W. A., Capturing novel mouse genes encoding chromosomal and other nuclear proteins, *J. Cell Sci.,* 111, 2575–2585, 1998.

11. Zambrowicz, B. P. et al., Disruption and sequence identification of 2,000 genes in mouse embryonic stem cells, *Nature,* 392, 608–611, 1998.

12. Skarnes, W. C., Gene trapping methods for the identification and functional analysis of cell surface proteins, *Meth. Enzymol.,* 328, 592–615, 2000.

13. Hrabe de Angelis, M. et al., Genome-wide, large-scale production of mutant mice by ENU mutagenesis, *Nat. Gen.,* 25, 444–447, 2000.

14. Kasarskis, A., Manova, K., and Anderson, K. V., A phenotype-based screen for embryonic lethal mutations in the mouse, *Proc. Natl. Acad. Sci. USA,* 95, 7485–7490, 1998.

15. Giger, R. J. et al., Neuropilin-2 is required *in vivo* for selective axon guidance responses to secreted semaphorins, *Neuron,* 25, 29–41, 2000.

16. Chen, H. et al., Neuropilin-2 regulates the development of selective cranial and sensory nerves and hippocampal mossy fiber projections, *Neuron,* 25, 43–56, 2000.

17. Gustincich, S., Feigenspan, A., Wu, D. K., Koopman, L. J., and Raviola, E., Control of dopamine release in the retina: A transgenic approach to neural networks, *Neuron,* 18, 723–736, 1997.

18. Fields-Berry, S. C., Halliday, A. L., and Cepko, C. L., A recombinant retrovirus encoding alkaline phosphatase confirms clonal boundary assignment in lineage analysis of murine retina, *Proc. Natl. Acad. Sci. USA,* 89, 693–697, 1992.

19. Mombaerts, P. et al., Visualizing an olfactory sensory map, *Cell,* 87, 675–686, 1996.

20. Mountford, P. S. and Smith, A. G., Internal ribosome entry sites and dicistronic RNAs in mammalian transgenesis, *Trends Gen.,* 11, 179–184, 1995.

21. Gunning, P., Leavitt, J., Muscat, G., Ng, S. Y., and Kedes, L., A human beta-actin expression vector system directs high-level accumulation of antisense transcripts, *Proc. Natl. Acad. Sci. USA,* 84, 4831–4835, 1987.

22. Schübeler, D., Maass, K., and Bode, J., Retargeting of retroviral integration sites for the predictable expression of transgenes and the analysis of cis-acting sequences, *Biochemistry,* 37, 11907–11914, 1998.

23. Seibler, J., Schübeler, D., Fiering, S., Groudine, M., and Bode, J., DNA cassette exchange in ES cells mediated by Flp recombinase: An efficient strategy for repeated modification of tagged loci by marker-free constructs, *Biochemistry,* 37, 6229–6234, 1998.

24. Berg, T. et al., Spectrum of mutations in alpha-mannosidosis, *Am. J. Human Gen.,* 64, 77–88, 1999.

25. Jones, M. Z. et al., Human mucopolysaccharidosis IIID: Clinical, biochemical, morphological and immunohistochemical characteristics, *J. Neurop. Exp. Neurol.,* 56, 1158–1167, 1997.

26. Pinson, K. I., Brennan, J., Monkley, S., Avery, B.J., and Skarnes, W.C., An LDL receptor-related protein mediates Wnt signaling in mice, *Nature,* 407, 535–538, 2000.

27. Serafini, T. et al., Netrin-1 is required for commissural axon guidance in the developing vertebrate nervous system, *Cell,* 87, 1001–1014, 1996.

28. Burgess, R. W., Skarnes, W. C., and Sanes, J. R., Agrin isoforms with distinct amino termini. Differential expression, localization, and function, *J. Cell Biol.,* 151, 41–52, 2000.

29. Yeo, T. T. et al., Deficient LAR expression decreases basal forebrain cholinergic neuronal size and hippocampal cholinergic innervation, *J. Neurosci. Res.,* 47, 348–360, 1997.

30. Tamai, K. et al., LDL Receptor-Related proteins in Wnt signal transduction, *Nature,* 407, 530–535, 2000.

31. Chiang, C. et al., Cyclopia and defective axial patterning in mice lacking Sonic hedgehog gene function, *Nature,* 383, 407–413, 1996.

32. St-Jacques, B., Hammerschmidt, M., and McMahon, A. P., Indian hedgehog signaling regulates proliferation and differentiation of chondrocytes and is essential for bone formation, *Genes Dev.,* 13, 2072–2086, 1999.

33. The, I., Bellaiche, Y., and Perrimon, N., Hedgehog movement is regulated through tout velu-dependent synthesis of a heparan sulfate proteoglycan, *Mol. Cell,* 4, 633–639, 1999.

34. Bellaiche, Y., The, I., and Perrimon, N., Tout-velu is a Drosophila homologue of the putative tumour suppressor EXT-1 and is needed for Hh diffusion [see comments], *Nature,* 394, 85–88, 1998.

35. Sagane, K., Yamazaki, K., Mizui, Y., and Tanaka, I., Cloning and chromosomal mapping of mouse ADAM11, ADAM22 and ADAM23, *Gene,* 236, 79–86, 1999.

36. Cal, S., Freije, J.M., Lopez, J.M., Takada, Y., and Lopez-Otin, C., ADAM23/MDC3, a human disintegrin that promotes cell adhesion via interaction with the alphavbeta3 integrin through an RGD-independent mechanism, *Mol. Biol. Cell.,* 4, 1457–1469, 2000.

37. Raper, J. A., Semaphorins and their receptors in vertebrates and invertebrates, *Curr. Op. Neuro.,* 10, 88–94, 2000.

38. Zhou, L. et al., Cloning and expression of a novel murine semaphorin with structural similarity to insect semaphorin I, *Mol. Cell. Neurosci.,* 9, 26–41, 1997.

39. Kikuchi, K. et al., Cloning and characterization of a novel class VI semaphorin, semaphorin Y, *Mol. Cell. Neurosci.,* 13, 9–23, 1999.

40. Xu, X. M. et al., The transmembrane protein semaphorin 6A repels embryonic sympathetic axons, *J. Neurosci.,* 20, 2638–48, 2000.

41. Klostermann, A., Lutz, B., Gertler, F., and Behl, C., The orthologous human and murine semaphorin 6A-1 proteins (SEMA6A-1/Sema6A-1) bind to the Enabled/Vasodilator-stimulated Phosphoprotein-like Protein (EVL) via a novel carboxyterminal Zyxin-like domain, *J. Biol. Chem.*, 275, 39647–39653, 2000.

42. Eckhardt, F. et al., A novel transmembrane semaphorin can bind c-src, *Mol. Cell. Neurosci.*, 9, 409–419, 1997.

43. Dottori, M. et al., EphA4 (Sek1) receptor tyrosine kinase is required for the development of the corticospinal tract, *Proc. Natl. Acad. Sci. USA*, 95, 13248–13253, 1998.

44. Helmbacher, F., Schneider-Maunoury, S., Topilko, P., Tiret, L., and Charnay, P., Targeting of the EphA4 tyrosine kinase receptor affects dorsal/ventral pathfinding of limb motor axons, *Development*, 127, 3313–3324, 2000.

44a. Kullander, K., Mather, N. K., Diella, F., Dottori, M., Boyd, A. W., and Klein, R., Kinase-dependent and kinase-independent functions of EphA4 receptors in major axon tract formation *in vivo*, *Neuron*, 29, 73–84, 2001.

45. Paine-Saunders, S., Viviano, B. L., Zupicich, J., Skarnes, W. C., and Saunders, S., Glypican-3 controls cellular responses to BMP4 in limb patterning and skeletal development, *Dev. Biol.*, 225, 179–187, 2000.

46. Lin, X. et al., Disruption of gastrulation and heparan sulfate biosynthesis in EXT1-deficient mice, *Dev. Biol.*, 224, 299–311, 2000.

47. Briscoe, J., Pierani, A., Jessell, T. M., and Ericson, J., A homeodomain protein code specifies progenitor cell identity and neuronal fate in the ventral neural tube, *Cell*, 101, 435–445, 2000.

48. Tsuchida, T. et al., Topographic organization of embryonic motor neurons defined by expression of LIM homeobox genes, *Cell*, 79, 957–970, 1994.

6 N-Ethyl-N-Nitrosourea Mutagenesis: Functional Analysis of the Mouse Nervous System and Behavior

Stephen J. Kanes, Gillian R. Leach, and Maja Bucan

CONTENTS

6.1 INTRODUCTION

As the human and mouse genome projects approach completion, the challenge will be to determine the function of the many genes being discovered. Often, newly identified genes can be assigned a function based on their homology with known genes, but, to fully understand their function, one must study the phenotypic consequences of mutations in these genes. This effort, often called functional genomics, has the potential to revolutionize our understanding of genes that affect behavior and perhaps provide critical clues about the biological underpinnings of neurological and psychiatric disorders. The mouse represents an important model organism for these studies; a variety of mutations can be engineered in individual genes and progeny of mutagenized mice can be examined to identify genetic variants with specific behavioral anomalies.

Large-scale mutagenesis screens have been successfully carried out for developmental anomalies and other phenotypic traits in a number of model organisms such as *Drosophila melanogaster*,[1,2] *Arabidopsis thaliana*,[3,4] and *Caenorhabditis elegans*.[5-9] A recent screen for a wide range of developmental mutants in the zebrafish, *Danio rerio*,[10-13] represents the first attempt of mutagenesis in a vertebrate organism that approaches saturation. In the mouse, large-scale mutagenesis studies employing the chemical mutagen N-ethyl-N-nitrosourea (ENU) are being carried out in several academic centers in the United States (Tennessee Mouse Genome Consortium/Oak Ridge National Laboratory [http://lsd.ornl.gov], Northwestern University [http://genome.northwestern.edu], and the Jackson Laboratory [www.jax.org]) and in Europe (Mammalian Genetics Research Center Harwell, U.K. [http://www.mgu.har.mrc.ac.uk/][14] and the Institute of Mammalian Genetics, Neuherberg, Germany [http://www.gsf.de/isg/institute.html][15]). These centers are assessing neurological and behavioral measures as part of a comprehensive phenotypic screen.

Because of the high mutation rate and a simple experimental procedure for generation of mutagenized mice, ENU screens can also be efficiently performed on a small scale, by individual laboratories. ENU is a highly effective chemical mutagen that has been used in a wide variety of model organisms, and in the mouse, ENU represents the mutagen of choice. ENU can be employed to produce mutations in a set of known genes, which can then be detected by DNA mutation analysis in progeny of mutagenized mice or in mutagen-treated embryonic stem (ES) cells (gene-based mutagenesis).[16] The unique power of this approach is in the ability to select novel mutations based on abnormal phenotypes (phenotype-based mutagenesis). To identify novel mutations with neurobiological defects, it is possible to screen for visible neurological symptoms (such as ataxia, tremor, circling) or specific pathological changes (cellular degeneration, loss of specific brain structures, altered morphology). Furthermore, in order to observe behavioral anomalies, it is usually necessary to utilize behavioral assays that can reveal subtle changes in mice that are apparently normal.

This chapter discusses the use of chemical mutagenesis and the way it has been, and can be, employed in the study of neurobiology and behavior, in particular, the study of behavioral models relevant to psychiatric illness. This approach involves (a) the identification of behavioral traits that can be modeled in rodents and their baseline analysis in several inbred strains,[17,18] (b) the identification of mouse mutants with anomalies in these simple behavioral traits in an ENU mutagenesis screen, and (c) the positional cloning of genes that underlie these traits. The goal of this chapter is to provide a framework for selecting appropriate phenotypic assays, breeding and mapping approaches, as well as provide guidance for the use of these mutants in the dissection of complex behavioral pathways.

6.2 ENU MUTAGENESIS TECHNIQUE

ENU is a potent chemical mutagen that induces mutations at a rate of approximately 1:750/locus/gamete.[19] In the mouse, the primary chemical lesion induced is a point mutation consisting of A-T to T-A transversions and A-T to G-C transitions.[20-22] Mutations induced by other mutagenesis techniques such as γ-irradiation or chlorambucil primarily cause large chromosomal rearrangements which may produce a phenotype due to the disruption of many genes. A useful feature of ENU-induced point mutations is that the mutant phenotype is caused by the mutation in a single gene, moreover, these mutations can result in either loss or gain of function as well as changes in the level of expression or function (hypomorphic or hypermorphic alleles).[22,23]

The action of ENU is most potent in germ cells, specifically the male spermatogonial cells.[21] Thus, the experimental approach to producing mutageneized mice involves treating males with ENU, most typically 300–400 mg/kg administered via intra-peritoneal injection once per week in divided doses over 3 to 4 weeks.[22,24-26] Starting a few weeks after treatment, mice undergo a sterile period of approximately 11–22 weeks. Those gametes that survive the treatment repopulate the testes and give rise to mutagenized sperm. The resulting gametes represent the clonal expansion of approximately 100 surviving spermatogonial stem cells.[24]

The optimal dose of ENU varies among inbred strains. Some inbred strains are more sensitive to ENU and may not recover fully after such treatment. The preferred ENU dose has been published for each of several commonly used inbred strains.[17,24,26] If these data are not available for a strain that represents the strain of choice for the studied phenotype, it is necessary, before initiating a new screen, to test several doses of ENU.[24,26]

6.3 SELECTION OF BEHAVIORAL PHENOTYPES

6.3.1 The Role of Endophenotypes in Modeling of Psychiatric Disorders

The initial step in a mutagenesis experiment is the selection of an appropriate assay that has the potential of uncovering anomalies in the phenotype of interest. In the case of animal models for psychiatric illnesses, the selection of phenotypes appropriate for a mutagenesis screen is made difficult by the way the disorders are defined. Psychiatric disorders are classified based on the presence or absence of empiric psychometrically validated criteria as set out in the *Diagnostic and Statistical Manual of Mental Disorders* of the American Psychiatric Association IV-TR.[27] This approach relies on the identification of symptoms but specifically avoids inclusion of biological markers or any other criteria based on etiology or pathophysiology.

An alternative to using DSM-IV diagnosis for genetic study involves the dissection of genetically complex psychiatric syndromes into measurable, easily modeled components known as endophenotypes. These endophenotypes may be more amenable to genetic analysis than the disorder itself. For example, in schizophrenia, abnormalities in sensorimotor gating,[28] smooth pursuit eye movement,[29-33] social interaction,[34-37] neurocognitive function,[38,39] olfaction,[40,41] structural[42-44] and functional brain imaging,[45-49] responses to acute pharmacologic challenge,[47,50] and medication sensitivity have all been suggested as potential endophenotypes for use in genetic linkage analysis.[51-57] Studying unaffected first-degree relatives of psychiatric patients for the presence of endophenotypic markers provides further evidence of a genetic basis for these phenotypes.[56,58] In addition, shifting the focus from DSM-IV criteria to endophenotypes will make the acquisition of multiplex families substantially easier as "affected" individuals are only required to carry the endophenotypic trait and not the full syndrome itself.[58] This approach has been successful in dissecting hypertension into genetically defined syndromes.[59-61] The modeling of endophenotypic traits in animal models may provide a way to harness the unique power of random mutagenesis to the dissection of complex neuropsychiatric illness.

To date, potential linkage for schizophrenia has been reported to chromosomes 1q21-22, 6p24, 8p, 13q32, 18p11.2, and 22q11.[62-68] Confirmed susceptibility loci for bipolar disorder have been detected on 4p16, 12q24, 13q32, 18p11.2, 21q21, 22q11-13, and Xq26.[69-76] Despite these numerous attempts to use standard genetic linkage approaches to the study of psychiatric illness, there have been no genes identified. Potential explanations for this difficulty include genetic complexity of the disease and ethnic diversity among the different populations being evaluated. The potential overlap between susceptibility loci for both bipolar disorder and schizophrenia on chromosomes 13, 18, and 22 is intriguing considering the increased risk of schizoaffective disorder and major depressive disorder in first degree relatives of probands with either schizophrenia or bipolar disorder. These data together may indicate the presence of shared genetic risk factors between these two disorders.[77,78] As Tsuang et al.[79] point out, despite the historic split between primary psychotic disorders (schizophrenia) and affective psychosis (bipolar disorder),[80] there may in fact be a shared genetic risk for some of the *dimensions* of psychiatric illness across categories, in this case the predisposition to experience hallucinations or delusions. Similarly, the presence of rest:activity abnormalities in mood disorders and schizophrenia, the presence of anxiety in mood and psychotic and anxiety disorders, or deficits in social interaction in both schizophrenia and autism may stem from similar genetic risk factors across disorder categories. It is important to note that recent linkage studies in schizophrenia demonstrate stronger linkage by broadening the disease phenotype to include schizoaffective disorder and bipolar disorder.[67,81]

The related concepts of endophenotypes and behavioral dimensions of psychiatric illness have direct relevance for the selection of behavioral paradigms in phenotype-based mutagenesis screens. It is the convergence of genetically validated endophenotypic measures of illness, and the ability to model those in mice, that should guide the selection of mouse models in random mutagenesis studies.

6.3.2 SELECTION OF BEHAVIORAL PARADIGM FOR ENU SCREENING

There are many dimensions of psychiatric disease such as psychosis and mood alterations that are not able to be modeled in the mouse. However, many others including anxiety, rest:activity abnormalities, sensorimotor gating, drug self-administration, and social interaction can be observed. As systematically complied by Crawley,[82] there are often several assays, some simple and fast, others complex and sophisticated, within each behavioral domain. Any behavioral mutation that is proven to be heritable will subsequently need to be evaluated as a model for a specific syndrome or endophenotype. The ideal assay should provide an unambiguous phenotype, be high-throughput, and have automated data acquisition and analysis.[83,84] The large number of mice that must be tested dictates the need for high-throughput and automated data acquisition/analysis. In order to mutagenize each gene at least one time with 95% confidence, 2000 mice (in a dominant screen) to 40,000 mice (in a recessive screen) mice must be tested[83,85] (and Chapter 4 of this volume). To address issues of uniformity of behavioral phenotyping, the analysis of novel mutations should be performed using a standard battery of tests that provide strong and consistent phenotypes. Behavioral differences observed in mutant animals should be replicated in multiple laboratories, using multiple tests of a behavioral domain.[86] In addition, Tarantino and Bucan[18] suggest that during behavioral assessment of potential novel mutations, one or two inbred strains (e.g., C57BL/6J or DBA/2J) should also be evaluated along with wild-type littermates to provide a running baseline by which to compare results at any given time point.[18]

Adequate performance in behavioral assays requires relatively intact neural and muscular function. Therefore, the first step in a screening panel is a comprehensive visual inspection for gross structural, developmental, musculoskeletal, craniofacial, sensory, and reflex abnormalities. Table 6.1 lists several well-described protocols for initial characterization of new offspring. In the absence of obvious abnormalities, selected behavioral and neurobiological phenotypes can then be characterized.

The screen for a novel behavioral mutation can be specifically aimed at the identification of a well-defined phenotype. For example, the circadian mutation *Clock* was isolated in a small-scale screen when mice were examined only in the wheel-running paradigm.[87] However, to make a small-scale screen more efficient, one can combine paradigms that encompass several behavioral domains. For example, a mutagenesis screen currently being performed in our laboratory involves examination of G_1 progeny of ENU-treated mice in a behavioral battery consisting of measure of anxiety, neuromuscular function, sensorimotor gating, olfaction, and rest:activity behavior.[87a] This experimental paradigm allows us to evaluate, prior to testing inheritance, how specific or pleiotropic the behavioral anomalies are in identified phenotypic deviants. Whereas mutations with specific phenotypes can be of particular value, combinations of phenotypes in single gene mutations can be even more informative in the search for models of psychiatric disorders where combinations of phenotypes are observed. For example, a mutant that shows abnormalities in both acoustic startle and anxiety-related measures would be an interesting model for schizophrenia.

TABLE 6.1
Comprehensive Mouse Observational Batteries

	Parameters
Irwin Observational Test Battery[140]	Physical appearance
	Locomotor activity
	Stereotyped behavior
	Seizure activity
	Arousal
	Cranial and peripheral reflexes
	Gait
	Aggression/passivity
	Grip strength
	Body tone
	Vocalization
	Respiratory rate
Moser Neurobehavioral Test Battery[141]	Autonomic tone
	Neuromuscular function
	Activity
	Sensorimotor
	Excitability
	Physiological measure
SHIRPA[142]	Behavioral observation
	Locomotor activity
	Food and water intake
	Balance and coordination
	Analgesia
	Histology
	Biochemistry
	Anxiety
	Learning and memory
	Prepulse inhibition
	Electromyography
	Electroencephalography
	Nerve conduction
	Magnetic resonance imaging

Based on phenotypic similarities between endophenotypes seen in major psychiatric illnesses, several robust behavioral paradigms have been selected as being particularly suitable for a random mutagenesis approach.

6.3.2.1 Sensorimotor Gating

Patients with schizophrenia, as well as other psychiatric disorders, have long been noted to exhibit deficiencies in concentration and attention.[80] Many schizophrenia patients complain of difficulty in learning new information and are often easily distracted by irrelevant sensory information. Some of these difficulties may in fact be related to impairments in sensory gating.[88-94] In humans, sensory gating can be measured by the degree of attenuation of a touch- or sound-induced startle response (eye blink or flinch) upon presentation of a nonstartling prepulse stimulus (prepulse inhibition of startle or PPI). PPI is one of the behaviors most easily translated from human to animal models.[91,92,94] In mice, it is measured as a whole-body response using a commercially available startle chamber. Inbred strains of mice vary widely in both their tactile and acoustic startle response as well as PPI.[95,96]

Studies of knockout mice indicate that single gene mutations are capable of measurably altering sensorimotor gating. For example, proline dehydrogenase *Prodh* (–/–) and NCAM-180 (–/–) mice both show abnormalities in startle,[97,98] and heterozygous *reeler* mice *rln* (–/+) show decreased PPI in addition to increased anxiety related measures. It is interesting to note that the human homolog of *Prodh* is located in 22q11, a candidate region for several psychiatric illnesses.[68,99] Both NCAM and the *reeler* gene product, reelin, are critical for neuronal migration. Postmortem studies demonstrated that NCAM is reduced in the hippocampus of schizophrenia patients[100,101] whereas reelin expression is reduced by 50% in prefrontal cortex, temporal cortex, hippocampus, caudate nucleus, and cerebellum.[102]

6.3.2.2 Anxiety

Anxiety, defined as unpleasant and unwarranted feelings of apprehension and fear, sometimes accompanied by physiological symptoms, is an important dimensional feature of many psychiatric disorders. Patients with psychotic disorders (schizophrenia, schizoaffective disorder), mood disorders (major depression and bipolar disorder), the "true" anxiety disorders (generalized anxiety disorder, panic disorder, specific and social phobia, obsessive compulsive disorder, and posttraumatic stress disorder), as well as many of the personality disorders, can exhibit significant levels of anxiety. In rodents, the most commonly used behavioral testing paradigm for assessment of anxiety is the brightly lit open field. Other assays include the elevated plus maze and elevated zero maze. All of these tests invoke a conflict between the rodent's drive to explore the environment and fear of being in an open exposed place.[103] The measures of anxious behavior include time spent in an open area (zero maze, plus maze), time spent near the wall (open field), defecation, and exploratory activity.[104,105] All of these paradigms have been shown to be predictive of the expected drug treatment response.[105] There is marked variation among inbred strains in their response to anxiety-related measures.[84,106]

There are several notable examples of induced mutations that demonstrate abnormalities in one or more measures of anxiety. Serotonin receptor 1a (*Htr1a*) knockout mice show elevated levels of anxiety related behaviors,[107-109] while Serotonin 1b (*Htr1b*)[110-112] and corticotropin-releasing factor receptor (*Crhr1*)[113,114] gene knockouts both show reduced levels of anxiety.

6.3.2.3 Learning and Memory

Learning and memory abnormalities are cardinal features of several psychiatric disorders including schizophrenia and major depression. The learning and memory problems in schizophrenia are particularly in the area of working memory and are thought to represent dysfunction of the prefrontal cortex.[115] There are several commonly used tasks to study learning and memory in mice. The hidden platform version of the Morris water maze is a test for spatial learning in which animals are trained to escape from a pool of water.[116,117] Mice are given multiple training sessions to learn the placement of the platform using visual cues in the room. The platform is then removed and the search pattern of the animal is recorded. Mice that have learned the location of the platform will spend more time in the area that had previously contained the platform and will make more crosses over the platform site. A second commonly used paradigm for studying learning and memory is the conditioned fear response.[117,118] Fear conditioning can evaluate two discrete forms of learning, cued and contextual, and there is evidence that different neural substrates support these two forms of learning. Other paradigms that have been used in mice for measuring learning and memory include operant conditioning, active and passive avoidance, go/no-go olfactory tasks, and social transmission food preparation tasks.[82]

Among the many learning and memory paradigms available, the conditioned fear response represents an ideal choice. With this test, several aspects of learning are measured simultaneously, data collection is for the most part automated, and there is low intrastrain variance in mice. After the behavioral phenotype is uncovered, further behavioral screening can then proceed in a nested fashion. Learning and memory mutants identified by abnormal fear conditioning response can be further characterized on more detailed measures such as Morris water maze, operant conditioning, avoidance, or spatial learning paradigms.

Several targeted mutations have been produced in genes known to be important in learning and memory. For example, mutations in calcium/calmodulin-dependent protein kinase II α,[119] CREB,[120] NMDA receptor ϵ1-subunit,[121] adenylate cyclase,[122,123] and mGluR5[124] display anomalies in different forms of learning and memory. In addition to the null mutations described above, conditional hypermorphic mutants have also been shown to display differences in learning and memory. For example, mice overexpressing the NMDA receptor 2b (Grin2b) in the forebrain show enhanced learning and memory in a variety of behavioral tasks including the Morris water maze, conditioned fear, a novel object recognition task, and fear extinction.[125] Deficits in learning and memory are also apparent in conditional knockouts for the neurotrophin tyrosine kinase receptor (TrkB).[126]

Many other screening paradigms are possible, including drug sensitivity, neuromuscular functioning, locomotor activity, social interaction, seizure susceptibility, olfaction, and circadian activity (see Chapter 4 of this volume for a description of the use of wheel-running activity in a mutagenesis screen).

6.4 SELECTION OF STRAIN

Inbred strains vary considerably in a wide variety of behavioral measures. Therefore, selection of the appropriate background strain is critical for mutagenesis and chromosomal localization of the mutant locus. Selection of a strain must enhance the possibility of detecting a mutation in a behavior of interest. For example, C57BL/6J (B6), a strain that performs well in learning tasks, is ideal for detecting mutations that adversely affect learning. However, if learning mutations in general are desired, a strain with a median phenotype should be selected to avoid both ceiling and floor effects. A second consideration in the selection of a strain for a mutagenesis experiment is the variability of the background strain in the phenotype of interest. In a reverse genetics experiment, the interpretation of a complex phenotype is facilitated by the ability to molecularly distinguish between mutant and normal mice using a transgene as a marker. However, in a forward genetic experiment, the abnormal phenotype is the starting point for the selection of a mutant and mapping and positional cloning of the mutated gene. Thus, if the background strain shows high variability for the targeted behavior, it can be difficult to reliably detect mutation-related variability. Therefore, for each phenotype, the most appropriate strain should be carefully considered. For many behavioral traits, extensive baseline data for a variety of inbred strains is available. See Crawley et al.[127] and Festing[128] for detailed information on the performance of many inbred strains in a variety of behavioral paradigms.

Selecting the appropriate strain to use in a mapping cross is as important as choosing the initial strain for mutagenesis. When deciding upon a strain for mapping, it is important to keep in mind that a host of additional genes in the mapping strain can severely affect the manifestation of the mutant phenotype. As previous studies have indicated, knowing the phenotypes of the two parental inbred strains does not necessarily predict the phenotype of F_1 or F_2 progeny between those strains. It is useful to know if major quantitative trait loci (QTL) segregate in the mapping cross, and if the phenotypic variability in the F_2 or N_2 backcross allows for identification of the mutagen-induced phenotype.

6.5 ENU BREEDING STRATEGIES

6.5.1 (SEMI) DOMINANT SCREEN

The simplest application of ENU mutagenesis is in the identification of dominant or semi-dominant mutations. Dominant mutations are ones that are fully expressed in heterozygotes, while semi-dominant mutations are those in which the heterozygote is intermediate between the wild-type and homozygous phenotype. Even though dominant mutations occur at approximately 1/10th the frequency of recessive alleles, the ease of this approach makes it an attractive one. Figure 6.1 details the steps needed for identification of dominant mutations. ENU-treated G_0 male mice are mated to wild-type females to produce a G_1 generation. The G_1 progeny are then screened for abnormalities in the phenotype(s) of interest.

The ability to identify abnormalities in heterozygotes clearly simplifies the maintenance of stocks, as well as the genetic and phenotypic characterization of the mutation. Two major limitations of this type of screen are that many genes may not show a dominant phenotype when mutated, and dominant phenotypes usually do not fully describe the function of a disrupted gene. Thus, after a dominant mutation is identified, its homozygous phenotype, as well as its loss of function phenotype, still needs to be characterized.

A well-known example of a behavioral gene identified using this method is the identification of *Clock* in a dominant ENU mutagenesis screen.[87] Three hundred four

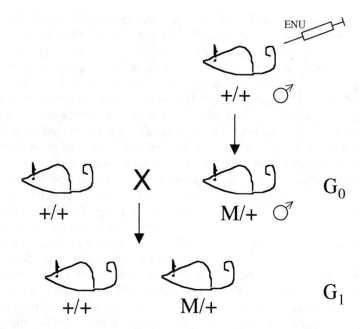

FIGURE 6.1 A dominant mutagenesis screen. Founder G_0 mice are created using ENU and mated to wild-type females. G_1 progeny are generated and screened for aberrant phenotypes.

G_1 offspring of ENU-treated C57BL/6J were assessed for circadian rhythm of wheel-running activity. Rhythms were obtained in both 14:10 h light:dark cycle (LD) and in total darkness (DD). Under these conditions, B6 mice show little variance in circadian period with a range of 23.3 to 23.8 h. A single G_1 mouse displayed a circadian period that lengthened over time in the DD condition. The inheritance pattern of this mutation was determined by backcrossing the mutant to wild-type B6 females, and subsequently outcrossing to both BALB/cJ (BALB) and C3H/HeJ (C3H). These crosses confirmed a fully penetrant, single-locus, dominant mutation. F_2 offspring of *Clock*/+ mice display three different phenotypes indicating that this mutation is semi-dominant. *Clock* was then mapped to the central portion of chromosome 5 and, subsequently, the disrupted gene was positionally cloned. *Clock* was revealed to be a key component in the circadian pathway, not only in mice, but also in lower species.[129,130] Several characteristics of this study make it an ideal case of a dominant screen. The mutant line displayed a robust semi-dominant phenotype, the background strain showed little variance in the measure, the data collection was both quantitative and automated, and the mapping strain was selected to highlight the presence of the mutation unambiguously.

6.5.2 RECESSIVE SCREEN

Many, perhaps most, mutations will not have either a dominant or semi-dominant phenotype. Figure 6.2 illustrates a breeding scheme for identifying recessive mutations in a random genome-wide mutagenesis screen. Mutagenized G_0 males are mated

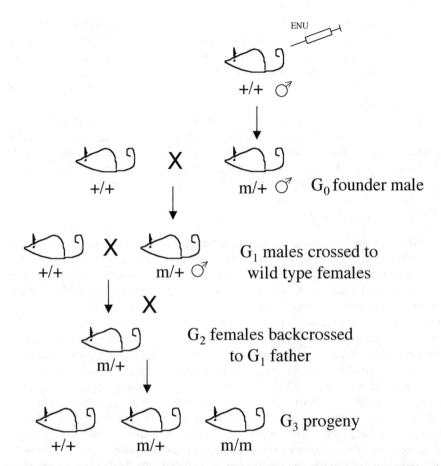

FIGURE 6.2 A recessive mutagenesis screen. Founder G_0 mice are created using ENU and mated to wild-type females. G_1 progeny are generated and G_1 males are mated to wild-type females to generate G_2 progeny. G_2 females are mated to their G_1 fathers to generate G_3 families that are screened for aberrant phenotypes.

with wild-type females followed by breeding of the F_1 progeny (G_1) to wild-type partners to establish families of siblings (G_2) sharing the same set of mutations. G_2 females are bred back to their G_1 fathers (backcross) to produce G_3 families that are then phenotyped. Since the rate of ENU-induced mutations is quite high, there is a good chance that several mutations will be manifested in the G_3 progeny, threatening the reliability of the phenotypic characterization and subsequent mapping.

To date, there is no example of a novel behavioral mutation identified using this strategy, although groups worldwide are incorporating behavioral assessment into their ongoing recessive screens. This approach was used successfully, however, to isolate developmental mutations.[131-133] Recessive screens have also been used to successfully dissect various enzymatic pathways in amino-acid metabolism.[134] For example, Harding et al.[135] described an ENU-induced mutation in the γ-glutamyl transpeptidase enzyme (GGT^{enu1}). These mice exhibit growth retardation and infertility. Interestingly, this mutation, while severe, differs from the GGT null mutation in several key features. Specifically, the GGT knockout exhibits bilateral cataracts, significantly higher levels of urinary glutathione, a shorter lifespan, and changes in coat color.[135] Thus, it is likely that the recessive ENU mutation retains some level of residual activity.

One of the key answers that large-scale recessive screens will provide to the field of developmental biology is the total number of essential genes or genes that are necessary for the survival of the embryo. In behavioral genetics, an important question is how many genes, when mutated, will give a "pure" behavioral phenotype, with no other developmental, visible, or neurological symptoms.

6.5.3 MODIFIER SCREEN

Modifier screens are a powerful technique wherein a novel mutation is detected by its ability to either *enhance* or *suppress* a dominant or recessive phenotype of a known mutation. These screens are aimed at the detection of both novel components of a behavioral pathway, as well as new alleles of the original (known) mutation. Figure 6.3 illustrates a breeding scheme that can be used in a modifier screen. ENU-G_1 males are bred to females homozygous for a gene known to affect the behavior of interest. All of the G_2 offspring carry one copy of the mutant gene null allele. If the ENU-G_1 male carries a new mutation, 50% of the G_2 offspring will inherit this ENU induced mutation; they will be "compound heterozygotes." If any of the random mutations interact with the known mutation, these compound heterozygotes will display a phenotype distinct from either the wild type or the heterozygote.

In the above described modifier screen, a previously characterized mutant line is used as a "sensitizer" when screening for novel mutations in interacting genes. Genetic mutations are not the only type of sensitizers that can be used. For instance, if a mutant phenotype is only expressed in a setting of a high temperature, sleep deprivation or a specific diet, the environmental modification acts as a sensitizer. For behavioral genetics, searching for novel phenotypes after pharmacologic treatment may be of particular importance. Although genes underlying the susceptibility of psychiatric disorders are still not known, the effect of a selected set of drugs on the course of the disease may be used to identify more models relevant to those diseases.

Figure 6.3 outlines a hypothetical sensitized behavioral screen aimed at the identification of novel genes involved in conferring sensitivity to psychostimulants. Norepinephrine (NE) is a catecholamine neurotransmitter whose actions include roles in learning and memory, reinforcement, affective illness, sleep–wake regulation, and anxiety. NE neurotransmission is terminated by a combination of enzymatic degradation and uptake via the norepinephrine transporter (NET). A recently described norepinephrine transporter NET (–/–) deficient mouse[136] displays supersensitivity to the action of the stimulants amphetamine and cocaine. Using the NET knockout, a two-step screen for novel random mutations that affect stimulant response can be devised. First, ENU-G_1 mice can be evaluated directly for amphetamine-induced locomotor activity (sensitized dominant screen). Those that display an increased (or decreased) activity as compared to baseline are treated as a dominant

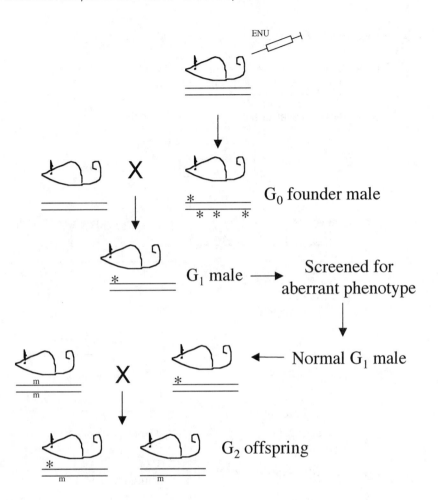

FIGURE 6.3 A modifier screen. Founder G_0 mice are created using ENU and mated to wild-type females. G_1 progeny are generated and G_1 males are phenotyped. Those that are normal are mated to females homozygous for the sensitizing mutation. The G_2 offspring from this cross are then phenotyped. * denotes location of ENU-induced point mutation(s); m denotes location of known mutation.

mutation whose phenotype will be further evaluated in the absence of drug treatment, as well as with different doses of drug. ENU-G_1 males that do not show altered activity are bred to NET (–/–) females. All the offspring from this cross carry one copy of the NET null allele and 50% of the offspring carry induced point mutations. Offspring that display stimulant hypersensitivity indicate that a random point mutation has been introduced that interacts with the NET null allele. Such a mutation can be either an allele of NET itself, or some other downstream, but related gene. Novel mutations identified in a pharmacological sensitized screen have to be further evaluated to determine if they affect a metabolic pathway or a primary neurobiological target.

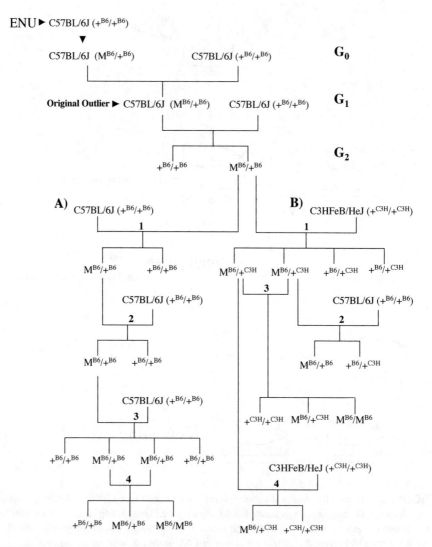

FIGURE 6.4 Breeding scheme for genetic characterization of dominant mutations. Section A of the figure illustrates maintaining the mutation on a homogeneous background. Section B of the figure illustrates outcrossing for mapping the mutant locus. See Section 6.6.1 for a full description of the numbered crosses.

6.6 GENETIC ANALYSIS

6.6.1 GENETIC CHARACTERIZATION
OF NOVEL DOMINANT MUTATIONS

An important and time-consuming step in the characterization of novel ENU muta-tions is the genetic characterization and chromosomal localization of the mutant locus. A recommended breeding scheme for characterization of novel mutations in a dominant screen is illustrated in Figure 6.4. This figure uses C57BL/6J as the original mutagenized strain and C3HFeB/HeJ as the strain introduced for mapping. Characterization of the mutant phenotype is initiated by mating the G_1 outlier to the original strain to generate G_2 progeny. The presence of affected progeny in the G_2 generation confirms transmission of the mutant locus. Typically, generation of two to three G_2 litters is needed to confirm inheritance of the new mutation. The ratio of affected to nonaffected progeny indicates the level of penetrance of the mutant trait. A 1:1 ratio of affected to nonaffected progeny indicates a fully penetrant mutation. If the number of affected progeny is below 50%, the mutant phenotype is not fully penetrant. Affected animals from the G_2 generation are used to establish two types of crosses: those that maintain the mutation on the homogenous genetic background (Figure 6.4A) and those that generate hybrid animals (F1 or F2) for genetic mapping of the mutant locus (Figure 6.4B).

To facilitate genetic analysis of behavioral mutations, especially those that are subtle, it is important to maintain them on a homogenous background. As illustrated in Figure 6.4A, three backcrosses (crosses 1–3) are performed by outcrossing the affected progeny back to the original strain, C57BL/6. This enables further progeny testing to provide additional confirmation of genetic inheritance and analysis of penetrance. These backcrosses also serve to eliminate the influence of other point mutations on the phenotype, including linked lethal mutations that would otherwise prevent generation of mice that are homozygous for the locus causing the mutant phenotype. In the case of mutations with a complex phenotype, these backcrosses aid in the dissection of a phenotype caused by two or more mutation events. The progeny generated by the backcrosses can be used to further define the behavioral anomaly through the use of additional behavioral assays. This analysis will indicate the level of specificity of the selected mutation and can uncover other behavioral traits that are more pronounced than the original phenotype. After three backcrosses, an intercross (cross 4) of two heterozygotes for the mutant locus can be attempted. This step allows the assessment of the recessive phenotype on the original genetic background. Heterozygote or homozygote progeny from the intercrosses can be used to initiate crosses to other genetic backgrounds.

Genetic mapping of the mutant locus requires crosses to an inbred strain different from the mutagenized strain. Ideally, the original phenotypic outlier should be outcrossed to several strains in order to identify a background that does not suppress the mutant phenotype. The "partner" strain for mapping can also be selected based on the available baseline data for the trait of interest in several inbred strains. The most suitable strain would have a phenotypic mean and variance similar to the strain used for mutagenesis. Additional information on the phenotypes of F_1 and F_2 progeny of these strains is useful in analyzing the segregating mutant locus.

Figure 6.4B demonstrates a series of mapping crosses. In cross 1, affected progeny from the G_2 population are mated with the mapping strain and affected F_1 progeny are used to establish a series of mapping crosses. A backcross (cross 2) to the original strain produce progeny for mapping. Both male and female F_1 are used to set up these crosses in order to evaluate whether or not parental origin affects phenotype. Intercrosses (cross 3) are set up subsequently, using F_1 males and females that were confirmed carriers of the mutant locus in the previous backcrosses. These crosses reveal the recessive phenotype on a mixed genetic background, and the nature of this phenotype is used to determine the optimal mapping strategy. For example, in some cases, dominant behavioral mutations can be associated with peri- or postnatal lethality. In the case of semi-dominant behavioral mutations, the phenotypic differences between three genotypes (+/+, M/+, and M/M) determines which mapping approach (backcross or intercross) is most suitable. Subsequent backcrosses (cross 4) to the mapping strain are particularly useful because they allow the application of heterozygosity mapping of partially congenic lines.[137] It is important that the mutant phenotype is monitored during these subsequent crosses because the phenotype can change when the new background is introduced.

6.6.2 GENETIC CHARACTERIZATION OF NOVEL RECESSIVE MUTATIONS

A recommended breeding scheme for the genetic characterization of novel mutations identified in a recessive screen is illustrated in Figure 6.5. Again, this figure uses C57BL/6J as the original mutagenized strain and C3HFeB/HeJ as the strain introduced for mapping. Affected animals from the G_3 generation are used to establish two types of crosses: those that will maintain the mutation on the homogenous genetic background (Figure 6.5A) and those that generate hybrid animals for genetic mapping of the mutant locus (Figure 6.5B).

Figure 6.5A illustrates a cross between a homozygote G_3 animal and the original background strain. After the backcross (cross 1), an intercross (cross 2) of two heterozygotes allows further assessment of the recessive phenotype on the original genetic background. This cross serves to eliminate the influence of other point mutations on the phenotype and to dissect a phenotype that was caused by two or more mutation events. These progeny also can be used to further define the behavioral anomaly through the use of additional behavioral assays. Homozygote progeny from the intercrosses can be used to initiate crosses to other genetic backgrounds.

Figure 6.5B illustrates a series of mapping crosses. In cross 1, homozygote progeny from the G_3 population are mated with the mapping strain and F_1 progeny are used to establish mapping crosses. Intercrosses (cross 2) are set up between the F_1 heterozygote progeny. These crosses reveal the recessive phenotype on a mixed genetic background. It is important that the mutant phenotype is monitored during these subsequent crosses because the phenotype can change when the new background is introduced.

6.7 CONCLUSION

Their small size, short gestation time, extensive homology with human genome, as well as the availability of sophisticated technology for genome analysis, make the laboratory mouse an ideal model organism for the study of human disease. A large collection of mutant lines, with anomalies in social and emotional profiles, drug sensitivity, learning and memory, liability to addiction and circadian rhythm activity, have established the mouse as a particularly useful model in the search for the genetic basis of behavior, as well as for genes underlying major psychiatric disorders.

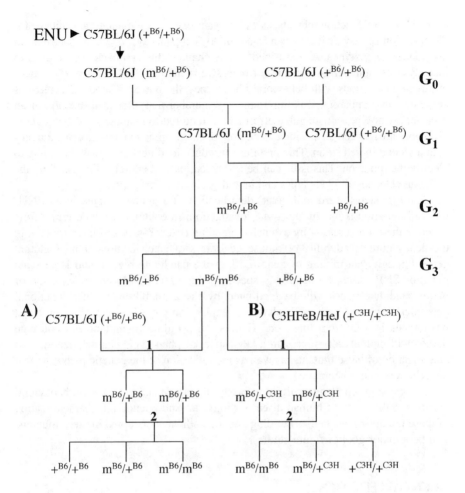

FIGURE 6.5 Breeding scheme for genetic characterization of recessive mutations. Section A of the figure illustrates maintaining the mutation on a homogeneous background. Section B of the figure illustrates outcrossing for mapping the mutant locus. The upper portion of the figure (G_1–G_3) illustrates the breeding scheme used to establish recessive mutant families.

With a growing number of genes identified through genome sequencing, analysis of behavior in mutant lines generated by targeted mutagenesis will continue to provide useful models. The ability to create tissue-specific and inducible mutant lines[138,139] will allow further dissection of developmental and neuroanatomical aspects of behavior. Random mutagenesis screens can be used as a complementary approach to these gene-driven searches to identify mutations with impaired behavior. In contrast to the traditional approaches to behavioral genetics, which exploit the phenotypic differences among inbred strains or spontaneous mutations, random mutagenesis involves screening for aberrant phenotypes in progeny of mice mutagenized by a chemical such as ENU.[22] ENU induces point mutations throughout the whole genome, thereby creating novel alleles of single genes. These complete or partial loss-of-function mutants, as well as gain-of-function mutants, are suitable for identification of new genes, their relevant biological pathways, and their contribution to complex phenotypes. As described in this chapter, this approach can be used to identify dominant, semi-dominant, or recessive mutations. With a growing number of targeted mutations with behavioral phenotypes, the power of modifier screens is now becoming apparent. It is important to remember that after identification of an aberrant phenotype, considerable effort is still required to map the underlying gene. Mutations are induced on an inbred background, but mapping requires crossing mutants with a second inbred strain. This step may introduce modifiers that mask the ability to identify the mutation. This issue can be addressed, but not entirely eliminated, by the judicious selection of phenotypes and inbred strain used for mapping.

The first step toward molecular identification of a gene disrupted in an ENU-induced mutation is its chromosomal localization. In contrast to the current efforts to refine the critical region by extensive breeding (backcrosses and intercrosses), in the near future, availability of mouse genome sequence, identification of mutant genes through data mining of genes/EST in the candidate region, and large-scale mutation/SNP analysis will greatly speed up this process. Moreover, selection of likely candidate genes will be facilitated by the availability of other functional genomics resources such as databases of expression patterns, interacting proteins, phenotypes in gene trap lines, etc. Expression profiling of mutant vs. wild-type tissues will undoubtedly represent a key aspect of phenotypic characterization and classification of these mutants, as well as identification of the genetic pathways that underlie a mutant phenotype.

A significant challenge will be to bring an understanding of mouse behavioral genetics to the study of human disease. Molecular and phenotypic characterization of these mutations, as well as building genetic pathways disrupted in these mutants, will be a major goal of future studies.

ACKNOWLEDGMENTS

The authors would like to thank their colleagues in the Center for Neurobiology and Behavior for numerous discussions and comments on this manuscript. This research is supported by NIH grants MH57855 (M.B.), AR45325-02 (M.B.). Dr. Kanes was partially supported by NIMH Training Grant MH19112, and by a Pfizer Scholar Award.

REFERENCES

1. Nusslein-Volhard, C., Kluding, H., and Jurgens, G., Genes affecting the segmental subdivision of the Drosophila embryo, *Cold Spring Harbor Symposia on Quantitative Biology,* 50, 145–154, 1985.
2. Nusslein-Volhard, C. and Wieschaus, E., Mutations affecting segment number and polarity in Drosophila, *Nature,* 287, 795–801, 1980.
3. Mayer, U., Torres-Ruiz, R.A., Berleth, T., Misera, S., and Jurgens, G., Mutations affecting body organization in the Arabidopsis embryo, *Nature,* 353, 402–407, 1991.

4. Hulskamp, M., Misra, S., and Jurgens, G., Genetic dissection of trichome cell development in Arabidopsis, *Cell,* 76, 555–566, 1994.
5. Brenner, S., The genetics of *Caenorhabditis elegans, Genetics,* 77, 71–94, 1974.
6. Hirsh, D. and Vanderslice, R., Temperature-sensitive developmental mutants of *Caenorhabditis elegans, Dev. Biol.,* 49, 220–235, 1976.
7. Anderson, P., Mutagenesis, *Meth. Cell Biol.,* 48, 31–58, 1995.
8. Barbazuk, W.B., Johnsen, R.C., and Baillie, D.L., The generation and genetic analysis of suppressors of lethal mutations in the *Caenorhabditis elegans* rol-3(V) gene, *Genetics,* 136, 129–143, 1994.
9. Williamson, V.M., Long, M., and Theodoris, G., Isolation of *Caenorhabditis elegans* mutants lacking alcohol dehydrogenase activity, *Biochem. Genet.,* 29, 313–323, 1991.
10. Driever, W., Solnica-Krezel, L., Schier, A.F., Neuhauss, S.C., Malicki, J., Stemple, D.L., Stainier, D.Y., Zwartkruis, F., Abdelilah, S., Rangini, Z., Belak, J., and Boggs, C., A genetic screen for mutations affecting embryogenesis in zebrafish, *Development,* 123, 37–46, 1996.
11. Haffter, P., Granato, M., Brand, M., Mullins, M.C., Hammerschmidt, M., Kane, D.A., Odenthal, J., van Eeden, F.J., Jiang, Y.J., Heisenberg, C.P., Kelsh, R.N., Furutani-Seiki, M., Vogelsang, E., Beuchle, D., Schach, U., Fabian, C., and Nusslein-Volhard, C., The identification of genes with unique and essential functions in the development of the zebrafish, *Danio rerio, Development,* 123, 1–36, 1996.
12. Mullins, M.C., Hammerschmidt M., Haffter P., and Nusslein-Volhard, C., Large-scale mutagenesis in the zebrafish: In search of genes controlling development in a vertebrate, *Curr. Biol.,* 4, 189–202, 1994.
13. Solnica-Krezel, L., Schier, A.F., and Driever, W., Efficient recovery of ENU-induced mutations from the zebrafish germline, *Genetics,* 136, 1401–1420, 1994.
14. Nolan, P.M., Peters, J., Vizor, L., Strivens, M., Washbourne, R., Hough, T., Wells, C., Glenister, P., Thornton, C., Martin, J., Fisher, E., Rogers, D., Hagan, J., Reavill, C., Gray, I., Wood, J., Spurr, N., Browne, M., Rastan, S., Hunter, J., and Brown, S.D., Implementation of a large-scale ENU mutagenesis program: Towards increasing the mouse mutant resource, *Mamm. Genome,* 11, 500–506, 2000.
15. Hrabe de Angelis, M.H., Flaswinkel, H., Fuchs, H., Rathkolb, B., Soewarto, D., Marschall, S., Heffner, S., Pargent, W., Wuensch, K., Jung, M., Reis, A., Richter, T., Alessandrini, F., Jakob, T., Fuchs, E., Kolb, H., Kremmer, E., Schaeble, K., Rollinski, B., Roscher, A., Peters, C., Meitinger, T., Strom, T., Steckler, T., Holsboer, F., Klopstock, T., Gekeler, F., Schindewolf, C., Jung, T., Avraham, K., Behrendt, H., Ring, J., Zimmer, A., Schughart, K., Pfeffer, K., Wolf, E., and Balling, R., Genome-wide, large-scale production of mutant mice by ENU mutagenesis, *Nat. Genet.,* 25, 444–447, 2000.
16. Chen, Y., Yee, D., Dains, K., Chatterjee, A., Cavalcoli, J., Schneider, E., Om, J., Woychik, R.P., and Magnuson, T., Genotype-based screen for ENU-induced mutations in mouse embryonic stem cells, *Nat. Genet.,* 24, 314–317, 2000.
17. Nolan, P.M., Kapfhamer, D., and Bucan, M., Random mutagenesis screen for dominant behavioral mutations in mice, *Methods,* 13, 379–395, 1997.
18. Tarantino, L.M. and Bucan, M., Dissection of behavior and psychiatric disorders using the mouse as a model, *Hum. Mol. Genet.,* 9, 953–965, 2000.
19. Russell, W.L., Kelly, E.M., Hunsicker, P.R., Bangham, J.W., Maddux, S.C., and Phipps, E.L., Specific-locus test shows ethylnitrosourea to be the most potent mutagen in the mouse, *Proc. Natl. Acad. Sci. USA,* 76, 5818–5819, 1979.

20. Vogel, E. and Natarajan, A.T., The relation between reaction kinetics and mutagenic action of mono-functional alkylating agents in higher eukaryotic systems. II. Total and partial sex-chromosome loss in Drosophila, *Mutat. Res.,* 62, 101–123, 1979.

21. Provost, G.S. and Short, J.M., Characterization of mutations induced by ethylnitrosourea in seminiferous tubule germ cells of transgenic B6C3F1 mice, *Proc. Natl. Acad. Sci. USA,* 91, 6564–6568, 1994.

22. Justice, M.J., Noveroske, J.K., Weber, J.S., Zeng, B., and Bradley, A., Mouse ENU mutagenesis, *Hum. Mol. Genet.,* 8, 1955–1963, 1999.

23. Davis, A.P. and Justice, M.J., Mouse alleles: If you've seen one, you haven't seen them all, *Trends Genet.,* 14, 438–441, 1998.

24. Justice, M.J., Zheng, B., Woychik, R.P., and Bradley, A., Using targeted large deletions and high-efficiency N-ethyl-N-nitrosourea mutagenesis for functional analyses of the mammalian genome, *Methods,* 13, 423–436, 1997.

25. Rinchik, E.M. and Russell, L.B., Germ line deletion mutations in the mouse: Tools for intensive functional and physical mapping of regions of the mammalian genome, in *Genome Analysis Volume I: Genetic Physical Mapping,* Cold Spring Harbor Laboratory Press, Cold Spring Harbor, NY, 1990.

26. Weber, J.S., Salinger,, A., and Justice, M.J., Optimal N-ethyl-N-nitrosourea (ENU) doses for inbred mouse strains, *Genesis,* 26, 230–233, 2000.

27. *Diagnostic Statistical Manual of Mental Disorders: DSM-IV-TR,* 4th ed., American Psychiatric Association, Washington, D.C., 2000.

28. Freedman, R., Coon, H., Myles-Worsley, M., Orr-Urtreger, A., Olincy, A., Davis, A., Polymeropoulos, M., Holik, J., Hopkins, J., Hoff, M., Rosenthal, J., Waldo, M.C., Reimherr, F., Wender, P., Yaw, J., Young, D.A., Breese, C.R., Adams, C., Patterson, D., Adler, L.E., Kruglyak, L., Leonard, S., and Byerley, W., Linkage of a neurophysiological deficit in schizophrenia to a chromosome 15 locus, *Proc. Natl. Acad. Sci. USA,* 94, 587–592, 1997.

29. Hutton, S.B., Crawford, T.J., Kennard, C., Barnes, T.R., and Joyce, E.M., Smooth pursuit eye tracking over a structured background in first-episode schizophrenic patients, *Eur. Arch. Psychiatry Clin. Neurosci.,* 250, 221–225, 2000.

30. Ross, R.G., Olincy, A., Harris, J.G., Sullivan, B., and Radant, A., Smooth pursuit eye movements in schizophrenia and attentional dysfunction: Adults with schizophrenia, ADHD, and a normal comparison group, *Biol. Psychiatry,* 48, 197–203, 2000.

31. Calkins, M.E. and Iacono, W.G., Eye movement dysfunction in schizophrenia: A heritable characteristic for enhancing phenotype definition, *Am. J. Med. Genet.,* 97, 72–76, 2000.

32. Lencer, R., Malchow, C.P., Trillenberg-Krecker, K., Schwinger, E., and Arolt, V., Eye-tracking dysfunction (ETD) in families with sporadic and familial schizophrenia, *Biol. Psychiatry,* 47, 391–401, 2000.

33. Waldo, M.C., Adler, L.E., Leonard, S., Olincy, A., Ross, R.G., Harris, J.G., and Freedman, R., Familial transmission of risk factors in the first-degree relatives of schizophrenic people, *Biol. Psychiatry,* 47, 231–239, 2000.

34. Salem, J.E. and Kring, A.M., Flat affect and social skills in schizophrenia: Evidence for their independence, *Psychiatry Res.,* 87, 159–67, 1999.

35. Manor, B.R., Gordon, E., Williams, L.M., Rennie, C.J., Bahramali, H., Latimer, C.R., Barry, R.J., and Meares, R.A., Eye movements reflect impaired face processing in patients with schizophrenia, *Biol. Psychiatry,* 46, 963–969, 1999.

36. Andreasen, N.C., Arndt, S., Alliger, R., Miller, D., and Flaum, M., Symptoms of schizophrenia. Methods, meanings, and mechanisms, *Arch. Gen. Psychiatry,* 52, 341–351, 1995.

37. Andreasen, N.C., Nopoulos, P., Schultz, S., Miller, D., Gupta, S., Swayze, V., and Flaum, M., Positive and negative symptoms of schizophrenia: Past, present, and future, *Acta Psychiatry Scand. Suppl.*, 384, 51–59, 1994.

38. Faraone, S.V., Seidman, J.J., Kremer, W.S., Pepple, J.R., Lyons, M., and Tsaung, M., Neuropsychological functioning among the non-psychotic relatives of schizophrenic patients: A diagnostic efficiency analysis, *J. Abnorm. Psychol.*, 104, 286–304, 1995.

39. Erwin, R.J., Turetsky, B.I., Moberg, P., Gur, R.C., and Gur, R.E., P50 abnormalities in schizophrenia: Relationship to clinical and neuropsychological indices of attention, *Schizophr. Res.*, 33, 157–167, 1998.

40. Moberg, P.J., Doty, R.L., Turetsky, B.I., Arnold, S.E., Mahr, R.N., Gur, R.C., Bilker, W., and Gur, R.E., Olfactory identification deficits in schizophrenia: Correlation with duration of illness, *Am. J. Psychiatry*, 154, 1016–1018, 1997.

41. Turetsky, B.I., Moberg, P.J., Yousem, D.M., Doty, R.L., Arnold, S.E., and Gur, R.E., Reduced olfactory bulb volume in patients with schizophrenia, *Am. J. Psychiatry*, 157, 828–830, 2000.

42. Gur, R.E., Turetsky, B.I., Cowell, P.E., Finkelman, C., Maany, V., Grossman, R.I., Arnold, S.E., Bilker, W.B., and Gur, R.C., Temporolimbic volume reductions in schizophrenia, *Arch. Gen. Psychiatry*, 57, 769–775, 2000.

43. Gur, R.E., Cowell, P.E., Latshaw, A., Turetsky, B.I., Grossman, R.I., Arnold, S.E., Bilker, W.B., and Gur, R.C., Reduced dorsal and orbital prefrontal gray matter volumes in schizophrenia, *Arch. Gen. Psychiatry*, 57, 761–768, 2000.

44. Gur, R.E., Maany, V., Mozley, P.D., Swanson, C., Bilker, W., and Gur, R.C., Subcortical MRI volumes in neuroleptic-naive and treated patients with schizophrenia, *Am. J. Psychiatry*, 155, 1711–1717, 1998.

45. Abi-Dargham, A., Rodenhiser, J., Printz, D., Zea-Ponce, Y., Gil, R., Kegeles, L.S., Weiss, R., Cooper, T.B., Mann, J.J., Van Heertum, R.L., Gorman, J.M., and Laruelle, M., From the cover: Increased baseline occupancy of D2 receptors by dopamine in schizophrenia, *Proc. Natl. Acad. Sci. USA*, 97, 8104–8109, 2000.

46. Laruelle, M., Abi-Dargham, A., Gil, R., Kegeles, L., and Innis, R., Increased dopamine transmission in schizophrenia: Relationship to illness phases, *Biol. Psychiatry*, 46, 56–72, 1999.

47. Laruelle, M., Imaging dopamine transmission in schizophrenia. A review and meta-analysis, *Q. J. Nucl. Med.*, 42, 211–221, 1998.

48. Laruelle, M., Gelernter, J., and Innis, R.B., D2 receptors binding potential is not affected by Taq1 polymorphism at the D2 receptor gene, *Mol. Psychiatry*, 3, 261–265, 1998.

49. Abi-Dargham, A., Gil, R., Krystal, J., Baldwin, R.M., Seibyl, J.P., Bowers, M., van Dyck, C.H., Charney, D.S., Innis, R.B., and Laruelle, M., Increased striatal dopamine transmission in schizophrenia: Confirmation in a second cohort, *Am. J. Psychiatry*, 155, 761–767, 1998.

50. Lander, E., Splitting schizophrenia, *Nature*, 336, 105–106, 1988.

51. Sams-Dodd, F., Phencyclidine in the social interaction test: An animal model of schizophrenia with face and predictive validity, *Rev. Neurosci.*, 10, 59–90, 1999.

52. Andersen, M.B., Zimmer, J., and Sams-Dodd, F., Specific behavioral effects related to age and cerebral ischemia in rats, *Pharmacol. Biochem. Behav.*, 62, 673–682, 1999.

53. Cornblatt, B., Obuchowski, M., Roberts, S., Pollack, S., and Erlenmeyer-Kimling, L., Cognitive and behavioral precursors of schizophrenia, *Dev. Psychopathol.*, 11, 487–508, 1999.

54. Lander, E.S., Splitting schizophrenia, *Nature*, 336, 105–106, 1988.

55. Paterson, A.H., Lander, E.S., Hewitt, J.D., Peterson, S., Lincoln, S.E., and Tanksley, S.D., Resolution of quantitative traits into Mendelian factors by using a complete linkage map of restriction fragment length polymorphisms, *Nature*, 335, 721–726, 1988.

56. Leboyer, M., Bellivier, F., Nosten-Bertrand, M., Jouvent, R., Pauls, D., and Mallet, J., Psychiatric genetics: Search for phenotypes, *Trends Neurosci.*, 21, 102–105, 1998.

57. Brown, V. and Smith, D., Mouse models of madness, *Mol. Psychiatry*, 4, 400–402, 1999.

58. Almasy, L., Porjesz, B., Blangero, J., Chorlian, D.B., O'Connor, S.J., Kuperman, S., Rohrbaugh, J., Bauer, L.O., Reich, T., Polich, J., and Begleiter, H., Heritability of event-related brain potentials in families with a history of alcoholism, *Am. J. Med. Genet. (Neuropsychiat. Genet.)*, 88, 383–390, 1999.

59. Mansfield, T.A., Simon, D.B., Farfel, Z., Bia, M., Tucci, J.R., Lebel, M., Gutkin, M., Vialettes, B., Christofilis, M.A., Kauppinen-Makelin, R., Mayan, H., Risch, N., and Lifton, R.P., Multilocus linkage of familial hyperkalaemia and hypertension, pseudo-hypoaldosteronism type II, to chromosomes 1q31-42 and 17p11-q21, *Nat. Genet.*, 16, 202–205, 1997.

60. Karet, F.E. and Lifton, R.P., Mutations contributing to human blood pressure variation, *Rec. Prog. Horm. Res.*, 52, 263–276, 1997.

61. Lifton, R.P., Molecular genetics of human blood pressure variation, *Science*, 272, 676–680, 1996.

62. Brzustowicz, L.M., Hodgkinson, K.A., Chow, E.W., Honer, W.G., and Bassett, A.S., Location of a major susceptibility locus for familial schizophrenia on chromosome 1q21-q22, *Science*, 288, 678–682, 2000.

63. Straub, R.E., MacLean, C.J., O'Neill, F.A., Burke, J., Murphy, B., Duke, F., Shinkwin, R., Webb, B.T., Zhang, J., Walsh, D., et al., A potential vulnerability locus for schizophrenia on chromosome 6p24-22: Evidence for genetic heterogeneity, *Nat. Genet.*, 11, 287–293, 1995.

64. Schwab, S.G., Albus, M., Hallmayer, J., Honig, S., Borrmann, M., Lichtermann, D., Ebstein, R.P., Ackenheil, M., Lerer, B., Risch, N., et al., Evaluation of a susceptibility gene for schizophrenia on chromosome 6p by multipoint affected sib-pair linkage analysis, *Nat. Genet.*, 11, 325–327, 1995.

65. Blouin, J.L., Dombroski, B.A., Nath, S.K., Lasseter, V.K., Wolyniec, P.S., Nestadt, G., Thornquist, M., Ullrich, G., McGrath, J., Kasch, L., Lamacz, M., Thomas, M.G., Gehrig, C., Radhakrishna, U., Snyder, S.E., Balk, K.G., Neufeld, K., Swartz, K.L., DeMarchi, N., Papadimitriou, G.N., Dikeos, D.G., Stefanis, C.N., Chakravarti, A., Childs, B., Pulver, A.E., et al., Schizophrenia susceptibility loci on chromosomes 13q32 and 8p21, *Nat. Genet.*, 20, 70–73, 1998.

66. Lin, M.W., Sham, P., Hwu, H.G., Collier, D., Murray, R., and Powell, J.F., Suggestive evidence for linkage of schizophrenia to markers on chromosome 13 in Caucasian but not Oriental populations, *Hum. Genet.*, 99, 417–420, 1997.

67. Schwab, S.G., Hallmayer, J., Lerer, B., Albus, M., Borrmann, M., Honig, S., Strauss, M., Segman, R., Lichtermann, D., Knapp, M., Trixler, M., Maier, W., and Wildenauer, D.B., Support for a chromosome 18p locus conferring susceptibility to functional psychoses in families with schizophrenia, by association and linkage analysis, *Am. J. Hum. Genet.*, 63, 1139–1152, 1998.

68. Pulver, A.E., Karayiorgou, M., Lasseter, V.K., Wolyniec, P., Kasch, L., Antonarakis, S., Housman, D., Kazazian, H.H., Meyers, D., Nestadt, G., et al., Follow-up of a report of a potential linkage for schizophrenia on chromosome 22q12-q13.1: Part 2, *Am. J. Med. Genet.*, 54, 44–50, 1994.

69. Blackwood, D.H., He, L., Morris, S.W., McLean, A., Whitton, C., Thomson, M., Walker, M.T., Woodburn, K., Sharp, C.M., Wright, A.F., Shibasaki, Y., St Clair, D.M., Porteous, D.J., and Muir, W.J., A locus for bipolar affective disorder on chromosome 4p, *Nat. Genet.,* 12, 427–430, 1996.

70. Detera-Wadleigh, S.D., Badner, J.A., Berrettini, W.H., Yoshikawa, T., Goldin, L.R., Turner, G., Rollins, D.Y., Moses, T., Sanders, A.R., Karkera, J.D., Esterling, L.E., Zeng, J., Ferraro, T.N., Guroff, J.J., Kazuba, D., Maxwell, M.E., Nurnberger, J.I., and Gershon, E.S., A high-density genome scan detects evidence for a bipolar-disorder susceptibility locus on 13q32 and other potential loci on 1q32 and 18p11.2, *Proc. Natl. Acad. Sci. USA,* 96, 5604–5609, 1999.

71. Berrettini, W.H., Ferraro, T.N., Goldin, L.R., Weeks, D.E., Detera-Wadleigh, S., Nurnberger, J.I., and Gershon, E.S., Chromosome 18 DNA markers and manic-depressive illness: Evidence for a susceptibility gene, *Proc. Natl. Acad. Sci. USA,* 91, 5918–5921, 1994.

72. Freimer, N.B., Reus, V.I., Escamilla, M., Spesny, M., Smith, L., Service, S., Gallegos, A., Meza, L., Batki, S., Vinogradov, S., Leon, P., and Sandkuijl, L.A., An approach to investigating linkage for bipolar disorder using large Costa Rican pedigrees, *Am. J. Med. Genet.,* 67, 254–263, 1996.

73. Escamilla, M.A., Spesny, M., Reus, V.I., Gallegos, A., Meza, L., Molina, J., Sandkuijl, L.A., Fournier, E., Leon, P.E., Smith, L.B., and Freimer, N.B., Use of linkage disequilibrium approaches to map genes for bipolar disorder in the Costa Rican population, *Am. J. Med. Genet.,* 67, 244–253, 1996.

74. Aita, V.M., Liu, J., Knowles, J.A., Terwilliger, J.D., Baltazar, R., Grunn, A., Loth, J.E., Kanyas, K., Lerer, B., Endicott, J., Wang, Z., Penchaszadeh, G., Gilliam, T.C., and Baron, M., A comprehensive linkage analysis of chromosome 21q22 supports prior evidence for a putative bipolar affective disorder locus, *Am. J. Hum. Genet.,* 64, 210–217, 1999.

75. Straub, R.E., Lehner, T., Luo, Y., Loth, J.E., Shao, W., Sharpe, L., Alexander, J.R., Das, K., Simon, R., Fieve, R.R., et al., A possible vulnerability locus for bipolar affective disorder on chromosome 21q22.3, *Nat. Genet.,* 8, 291–296, 1994.

76. Baron, M., Straub, R.E., Lehner, T., Endicott, J., Ott, J., Gilliam, T.C., and Lerer, B., Bipolar disorder and linkage to Xq28, *Nat. Genet.,* 7, 461–462, 1994.

77. Berrettini, W.H., Susceptibility loci for bipolar disorder: Overlap with inherited vulnerability to schizophrenia, *Biol. Psychiatry,* 47, 245–251, 2000.

78. Blehar, M.C., Weissman, M.M., Gershon, E.S., and Hirschfeld, R.M., Family and genetic studies of affective disorders, *Arch. Gen. Psychiatry,* 45, 289–292, 1988.

79. Tsuang, M.T., Stone, W.S., and Faraone, S.V., Toward reformulating the diagnosis of schizophrenia, *Am. J. Psychiatry,* 157, 1041–1050, 2000.

80. Kraepelin, E., *Dementia Praecox Paraphrenia*, 1971 ed., Robert E. Krieger, New York, 1919.

81. Wildenauer, D.B., Schwab, S.G., Maier, W., and Detera-Wadleigh, S.D., Do schizophrenia and affective disorder share susceptibility genes?, *Schizophr. Res.,* 39, 107–111, 1999.

82. Crawley, J.N., *What's Wrong with My Mouse?* 1st ed., John Wiley & Sons, New York, 2000.

83. Pinto, L.H. and Takahashi, J.S., Genetic dissection of mouse behavior using induced mutagenesis, in *Handbook of Molecular-Genetic Techniques for Brain Behavior Research*, Crusio, W.E., Gerlai, R.T., Eds., Elsevier, Amsterdam, 1999.

84. Tarantino, L.M., Gould, T.J., Druhan, J.P., and Bucan, M., Behavior and mutagenesis screens: The importance of baseline analysis of inbred strains, *Mammal. Genome,* 11, 555–564, 1999.

85. Schimenti, J. and Bucan, M., Functional genomics in the mouse: Phenotype-based mutagenesis screens, *Genome Res.,* 8, 698–710, 1998.

86. Crabbe, J.C., Wahlsten, D., and Dudek, B.C., Genetics of mouse behavior: Interactions with laboratory environment, *Science,* 284, 1670–1672, 1999.

87. Vitaterna, M.H., King, D.P., Chang, A.M., Kornhauser, J.M., Lowrey, P.L., McDonald, J.D., Dove, W.F., Pinto, L.H., Turek, F.W., and Takahashi, J.S., Mutagenesis and mapping of a mouse gene, Clock, essential for circadian behavior, *Science,* 264, 719–725, 1994.

87a. Kanes, S.J., Leach, G.R., Tarantino, L.M., Schimenti, J., and Bucan, M., unpublished data.

88. Grillon, C., Ameli, R., Charney, D.S., Krystal, J., and Braff, D., Startle gating deficits occur across prepulse intensities in schizophrenic patients, *Biol. Psychiatry,* 32, 939–943, 1992.

89. Zisook, S., Heaton, R., Moranville, J., Kuck, J., Jernigan, T., and Braff, D., Past substance abuse and clinical course of schizophrenia, *Am. J. Psychiatry,* 149, 552–553, 1992.

90. Judd, L.L., McAdams, L., Budnick, B., and Braff, D.L., Sensory gating deficits in schizophrenia: New results, *Am. J. Psychiatry,* 149, 488–493, 1992.

91. Braff, D.L., Grillon, C., and Geyer, M.A., Gating and habituation of the startle reflex in schizophrenic patients, *Arch. Gen. Psychiatry,* 49, 206–215, 1992.

92. Braff, D.L. and Geyer, M.A., Sensorimotor gating and schizophrenia. Human and animal model studies, *Arch. Gen. Psychiatry,* 47, 181–188, 1990.

93. Grillon, C., Courchesne, E., Ameli, R., Geyer, M.A., and Braff, D.L., Increased distractibility in schizophrenic patients. Electrophysiologic and behavioral evidence, *Arch. Gen. Psychiatry,* 47, 171–179, 1990.

94. Geyer, M.A., Swerdlow, N.R., Mansbach, R.S., and Braff, D.L., Startle response models of sensorimotor gating and habituation deficits in schizophrenia, *Brain Res. Bull.,* 25, 485–498, 1990.

95. Paylor, R. and Crawley, J.N., Inbred strain differences in prepulse inhibition of the mouse startle response, *Psychopharmacology,* 132, 169–180, 1997.

96. Dulawa, S.C. and Geyer, M.A., Psychopharmacology of prepulse inhibition in mice, *Chin. J. Physiol,* 39, 139–146, 1996.

97. Gogos, J.A., Santha, M., Takacs, Z., Beck, K.D., Luine, V., Lucas, L.R., Nadler, J.V., and Karayiorgou, M., The gene encoding proline dehydrogenase modulates sensorimotor gating in mice, *Nat. Genet.,* 21, 434–439, 1999.

98. Wood, G.K., Tomasiewicz, H., Rutishauser, U., Magnuson, T., Quirion, R., Rochford, J., and Srivastava, L.K., NCAM-180 knockout mice display increased lateral ventricle size and reduced prepulse inhibition of startle, *Neuroreport,* 9, 461–466, 1998.

99. Karayiorgou, M. and Gogos, J., Dissecting the genetic complexity of schizophrenia, *Mol. Psychiatry,* 2, 211–223, 1997.

100. Barbeau, D., Liang, J.J., Robitalille, Y., Quirion, R., and Srivastava, L.K., Decreased expression of the embryonic form of the neural cell adhesion molecule in schizophrenic brains, *Proc. Natl. Acad. Sci. USA,* 92, 2785–2789, 1995.

101. Weickert, C.S. and Weinberger, D.R., A candidate molecule approach to defining developmental pathology in schizophrenia, *Schizophr. Bull.,* 24, 303–316, 1998.

102. Impagnatiello, F., Guidotti, A.R., Pesold, C., Dwivedi, Y., Caruncho, H., Pisu, M.G., Uzunov, D.P., Smalheiser, N.R., Davis, J.M., Pandey, G.N., Pappas, G.D., Tueting, P., Sharma, R.P., and Costa, E., A decrease of reelin expression as a putative vulnerability factor in schizophrenia, *Proc. Natl. Acad. Sci. USA*, 95, 15718–15723, 1998.

103. Hall, C.S., Emotional behavior in the rat. I. Defaecation and urination as measures of individual differences in emotionality, *J. Comp. Psychol.*, 18, 385–403, 1934.

104. Lister, R.G., The use of a plus-maze to measure anxiety in the mouse, *Psychopharmacology*, 92, 180–185, 1987.

105. Shepherd, J.K., Grewal, S.S., Fletcher, A., Bill, D.J., and Dourish, C.T., Behavioural and pharmacological characterisation of the elevated "zero-maze" as an animal model of anxiety, *Psychopharmacology (Berl.)*, 116, 56–64, 1994.

106. Mathis, C., Paul, S.M., and Crawley, J.N., Characterization of benzodiazepine-sensitive behaviors in the A/J and C57BL/6J inbred strains of mice, *Behav. Genet.*, 24, 171–180, 1994.

107. Heisler, L.K., Chu, H.M., Brennan, T.J., Danao, J.A., Bajwa, P., Parsons, L.H., and Tecott, L.H., Elevated anxiety and antidepressant-like responses in serotonin 5-HT1A receptor mutant mice, *Proc. Natl. Acad. Sci. USA*, 95, 15049–15054, 1998.

108. Parks, C.L., Robinson, P.S., Sibille, E., Shenk, T., and Toth, M., Increased anxiety of mice lacking the serotonin1A receptor, *Proc. Natl. Acad. Sci. USA*, 95, 10734–10739, 1998.

109. Ramboz, S., Oosting, R., Amara, D.A., Kung, H.F., Blier, P., Mendelsohn, M., Mann, J.J., Brunner, D., and Hen, R., Serotonin receptor 1A knockout: An animal model of anxiety-related disorder, *Proc. Natl. Acad. Sci. USA*, 95, 14476–14481, 1998.

110. Brunner, D., Buhot, M.C., Hen, R., and Hofer, M., Anxiety, motor activation, and maternal-infant interactions in 5HT1B knockout mice, *Behav. Neurosci.*, 113, 587–601, 1999.

111. Malleret, G., Hen, R., Guillou, J.L., Segu, L., and Buhot, M.C., 5-HT1B receptor knock-out mice exhibit increased exploratory activity and enhanced spatial memory performance in the Morris water maze, *J. Neurosci.*, 19, 6157–6168, 1999.

112. Zhuang, X., Gross, C., Santarelli, L., Compan, V., Trillat, A.C., and Hen, R., Altered emotional states in knockout mice lacking 5-HT1A or 5-HT1B receptors, *Neuropsychopharmacology*, 21, 52S-60S, 1999.

113. Smith, G.W., Aubry, J.M., Dellu, F., Contarino, A., Bilezikjian, L.M., Gold, L.H., Chen, R., Marchuk, Y., Hauser, C., Bentley, C.A., Sawchenko, P.E., Koob, G.F., Vale, W., and Lee, K.F., Corticotropin releasing factor receptor 1-deficient mice display decreased anxiety, impaired stress response, and aberrant neuroendocrine development, *Neuron*, 20, 1093–1102, 1998.

114. Timpl, P., Spanagel, R., Sillaber, I., Kresse, A., Reul, J.M., Stalla, G.K., Blanquet, V., Steckler, T., Holsboer, F., and Wurst, W., Impaired stress response and reduced anxiety in mice lacking a functional corticotropin-releasing hormone receptor, *Nat. Genet.*, 19, 162–166, 1998.

115. Callicott, J.H., Bertolino, A., Mattay, V.S., Langheim, F.J., Duyn, J., Coppola, R., Goldberg, T.E., and Weinberger, D.R., Physiological dysfunction of the dorsolateral prefrontal cortex in schizophrenia revisited, *Cereb. Cortex*, 10, 1078–1092, 2000.

116. Klapdor, K. and van der staay, F.J., The Morris water-escape task in mice: Strain differences and effects of intra-maze contrast and brightness, *Physiol. Behav.*, 60, 1247–1254, 1996.

117. Logue, S.F., Paylor, R., and Wehner, J.M., Hippocampal lesions cause learning deficits in inbred mice in the Morris water maze and conditioned-fear task, *Behav. Neurosci.,* 111, 104–113, 1997.

118. Owen, E.H., Logue, S.F., Rasmussen, D.L., and Wehner, J.M., Assessment of learning by the Morris water task and fear conditioning in inbred mouse strains and F1 hybrids: Implications of genetic background for single gene mutations and quantitative trait loci analyses, *Neuroscience,* 80, 1087–1099, 1997.

119. Giese, K.P., Fedorov, N.B., Filipkowski, R.K., and Silva, A.J., Autophosphorylation at Thr286 of the alpha calcium-calmodulin kinase II in LTP and learning, *Science,* 279, 870–873, 1998.

120. Gass, P., Wolfer, D.P., Balschun, D., Rudolph, D., Frey, U., Lipp, H.P., and Schutz, G., Deficits in memory tasks of mice with CREB mutations depend on gene dosage, *Learn. Mem.,* 5, 274–288, 1998.

121. Sakimura, K., Kutsuwada, T., Ito, I., Manabe, T., Takayama, C., Kushiya, E., Yagi, T., Aizawa, S., Inoue, Y., Sugiyama, H., et al., Reduced hippocampal LTP and spatial learning in mice lacking NMDA receptor epsilon 1 subunit, *Nature,* 373, 151–155, 1995.

122. Wu, Z.L., Thomas, S.A., Villacres, E.C., Xia, Z., Simmons, M.L., Chavkin, C., Palmiter, R.D., and Storm, D.R., Altered behavior and long-term potentiation in type I adenylyl cyclase mutant mice, *Proc. Natl. Acad. Sci. USA,* 92, 220–224, 1995.

123. Storm, D.R., Hansel, C., Hacker, B., Parent, A., and Linden, D.J., Impaired cerebellar long-term potentiation in type I adenylyl cyclase mutant mice, *Neuron,* 20, 1199–1210, 1998.

124. Lu, Y.M., Jia, Z., Janus, C., Henderson, J.T., Gerlai, R., Wojtowicz, J.M., and Roder, J.C., Mice lacking metabotropic glutamate receptor 5 show impaired learning and reduced CA1 long-term potentiation (LTP) but normal CA3 LTP, *J. Neurosci.,* 17, 5196–5205, 1997.

125. Tang, Y.P., Shimizu, E., Dube, G.R., Rampon, C., Kerchner, G.A., Zhuo, M., Liu, G., and Tsien, J.Z., Genetic enhancement of learning and memory in mice, *Nature,* 401, 63–69, 1999.

126. Minichiello, L., Korte, M., Wolfer, D., Kuhn, R., Unsicker, K., Cestari, V., Rossi-Arnaud, C., Lipp, H.P., Bonhoeffer, T., and Klein, R., Essential role for TrkB receptors in hippocampus-mediated learning, *Neuron,* 24, 401–414, 1999.

127. Crawley, J.N., Belknap, J.K., Collings, A., Crabbe, J.C., Frankel, W., Henderson, N., Hitzemann, R.J., Maxson, S.C., Miner, L.L., Silva, A.J., Wehner, J.M., Wynshaw-Boris, A., and Paylor, R., Behavioral phenotypes of inbred mouse strains: Implications and recommendations for molecular studies, *Psychopharmacology,* 132, 107–124, 1997.

128. Festing, M.F.W., *Inbred Strains Biomedical Research*, Oxford University Press, New York, 1979.

129. Antoch, M.P., Song, E.J., Chang, A.M., Vitaterna, M.H., Zhao, Y., Wilsbacher, L.D., Sangoram, A.M., King, D.P., Pinto, L.H., and Takahashi, J.S., Functional identification of the mouse circadian Clock gene by transgenic BAC rescue, *Cell,* 89, 655–667, 1997.

130. King, D.P., Zhao, Y., Sangoram, A.M., Wilsbacher, L.D., Tanaka, M., Antoch, M.P., Steeves, T.D., Vitaterna, M.H., Kornhauser, J.M., Lowrey, P.L., Turek, F.W., and Takahashi, J.S., Positional cloning of the mouse circadian clock gene, *Cell,* 89, 641–653, 1997.

131. Bode, V.C., Ethylnitrosourea mutagenesis and the isolation of mutant alleles for specific genes located in the T region of mouse chromosome 17, *Genetics*, 108, 457–470, 1984.

132. Bode, V.C., McDonald, J.D., Guenet, J.L., and Simon, D., hph-1: A mouse mutant with hereditary hyperphenylalaninemia induced by ethylnitrosourea mutagenesis, *Genetics*, 118, 299–305, 1988.

133. Kasarskis, A., Manova, K., and Anderson, K.V., A phenotype-based screen for embryonic lethal mutations in the mouse, *Proc. Natl. Acad. Sci. USA*, 95, 7485–7490, 1998.

134. McDonald, J.D., Bode, V.C., Dove, W.F., and Shedlovsky, A., Pahhph-5: A mouse mutant deficient in phenylalanine hydroxylase, *Proc. Natl. Acad. Sci. USA*, 87, 1965–1967, 1990.

135. Harding, C.O., Williams, P., Wagner, E., Chang, D.S., Wild, K., Colwell, R.E., and Wolff, J.A., Mice with genetic gamma-glutamyl transpeptidase deficiency exhibit glutathionuria, severe growth failure, reduced life spans, and infertility, *J. Biol. Chem.*, 272, 12560–12567, 1997.

136. Xu, F., Gainetdinov, R.R., Wetsel, W.C., Jones, S.R., Bohn, L.M., Miller, G.W., Wang, Y.M., and Caron, M.G., Mice lacking the norepinephrine transporter are supersensitive to psychostimulants, *Nat. Neurosci.*, 3, 465–471, 2000.

137. Nolan, P.M., Sollars, P.J., Bohne, B.A., Ewens, W.J., Pickard, G.E., and Bucan, M., Heterozygosity mapping of partially congenic lines: Mapping of a semidominant neurological mutation, Wheels (Whl), on mouse chromosome 4, *Genetics*, 140, 245–254, 1995.

138. Kuhn, R., Schwenk, F., Aguet, M., and Rajewsky, K., Inducible gene targeting in mice, *Science*, 269, 1427–1429, 1995.

139. Sauer, B., Inducible gene targeting in mice using the Cre/lox system, *Methods*, 14, 381–392, 1998.

140. Irwin, S., Comprehensive behavioral assessment: Ia. A systematic, quantitative procedure for assessing the behavioral and physiologic state of the mouse, *Psychopharmacologia*, 13, 222–257, 1968.

141. Moser, V.C., Tilson, H.A., MacPhail, R.C., Becking, G.C., Cuomo, V., Frantik, E., Kulig, B.M., and Winneke, G., The IPCS collaborative study on neurobehavioral screening methods: II. Protocol design and testing procedures, *Neurotoxicology*, 18, 929–938, 1997.

142. Rogers, D.C., Fisher, E.M., Brown, S.D., Peters, J., Hunter, A.J., and Martin, J.E., Behavioral and functional analysis of mouse phenotype: SHIRPA, a proposed protocol for comprehensive phenotype assessment, *Mamm. Genome*, 8, 711–713, 1997.

Knockouts

A.

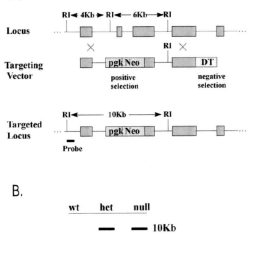

FIGURE 3.2 Standard gene targeting approach. (a) Example of a targeting vector using both positive and negative selection. Positive selection is with a neomycin resistance gene carrying its own weak promoter (pgkNeo). Negative selection, in this case, uses the diphtheria toxin gene (DT). A homologous recombination event removes the DT gene and confers Neo resistance. (b) The targeting event can be detected using Southern blot hybridization using a probe (shown in a) that is external to the targeting construct.

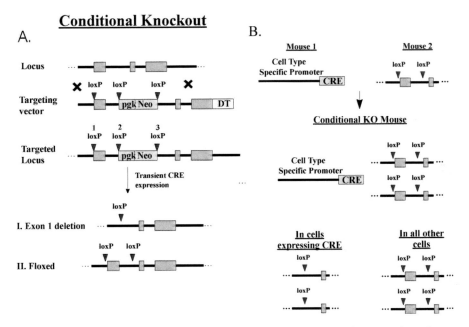

FIGURE 3.3 Conditional knockouts. (a) Two step targeting approach for generating a locus containing flanking loxP sites (floxed locus). The targeting vector is designed with 3 loxP sites flanking the NeoR gene and a critical exon of the target locus. Following successful targeting in ES cells, the targeted line is transiently transfected with a vector expressing CRE recombinase. Recombination between loxP sites 1 and 3 removes both the NeoR gene and the critical exon, resulting in what should be a standard null allele. Recombination between loxP sites 2 and 3 removes the NeoR gene and leaves 2 loxP sites in introns flanking the critical exon. These ES cell lines can then be used to generate a null and floxed locus, respectively. (b) To obtain a conditional knockout, a mouse is generated that is homozygous for the floxed locus and in addition carries a transgene expressing CRE recombinase in the cell types in which a knockout is desired. Expression of the recombinase in those cells should result in deletion of the critical exon while sparing the remaining tissues in the animal.

Inducible Transgenic

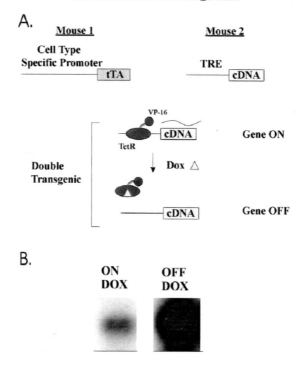

A.

Mouse 1   Mouse 2

Cell Type Specific Promoter   TRE

tTA   cDNA

VP-16

cDNA   Gene ON

TetR

Double Transgenic   Dox △

cDNA   Gene OFF

B.

ON DOX   OFF DOX

FIGURE 3.4 Regulated transgene expression. (a) Two transgenes are introduced into the same mouse to obtain doxycycline regulated transgene expression. In mouse 1 a cell type specific promoter is used to drive expression of the tTA transcription factor. In mouse 2 the gene of interest is fused to a TRE (Tetracycline Responsive Element). When both transgenes are introduced into the same mouse, expression of the TRE linked gene is activated only in those cells that express tTA. (b) Blot showing the regulation of a TRE linked gene by doxycycline (DOX).

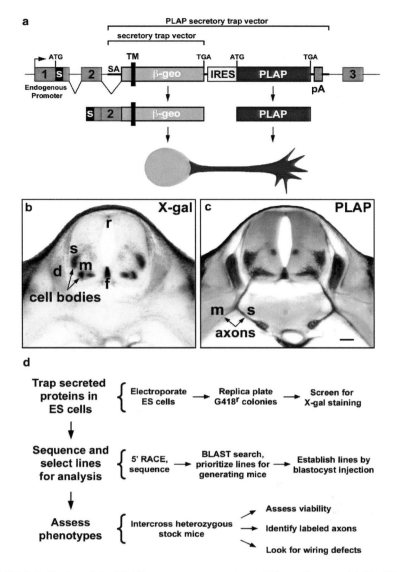

FIGURE 5.1 Design of the PLAP secretory trap vector and flow of screen. (a) The PLAP secretory trap vector is a modification of the original secretory trap vector [including splice acceptor site (SA), CD4 transmembrane domain (TM), and β-geo sequences[9]] to which has been added an internal ribosome entry site (IRES) and PLAP sequences. Integration into an endogenous gene drives expression of a bicistronic transcript beginning with the ATG in exon 1 and ending with the polyadenylation signal (pA) in the vector. Two proteins are made from the transcript (middle line): (1) a fusion between upstream exons of the trapped gene and the transmembrane β-geo protein and (2) a wild type PLAP protein, translated independently from the ATG in the IRES (s, signal sequence). The β-geo fusion protein is localized to neuronal cell bodies and PLAP is localized to axons (bottom line). (b–c) Adjacent sections of embryonic spinal cord from an E12.5 embryo heterozygous for an insertion in *KST37*. In (b), blue X-gal staining labels the cell bodies of dorsal root ganglia (d), sympathetic preganglionic (s), and motor (m) neurons, as well as the floor plate (f) and roof plate (r). In c, purple PLAP staining labels the corresponding axons as they project out of the spinal cord into the periphery. Scale bar, 100 μm. (d) The screen consists of three major stages, each involving several steps. (All panels from Chapter Reference 3.)

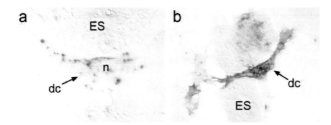

FIGURE 5.3 Secretory X-gal staining pattern. (a) X-gal staining is evident in a single large differentiated cell (dc) while neighboring clumps of undifferentiated embryonic stem cells (ES) do not exhibit detectable levels of staining in this line. The punctate pattern characteristic of genuine secretory trap insertions is easy to see in this large spread-out cell, with discrete dots of various sizes around the nucleus (n) and throughout the cytoplasm. (b) PLAP staining outlines the surface morphology of a large differentiated cell (dc). Again, ES cells in the culture are largely unstained. (Note that different lines can show a different distribution of staining throughout the culture.)

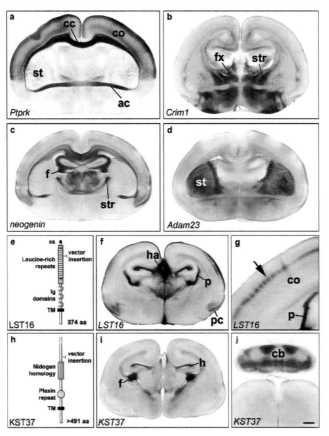

FIGURE 5.4
A survey of axonal populations labeled by PLAP in different mouse lines. (a–d) P0 brains, coronal sections, dorsal is up. ac, anterior commissure; co, cortex; cc, corpus callosum; f, fimbria; fx, fornix; st, striatum; str, stria terminalis. (a) *Ptprk* homozygote. (b) *Crim1* heterozygote. (c) *Neogenin* homozygote. (d) *Adam23* homozygote. (e) *LST16* predicted protein structure (Ig, immunoglobulin domains). (f) *LST16* homozygote, coronal section. ha, habenula; pc, piriform cortex; p, pia. (g) *LST16* heterozygote, coronal section. co, somatosensory cortex; arrow indicates barrels. (h) *KST37* predicted protein structure. (i) *KST37* homozygote, coronal section. h, hippocampus; f, fimbria j, *KST37* homozygote, transverse section. cb, cerebellum. Scale bar, 125 μm (a–d); 140 μm (f, i), 205 μm (g), 85 μm (j). (All panels from Chapter Reference 3.)

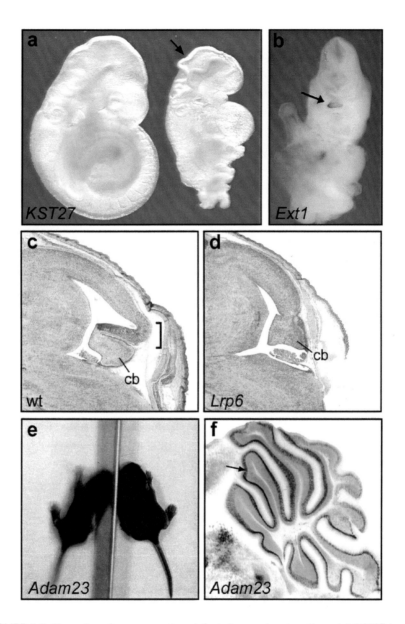

FIGURE 5.5 Examples of nervous system defects in secretory trap lines. (a) *KST27* (novel, predicted multi-TM protein). Mutant embryos (right; compare to wild-type, left) fail to turn properly and show severe defects in the developing CNS (arrow shows open neural tube at forebrain level). (b) *Ext1*. Mutant embryos show severe limb defects and variable loss of ventral midline structures, often leading to cyclopia (arrow shows single eye). (c–d) *Lrp6*. Saggital sections through the brain of wild-type (c) and mutant (d) neonates show a loss of the inferior colliculus (bracket in c) and malformation of the cerebellum (cb) in the mutant. (e) *Adam23*. Wild-type (right) and mutant (left) pups at P12. The mutant animal is noticeably smaller than its littermate, is ataxic, and displays a strong tremor (resulting in a blurry outline in this one-second exposure). (f) *Adam23* is strongly expressed in Purkinje cells in the cerebellum (arrow; 200 μm section stained for β-gal activity). (Panels a, b, e, and f from Mitchell et al., 2001; panels c and d from Pinson et al., 2000.)

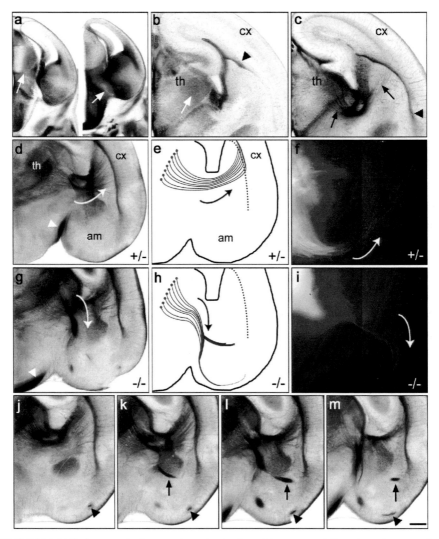

FIGURE 5.6 Thalamocortical axon misrouting in *Sema6A* mutants. All sections shown are in the coronal plane with dorsal to the top. (a) *Sema6A* homozygote, E14.5, X-gal (hemisection on left; midline to left) and PLAP (right) staining. (b–c) *Sema6A* heterozygote, P0, X-gal (b) and PLAP (c) staining in adjacent sections. cx, cortex; th, thalamus; white arrow, cell bodies in dorsal thalamus. Arrowheads point to staining in a glial palisade. (d–f) In *Sema6A* heterozygotes (P0), PLAP staining (d) reveals thalamocortical axons projecting normally through the internal capsule and up into the cortex (arrow; in schematic form in e). (f) Injection of DiI into the dorsal thalamus labels thalamocortical axons and confirms the accuracy of the PLAP staining pattern. (g–i) In P0 homozygotes, thalamocortical axons project abnormally down into the amygdala instead of up into the cortex [arrow in g (PLAP); in schematic form in (h)]. The same phenotype is observed by DiI labeling (i). [No difference is observed in the projections of the optic tract in homozygous mutants (arrowheads in d and g)]. (j–m) A series of adjacent sections through a homozygote at P0 progressing from rostral (left) to caudal (right). At rostral levels thalamocortical projections appear normal but a greater percentage become disrupted in progressively more caudal sections. Individual branches within the amygdala can be followed through several sections (arrows, arrowheads). Scale bar, 350 μm in a–c, j–m; 400 μm in d and g; 200 μm in f and i. th, thalamus; cx, cortex; am, amygdala. (Panels a–i from Chapter Reference 3.)

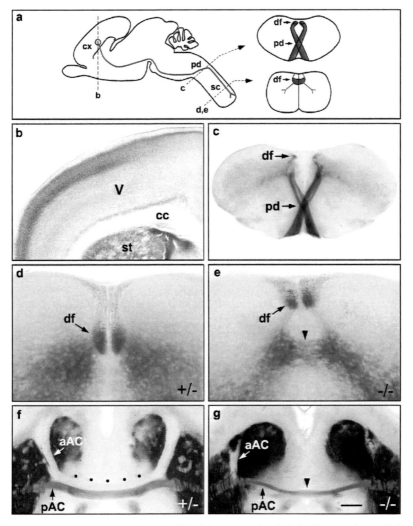

FIGURE 5.7 Axon guidance defects in EphA4 mutants. (a–e) *EphA4* expression and pheno-types in the corticospinal tract (CST). (a) Diagram of the mouse corticospinal tract: sagittal view on the left, transverse sections on the right. Cell bodies of corticospinal neurons (blue circle) reside in the motor cortex (cx) and send their axons (purple line) through the brain and toward the spinal cord (sc). CST axons cross the midline in the pyramidal decussation (pd, upper right), travel dorsally, and then project down the spinal cord in the dorsal funiculus (df, lower right). In the spinal cord, CST axons send collateral branches into the gray matter. Dashed lines indicate coronal (b) and transverse (c–e) planes of section in the following pan-els. b, X-gal staining of an *EphA4* heterozygote brain at P4 reveals expression in cell bodies of layer V motor cortex (V), as well as other layers. cc, corpus callosum; st, striatum. (c and d) PLAP staining of an *EphA4* heterozygote at P4 shows strong staining of corticospinal axons crossing in the pyramidal decussation (c) and continuing in the dorsal funiculus (df) of the spinal cord (d). (e) In the homozygous mutant spinal cord at the same age, the dorsal funicu-lus is shifted dorsally and some axons cross the midline (arrowhead). (f and g) In the anterior commissure at P6, PLAP histochemistry reveals specific staining in the posterior limb (pAC) but not the anterior limb (aAC and outlined by dots) in the heterozygote (f). In the mutant (g), the anterior limb fails to cross the midline (arrowhead), the posterior limb is thinner than nor-mal, and both are shifted ventrally. Scale bar, 315 μm in b, 360 μm in c, 60 μm in d and e, and 400 μm in f and g. (Panels a–e from Chapter Reference 3.)

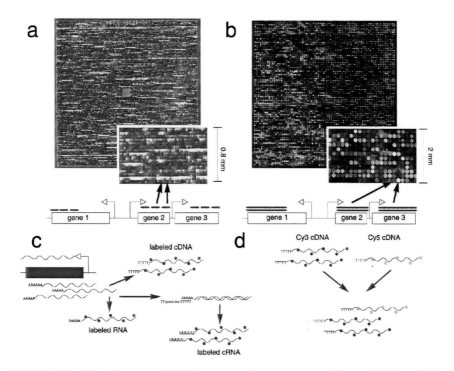

FIGURE 7.1 Principle types of DNA arrays used in gene expression monitoring. (a) An oligonucleotide array from Affymetrix and (b) a cDNA array are shown. Images are shown following hybridization of labeled samples and fluorescence detection. In the case of photolithographically synthesized arrays, $\sim 10^7$ copies of each selected oligonucleotide (usually 25 nucleotides in length) are synthesized base by base in a highly parallel, combinatorial synthesis strategy in hundreds of thousands of spatially separated different 24×24 μm areas on a 1.28 $\times$ 1.28 cm glass surface (some current arrays are even higher density, using 20×20 micron features with over 400,000 different probes). For robotic deposition, approximately one nanogram of material is deposited at intervals of 100 to 300 μm. Although oligonucleotide probes vary in their hybridization efficiency, quantitative estimates of the number of transcripts per cell can be obtained directly by averaging the signal from multiple probes. For technical reasons, the information obtained from spotted cDNA arrays gives the relative concentration (ratio) of a given transcript in two different samples derived from competitive, 2-color hybridization reactions. (c) Different methods for preparing labeled material for measurements of gene expression (mRNA abundance) levels. RNA can be labeled directly using chemical or enzymatic methods, DNA can be end-labeled using terminal transferase and biotinylated nucleotides, and labeled nucleotides can be incorporated into cDNA during or after reverse transcription of polyadenylated RNA. In the protocol used most frequently for oligonucleotide arrays, cDNA is generated from cellular mRNA using an oligo dT primer that carries a T7 promoter at its 5' end. The double-stranded cDNA intermediate serves as template for a reverse transcription reaction in which labeled nucleotides are incorporated into cRNA. The advantage of this approach is that the original, cellular mRNA is effectively amplified (typically by a factor of 50 to 200) in a linear, unbiased, and reproducible fashion. Commonly used labeling groups include the fluorophores fluorescein, Cy3 (or Cy5 and other Cy dyes), and nonfluorescent biotin, which is subsequently made fluorescent by staining with a streptavidin-phycoerythrin conjugate. (d) Two-color hybridization strategy often used with cDNA microarrays. cDNA from two different conditions is labeled with two different fluorescent dyes (e.g., Cy3 and Cy5), and the two samples are co-hybridized to an array. After washing, the array is scanned at two different wavelengths to quantitate the relative transcript abundance for each condition. cDNA array image courtesy of J. DeRisi and P.O. Brown. (Figure reprinted from Lockhart and Winzeler[32], with permission from *Nature,* copyright 2000, Macmillan Magazines Limited.)

a

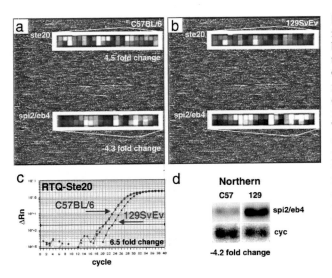

mRNA reference sequence

5' ———————————————————————————————— 3'

Reference sequence

Spaced DNA probe pairs

··· TGTGATGGTGGGAATGGGTCAGAAGGACTCCTATGTGGGTGACGAGGCC ···

TTACCCAGTCTTCCTGAGGATACACCCAC Perfect Match Oligo

TTACCCAGTCTTGCTGAGGATACACCCAC Mismatch Oligo

Perfect match probe cells

Fluorescence Intensity Image

Mismatch probe cells

FIGURE 7.2 (a) Gene expression probe set layout for oligonucleotide arrays. Multiple oligonucleotide probes that are complementary to the mRNA of interest are chosen, typically from the region near the 3′ end for each transcript of interest. In a position physically adjacent (below in the schematic) to each perfect match (PM) probe, is a probe that has a single base difference in the middle (the mismatch, or MM probe). The MM probes serve as specificity controls and allow the discrimination between signals that are due to the specific RNA of interest and those that may be due to cross-hybridization. The PM probes are chosen based on a measure of sequence uniqueness, expected absence of secondary structure, and a set of sequence-based selection rules. The probes are not necessarily chosen from equally spaced regions of the transcript, and they may have some degree of sequence overlap. For quantitation, the PM minus MM differences are used because subtracting the MM signals helps reduce contributions due to background and cross-hybridization. The patterns of hybridization (i.e., the consistency of PM signals that are larger than MM signals, as expected from specific hybridization of the RNA for which the probes were designed) are used to make a qualitative assessment of "Present" or "Absent" for each probe set (more precisely, "detectable" and "not detectable"). The average of the PM–MM values (referred to as the "average difference"), after discarding outliers, is used to make a quantitative assessment of RNA abundance. In some newer designs, the multiple PM/MM pairs for each gene are not located next to each other, but are distributed around the array to minimize effects of spatial variation.

a C57BL/6
ste20
4.5 fold change
spi2/eb4
-4.3 fold change

b 129SvEv
ste20
spi2/eb4

c RTQ-Ste20
C57BL/6
129SvEv
6.5 fold change
ΔRn
cycle

d Northern
C57 129
spi2/eb4
cyc
-4.2 fold change

FIGURE 7.4 Array images following hybridization of RNA from the hippocampus of a C57BL/6 and a 129SvEv mouse. The array shown covers more than 6000 murine genes and ESTs. The expanded images show the raw data for the probe sets for two genes (ste20 and spi2/eb4) found to be differentially expressed between the two strains, along with the magnitude of the expression changes. The array results were confirmed using quantitative RT-PCR (Taqman) and Northern blots. The expression results for these genes are in qualitative and quantitative agreement between the different methods; for ste20, the array indicated a 4.5-fold and RT-PCR a 6.5-fold change; for spi2/eb4, the array indicated a change of −4.3-fold while the Northern blot results indicated a −4.2-fold change.

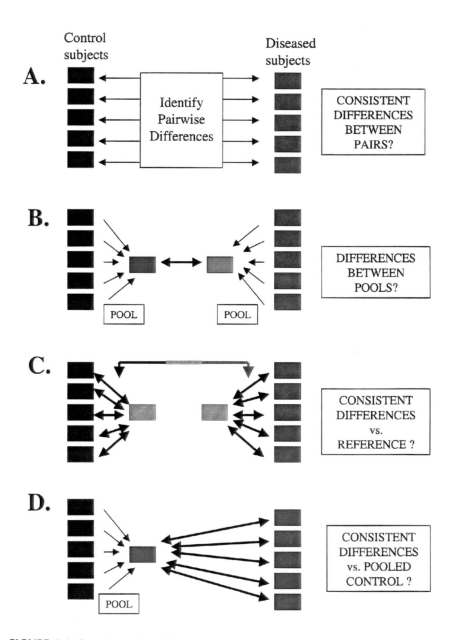

FIGURE 8.4 Control-sample pairing paradigms. (a) Pairwise comparisons to matched controls on different microarrays. (b) Pooling of controls and experimental samples, and comparing the pools on the same microarrays. (c) Comparing individual controls and samples to the same "reference," using different microarrays. This is followed by an indirect comparison of data that are standardized to the reference pool. (d) Control subjects are pooled, and the experimental samples are hybridized against the same pool in independent reactions.

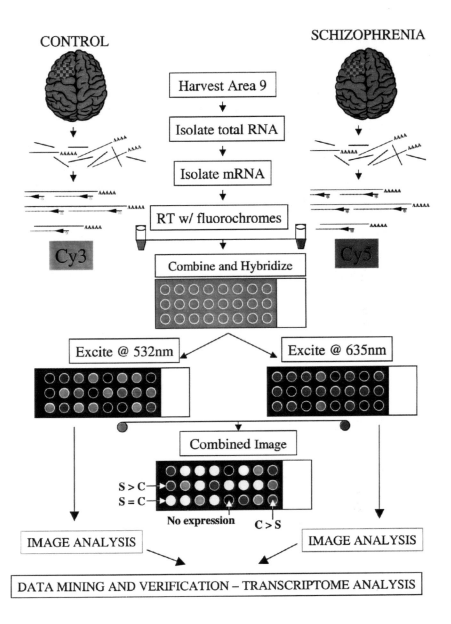

FIGURE 8.5 Experimental procedure for cDNA microarrays. After matching the subjects and the anatomical/histological identification of the brain area of interest, total RNA and/or mRNA are isolated. Samples are separately labeled with Cy3- or Cy5-modified oligonucleotides, primers, or couplers in a standard reverse transcription reaction. The fluorochrome-labeled cDNA is combined and hybridized onto the same microarray. After a series of stringent washes, the microarrays are scanned under the appropriate excitation wavelength. The obtained images are adjusted and analyzed, and the fluorescent intensities are compared between the red and green channels. Note that the combined image, while esthetically pleasing, has no significant value for data analysis.

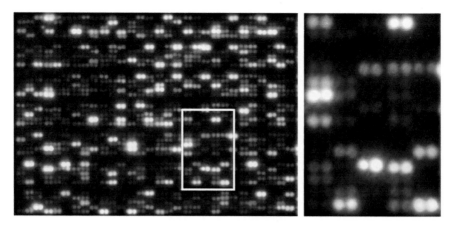

FIGURE 8.6 Membrane arrays are a viable alternative to microarray expression analyses. Two sequential ^{32}P radioisotope-labeled hybridizations were performed on the same nitrocellulose membrane. One of the resulting images was pseudo-colored with red, the other with green, and they were superimposed creating a microarray-like image. The overlaid image shows significant expression differences[102] (Courtesy of Drs. Z. Korade-Mirnics, S. Corey, and J. Burnside.)

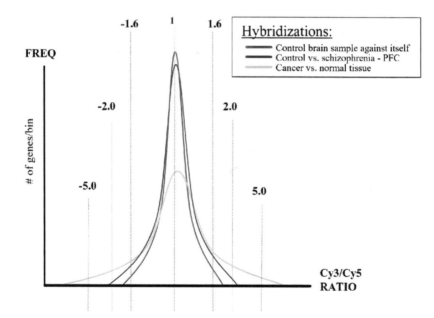

FIGURE 8.8 Cy3 vs. Cy5 ratio distribution of all genes in three typical hybridizations. On the X axis, "1" bin represents equal Cy3/Cy5 intensity; Y axis reports the number of genes in the corresponding bin for each of the three experiments. Note that a prefrontal comparison between a subject with schizophrenia and its matched control (blue line) exceeds experimental noise (red line) in only about 5% of the observations, with virtually no expression differences reported over 5-fold. In contrast, when a malignant to normal tissue comparison is made (green line), gene expression differences are more numerous and robust. In microarray experiments involving brain tissue, many "true" gene expression differences are difficult to discern from the combined noise of the microarray assay (labeling, hybridization, microarray-related noises).

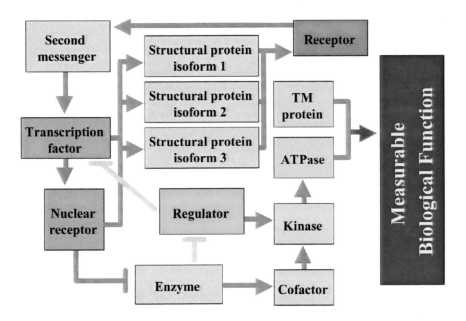

FIGURE 8.9 A functional gene group. In a "gene group" analysis, structurally unrelated genes, whose gene products collaborate to perform a biologically measurable function, are grouped into a *functional* cascade. Expression differences are assessed at the gene group level. Note that impairment in *different* genes of the cascade may lead to deficits in the *same biological* function. Furthermore, deficits in a critical gene may lead to adaptive expression changes in other genes of the same pathway.

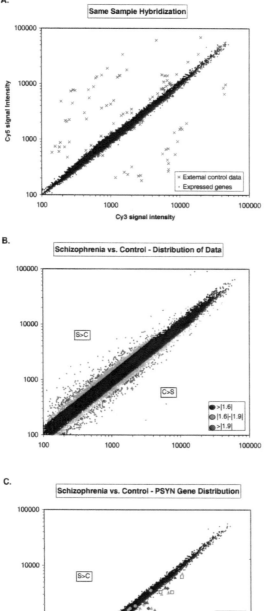

FIGURE 8.10 cDNA microarray hybridization results. Each panel represents a log/log plot of adjusted Cy3 (X axis) and Cy5 (Y axis) signal intensities. (a) Two aliquots of the same control PFC sample were primer-labeled with Cy3 and Cy5, combined, and hybridized onto the same UniGEM-V2 microarray. Out of 5400 expressed genes, only 2 showed a Cy3/Cy5 ratio >|1.6| (<0.01%). Gray × markers represent different spiked-in controls in various ratios and amounts that were labeled with the samples, revealing sufficient array sensitivity and resolution to detect expression differences. (b) Four experiments are plotted together, each comparing a PFC of a different subject with schizophrenia to the same brain region of a matched control. Cy3/Cy5 signal differences |1.6| through |1.8| were defined as "probably true" expression differences (green spots), while signal differences >|1.9| were considered "true" expression differences (red spots) between the samples. (c) RGS4, NSF, SYN2, GAD67, and AMPA2 expression is decreased in schizophrenic subjects across most of the eleven pairwise comparisons. Note that these data overlap minimally with the observations obtained in the control experiment (Panel A — Noise).

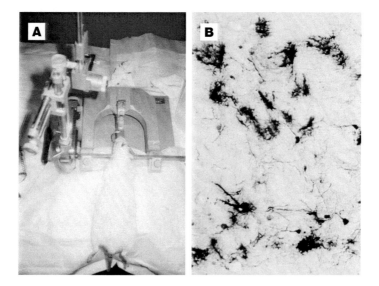

FIGURE 10.3 Direct injection of vectors into rodent brain. (a) Stereotactic apparatus. Anesthetized rat is positioned in Kopf apparatus for stereotactic injection. (b) Labeled neurons and glia. Neurons and glia within the globus pallidus express GFP after injection with a hybrid HSV amplicon vector (packaged helper virus-free[267]) carrying this gene, CD-1 mice were injected with 2 ul of vector at a concentration of 3.5×10^3 t.u. in 1 μl into the striatum. Four days after injection the animals were sacrificed and standard immunocytochemistry was used on brain sections to detect GFP. Neurons are characterized by a large cell body and a single axonal projection, while astrocytes have very dense arborization. Section thickness = 30 μm, magnification = 200×. (Figure provided by Dr. Joanna Bakowska, Massachusetts General Hospital.)

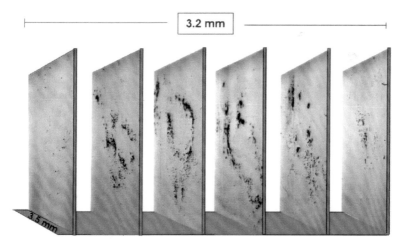

FIGURE 10.4 Multiple direct injections of vector into tumor at single time. Twenty-five μl (total vol.) lacZ-bearing HSV amplicon vector (titer: 3.8×10^8 tu/ml; helper-virus-free) were injected at multiple sites in a 50 mm³ subcutaneous human glioma tumor (Gli 36 3A). Transduction of tumor cells (β-galactosidase expressing) were visualized 48 hr post injection by X-gal staining. Magnification 40×; serial; 50 μm frozen section.

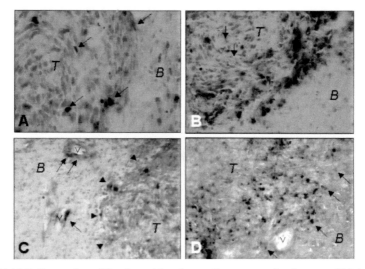

FIGURE 10.5 Expression of β-galactosidase in rat gliosarcoma after intracarotid injection of replication-deficient adenovirus vectors and liposome-DNA complexes (lipoplexes). (a) Photomicrograph of a tumor 48 hr after ipsilateral intra-carotid vector injection in the absence of BK. Note the relatively high number of transduced cells (arrows) in the tumor periphery and, to a certain extent, in the tumor center (T). B = brain adjacent to tumor (magnification 300×, 20 μm frozen section, counterstained with hematoxylin). (b) Photomicrograph of a tumor 48 hr after intra-carotid BK infusion and vector injection. Increased numbers of stained cells are distributed throughout the tumor (T) (compared to A). B = normal brain (magnification 200×, 20 μm frozen section, counterstained with hematoxylin). (c) Photomicrograph of a tumor 48 hr after injection of lipoplex in the absence of BK. Note the high number of transduced cells (arrowheads) throughout the tumor (T). Endothelial cells in capillaries (V) near the tumor/brain border also stain positively (arrows). B = normal brain (magnification 200×, 20 μm frozen section, counterstained with Neutral Red). (d) Photomicrograph of a tumor 48 hr after intra-carotid BK infusion and injection of lipoplex. The number of transduced cells (arrows) throughout the tumor (T) is higher than in the absence of BK, V = tumor vessel, B = normal brain (magnification 200×, 20 μm frozen section, counterstained with Neutral Red).

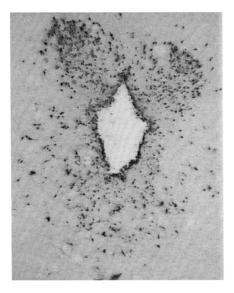

FIGURE 10.7 Intrathecal delivery of transgene to ependyma and associated brain cells. Ten μl of a replication-conditional rHSV vector (10^8 t.u.) bearing the lacZ gene was injected intrathecally through the cisterna magna of rats (see Chapter Reference 238). Two days later animals were sacrificed and brain sections were analyzed by immunocytochemistry for expression of beta-galactosidase. Extensive staining of the ependymal layer and neural cells projecting to that layer was observed. Section thickness = 30 μm. (Figure provided by Dr. Christof Kramm, Massachusetts General Hospital.)

Section 3

7 DNA Arrays and Gene Expression Analysis in the Brain

David J. Lockhart and Carrolee Barlow

CONTENTS

7.1 INTRODUCTION

Two factors are driving a significant change in the way biological and biomedical research is done: the massive increase in the amount of DNA sequence information and the development of technologies to exploit its use. A large amount of sequence information is available for a host of organisms, and new experimental methods have made it possible to gain a global view of molecular and cellular events involving many genes without sacrificing the ability to obtain specific and quantitative information about individual genes. Among the most useful and versatile tools developed for molecular and cellular studies are high-density DNA arrays that allow complex mixtures of RNA and DNA to be interrogated in a highly parallel fashion. DNA arrays can be employed for many different purposes, and they have been put to greatest use to measure gene expression levels (messenger RNA abundance) for tens of thousands of genes simultaneously. The goal of these methods is to understand the underlying workings of the cell, and how all the various components work together to comprise properly functioning cells, tissues, organs, and organisms.

Obviously, the brain is a complex and inhomogeneous organ containing a large number of different regions and cell types. This does not mean, however, that the brain is too complex to be studied using these new tools. Instead, what is clear is that extra care must be taken, experiments need to be designed with the unique features of the brain in mind, and that array-based measurements need to be applied in combination with other methods. Systematic and large-scale measurements of gene expression in multiple brain regions are helping to lay the foundation for asking system-wide questions concerning brain structure and function at the molecular and genetic level. There is little doubt that continuing advances in gene-targeting technology combined with robust behavioral analysis and global gene expression methods will provide new avenues for studying the brain and further our ability to understand the functions of genes, the interactions between the encoded proteins, the relationship between genotype and phenotype, and the unique functions of specific brain regions. In this chapter we discuss how to apply array-based methods to the study of cells and complex tissue, and describe some special considerations for applying these methods to the study of the brain.

7.1.1 GLOBAL GENE EXPRESSION EXPERIMENTS — AN OVERVIEW

The collection of genes that are expressed or transcribed from genomic DNA, sometimes referred to as the expression profile or the "transcriptome," is a major determinant of cellular phenotype and function. Differences in gene expression are both responsible for morphological and phenotypic differences as well as indicative of cellular responses to environmental stimuli and perturbations. Unlike the genome, the transcriptome is highly dynamic and changes rapidly and dramatically in response to perturbations or even during normal cellular events such as DNA replication and cell division.[1,2] In terms of understanding gene function, knowing when, where, and to what extent a gene is expressed is central to understanding the activity and biological roles of its encoded protein. Expression patterns can also help identify genes that are important for a process, function, or phenotype, even in the absence of functional or positional information that might have implicated the gene as a "candidate." In addition, changes in the multigene patterns of expression can provide clues about regulatory mechanisms and broader cellular functions and biochemical pathways.

7.1.2 DNA Arrays

Nucleic acid arrays have been constructed for a wide variety of different organisms[3-7] and have been used successfully to measure transcript abundance in many different experiments (see Color Figure 7.1).* The arrays are passive devices that work by hybridization of labeled RNA or DNA samples to DNA molecules attached at specific locations on a surface. The DNA probes on the surface effectively "count" the number of molecules of each type by binding to molecules that contain their complementary sequence. The sequence of the oligonucleotide or cDNA at each physical location (or address) is generally known or can be determined, the recognition rules that govern hybridization are fairly well understood, and the reactions can be made sufficiently specific. Because of these factors, the signal intensity at each position following a hybridization reaction yields a measure of the number of molecules bound and also their identity.

DNA arrays are generally produced in one of two basic ways: by deposition of nucleic acids (PCR products, plasmids, or oligonucleotides) onto a glass slide,[8] or by *in situ* synthesis of oligonucleotides using photolithography.[9] It is also possible to make arrays by depositing and attaching pre-made oligonucleotides or synthesizing them on a solid support by the spatially specific application of reactants. In the case of photolithographically synthesized arrays, ~10^7 copies of each selected oligonucleotide (usually 25 nucleotides in length) are synthesized base by base in a highly parallel, combinatorial synthesis strategy to make hundreds of thousands of different probes in distinct regions on a flat glass surface (Color Figure 7.1). Regardless of how they are made, DNA arrays are simply large collections of oligonucleotides or cDNA at distinct positions on glass, and most of their uses amount to the counting of different molecules. Typically, for oligonucleotide arrays, multiple probes per gene are placed on the array (often 15 to 20 pairs, but the number can be reduced to further increase experimental efficiency), while in the case of robotic deposition of cDNA, a single, longer (100 to 1000 bp), typically double-stranded DNA probe is used for each gene or EST. In both cases, surface-bound probes are usually chosen from sequence located nearer to the 3' end of the gene (near the poly-A tail in eukaryotic mRNA), and different probes can be used for different

---

* Color figures follow page 140.

exons to enable the detection of variant splice forms. The monitored genes can be of known or unknown function; all that is needed to design probes for an array is at least a couple hundred bases of sequence information to design either PCR primers to make cDNA, or from which to choose appropriate complementary oligonucleotides.

Following preparation of samples and hybridization of labeled samples (typically overnight), the arrays are scanned, and the quantitative fluorescence image along with the known identity of the probes is used to assess the "presence" or "absence" (more precisely, the detectability above thresholds based on background and noise levels) of particular transcripts, and their relative abundance in one or more samples. Messenger RNA present at one to a few copies (relative abundance of ~1:300,000) to thousands of copies per mammalian cell can be detected,[4,8,10] and changes as subtle as a factor of 1.1 to 2.0 can be reliably detected (although changes of at least a factor of 1.5 are more routinely trustworthy) if data quality is high and replicate experiments are performed.

There are a number of possible variations on the basic experimental approach, but the key elements of parallel hybridization to localized, surface-bound nucleic acid probes and subsequent detection and quantification of bound molecules are ubiquitous. In this chapter we will emphasize the use of high-density arrays of relatively short, specifically chosen, *in situ* synthesized DNA oligonucleotides on glass (often called DNA microarrays, oligonucleotide arrays, Affymetrix GeneChip arrays, or simply "chips") and their use for parallel gene expression (mRNA abundance) measurements.

7.2 DETAILED METHODS

7.2.1 OLIGONUCLEOTIDE ARRAY DESIGN

Oligonucleotide arrays for gene expression measurements are designed directly from gene, gene fragment, or EST sequence information (see Color Figure 7.2a).* Fortunately, the amount of sequence information, the "raw material" needed to implement the new genomics technologies, has grown rapidly. At least partial sequence has been obtained for tens of thousands of mouse and rat genes, and a working draft of the entire human genome has been assembled. For each RNA, multiple different oligonucleotide probes are chosen that are complementary in sequence to each monitored mRNA. The advantage of having multiple, different oligonucleotide probes (nonoverlapping if possible, but minimally overlapping if necessary) is that they serve as independent detectors for the same gene. The use of redundant, but different, probes for each mRNA also increases both the qualitative and quantitative accuracy of the results because consistent patterns of hybridization (or hybridization differences) across probes of different sequence for the same gene can be recognized, and the average behavior across the probe set can be used for quantitation rather than relying on the intensity of only a single detector or "spot."

---

* Color figures follow page 140.

One advantage of using shorter, specifically chosen oligonucleotide probes rather than longer cDNA is a greater ability to discriminate between genes with similar sequences. Improved specificity results because the probes can be chosen from the regions of each gene that are the most unique relative to other related family members and all other known expressed sequences in the particular organism. The choice of probes is further guided by a set of sequence-based, empirically determined rules (based on the observed hybridization behavior of many oligonucleotides on arrays), which increase the odds of choosing probes that produce strong and specific hybridization signals in the presence of the complementary sequence. The process of probe selection is not unlike that used to select specific primers for PCR: the desired outcome is oligonucleotides that specifically target the gene or region of interest. But because this selection process is imperfect, and it is still difficult to predict *a priori* the behavior of oligonucleotides on surfaces in the presence of complex mixtures of DNA and RNA, it is helpful to have multiple probes per gene. Gene-specific probes can be chosen from any region of the transcript, but in order to decrease the effects of RNA degradation on the data quality, the oligonucleotide probes for eukaryotic arrays are typically chosen from the 200 to 600 bases nearest the poly-A tail at the 3' end of the transcript. The selected probes usually come in pairs, and are arranged on the arrays as physically adjacent perfect match (PM) and mismatch (MM) partners (the PM probes are designed to be perfectly complementary to the mRNA sequence, and the MM probes are identical except for a single base difference in a central position as shown in Color Figure 7.2a). The MM probes serve as internal controls for hybridization specificity and enable the effective subtraction of local background and cross-hybridization signals. In addition to the sets of probes for each gene or EST, there are a host of additional control probes that are used for grid alignment, spatial normalization, array identification, overall detection sensitivity and specificity, and for assessments of RNA, array, and data quality.

7.2.2 PROCEDURAL OVERVIEW

For these approaches to be most useful, the experimental steps must be relatively easy to perform so that it does not require a heroic effort to obtain an expression profile. The basic steps, prior to hybridization, for performing an array-based expression measurement are similar to those necessary for any mRNA measurement (e.g., northerns, RT-PCR), and involve handling animals, tissues, cells, and then RNA. The exact procedures following extraction of total RNA tend to be more specific to arrays, and even the specific type of array used (e.g., cDNA arrays vs. oligonucleotide GeneChip arrays available from Affymetrix), but the protocols all employ basic molecular biological techniques and reagents. For the Affymetrix arrays, the cellular mRNA is usually amplified (by a factor of 50 to 200) using a linear *in vitro* transcription (IVT) reaction. The IVT reaction is run in the presence of labeled ribonucleotides to produce labeled, complementary RNA (cRNA). The single-stranded cRNA is fragmented randomly to an average size of 30–50 bases prior to hybridization to minimize the possible effects of RNA secondary structure and to enhance hybridization specificity. Following hybridization (typically overnight at a temperature of 40–50°C), the sample is recovered from the hybridization cartridge and saved for future use (samples can be rehybridized multiple times). The arrays are washed to remove weakly bound molecules and to reduce background signals and then they are "read" using a specially designed laser confocal scanner that scans the entire array at a spatial resolution of 3 microns in only 5 to 10 min. This scan produces the raw data file (the "image") that is then quantitatively analyzed and interpreted, as discussed below. The raw data file (the .dat file) for a 1.28 × 1.28 cm array read at a resolution of 3 microns per pixel is approximately 44 megabytes in size, and the subsequent processed files (the .cel and the .chp files) are both about 10 megabytes in size. That means the basic data from a single experiment requires at least 64 megabytes of storage, and such data can be collected by a single person with a single scanner as often as every 10 to 15 min.

7.2.3 TISSUE DISSECTION AND RNA PREPARATION

Several methods exist for obtaining high-quality RNA from brain tissue. Based on our experience in studies of the mouse brain, we have found that animal to animal variation and variation due to dissection can be readily minimized. When using animals, it is important that they be handled in a systematic and consistent manner prior to obtaining tissue. All animals are singly housed for 7 days prior to sacrifice. All euthanasia is performed using cervical dislocation. We also go so far as to perform all dissections at specified hours of the day. Dissections are carried out on petri dishes filled with wet ice. Samples are dissected and frozen in dry ice and stored at –80°C until RNA is extracted. To prepare total RNA from the tissue, TRIzol (GIBCO-BRL) is added at approximately 1 ml per 100 mg tissue to the frozen tissues and then homogenized (Polytron, Kinematica) at maximum speed for 90–120 s. RNA is resuspended in RNase-free water at a concentration of 1 mg/ml. The quality of the total RNA is checked on an agarose gel (to check the distribution of RNA lengths) and by an absorption measurement of the RNA in TE and H_2O (see below for more details).

7.2.4 SAMPLE PREPARATION

In most current implementations of array-based approaches, the RNA or DNA to be hybridized must be labeled prior to the hybridization reaction so that surface-bound molecules can be fluorescently detected and quantitated (see Color Figure 7.1). Either RNA or DNA can be hybridized to arrays, and different methods can be used to prepare labeled material. Commonly used labels include the fluorophores fluorescein, Cy3, or Cy5. Labeling with biotin is most common for use with oligonucleotide arrays. The biotin-containing molecules (biotin is nonfluorescent) following hybridization are made fluorescent by staining with a streptavidin-phycoerythrin conjugate (phycoerythrin is a naturally occurring phycobiliprotein that contains multiple fluorescent molecules in a protein-chromophore complex). We have found that biotin-phycoerythrin labeling yields approximately 10 times more signal per bound molecule than straight incorporation of fluorescein. With oligonucleotide arrays, each sample is labeled identically and hybridized independently to different arrays. Signal intensities can then be compared directly between any two or more array experiments. In the two-color hybridization strategy often used with cDNA microarrays, reverse-transcribed cDNA from two different samples is labeled with two different fluorescent dyes, and the two samples are co-hybridized to an array (see Color Figure 7.1).

7.2.5 STANDARD SAMPLE PREPARATION FOR USE WITH OLIGONUCLEOTIDE ARRAYS

The standard procedure for mRNA amplification and labeling has been described.[4,8,10] Briefly, starting with 1 to 10 μg of total RNA, the mRNA is converted to double-stranded cDNA (in the presence of total RNA, without a poly-A pre-purification step) using a cDNA synthesis kit with an oligo dT primer that contains a T7 RNA polymerase promoter site. From the cDNA is made many copies (typically 50 to 200 copies of cRNA per cDNA molecule) of labeled anti-sense RNA using a linear *in vitro* transcription (IVT) reaction in the presence of labeled nucleotides. This reaction usually yields approximately 40 to 100 μg of labeled RNA, which is more than needed as only 10 to 30 μg of labeled material is used for array hybridization. This amplification and labeling procedure is highly reproducible and introduces very little bias in the mRNA population (i.e., it produces a faithful and quantitative representation of the original mRNA population).

7.2.6 ARRAY DATA ANALYSIS

Following a quantitative fluorescence scan of a typical, photolithographically synthesized oligonucleotide array, a grid is aligned to the image using the known dimensions of the array and the corner and edge controls (laid out in specific patterns on every array) as markers. The individual pixels (typically 50–60 per 24×24 μm synthesis feature — newer designs available from Affymetrix use 20×20 μm features) within each region are averaged (or most commonly, the 75th percentile pixel intensity value is used) after systematically ignoring those at the border and discarding outliers. The qualitative assessment of "present" or "absent" (more correctly, "detected" or below the threshold of detection, respectively) is based on a "voting scheme," with the number of instances in which the PM signal is significantly larger than the MM signal across the redundant set of probes calculated for each gene (see Color Figure 7.2a). In the analysis for each individual probe set, a call of "present" requires a reasonably consistent, but not unanimous, vote across the PM/MM pairs. (The meaning of "significantly larger" in this analysis is determined by a measurement of the noise, with thresholds set as a multiple of the minimum background noise in any given array image.) If the requirements for making a present call are not met, then the gene is called "absent," but it is more correct to call it "not clearly detected." In other words, it is not a good idea to interpret a call of "absent" as an affirmative statement that the RNA for which the probe set was designed is not in the sample, but simply that the signal was not sufficiently strong and consistent to make a clear call of present. Using standard analysis methods, it has been shown that low-abundance RNA can be detected and that the hybridization signal is proportional to RNA concentration over a very wide dynamic range (see Color Figure 7.2b).

A voting scheme that requires some level of consistency across a set of different probes for each gene is analogous to the requirements for a jury in a civil trial (preponderance of evidence, with most jurors in agreement), as opposed to a criminal trial (beyond a reasonable doubt, unanimous jury vote). With this approach, everything does not have to be perfect, and all oligonucleotide probes do not have to behave identically. This is important, because it is not possible to predict the specific hybridization and cross-hybridization properties of oligo probes in advance. Also, with multiple probes, it is possible to recognize weak but real patterns in the presence of much larger signals, and the use of multiprobe patterns increases sensitivity (because not as much is required of each individual probe) and also confidence that the signal is primarily due to specific hybridization of the RNA for which the probe set was designed. In addition, the ability to average across multiple probes per gene improves the quantitative accuracy and reproducibility of the quantitative results because average behavior is more impervious to both systematic (e.g., cross-hybridization, poor probe performance, the effects of polymorphisms or sequencing errors) and random (e.g., noise, physical defects on the arrays) errors.

The details of the way the noise is calculated from each array image, the way thresholds are set, and the logic of the voting scheme were established based on extensive quantitative spiking and reconstruction experiments, and on an assessment of an acceptable false-positive rate (the false-positive rate for "present" calls in any single measurement on an array is generally less than 1% of all genes monitored when using standard conditions and default analysis parameters).[4,9,10] Along with the qualitative assessment of the pattern to make a call of present or absent, there is a quantitative assessment to estimate the RNA concentration or abundance. The determination of quantitative RNA abundance is determined from the average of the pairwise PM minus MM differences (referred to as the "average difference," the quantity shown to be proportional to RNA concentration) across the set of probes for each RNA. When assessing the differences between two different RNA samples (hybridized independently to two different arrays), similar logic and criteria are used, except the primary determinants in this case are the **changes** in the individual PM-MM values across the probe set. Prior to comparing any two or more measurements, all signal intensities on an array are multiplied by a factor (a linear "scaling factor" in the simplest case) that makes the mean PM-MM value for any array measurement equal to a preset value. This simple global scaling process is designed to correct for any interarray differences, or small differences in sample concentration, labeling efficiency or fluorescence detection, and it appears to work rather well when experiments are performed in a consistent fashion (e.g., identical hybridization and washing conditions, and the same amount of labeled material hybridized). In the case of a pair-wise comparison of array results, the patterns of change (with consistent "voting") and the magnitude of the changes are used to make both qualitative calls of "increase" or "decrease," and quantitative assessments of the absolute size (differences in signal, related to changes in the number of copies per cell) and the relative size (ratio or "fold change") of any differences. These methods for qualitative and quantitative assessments of mRNA abundance and differential expression are codified in the standard, commercially available Affymetrix GeneChip analysis software. Because of the richness of the data and the built-in redundancy (at the level of having both multiple pixels per feature and multiple features per gene or EST), there are of course a number of alternative ways in which data of this type could be assessed. New approaches that may help improve sensitivity and quantitative accuracy while using fewer probes per gene (and thus more genes per array) appear promising. These issues are being explored by many groups, with the attainable goal of a significant increase in the information content per array and data quality without an increase in the difficulty, expense, or time required for an experiment.

7.2.7 FOLD CHANGES AND EXPRESSION DIFFERENCES

The extent of change in expression level for any gene is commonly given as the "fold change." For example, if the expression level went from 5 to 10 copies per cell, this would be a 2-fold change; 5 to 15 copies per cell, a 3-fold change, and so on. Often we care most about the relative size of a change rather than exactly how many copies of mRNA per cell are found for a given gene, and we generally would not interpret a change from 5 to 10 copies per cell any differently than we would a change from 20 to 40 copies. Another reason the ratio or fold change is used is that with spotted cDNA arrays, the readout is a ratio of two intensities at each "spot" after a competitive hybridization of two samples labeled with different fluorophores that emit at different wavelengths (i.e., only the ratio is interpreted). But a problem arises no matter which type of array is being used if in one of the two cases, the mRNA is absent or the level is extremely low. For example, if the abundance of a transcript really goes from 0 to 10 copies per cell, the fold change is infinite, and the *difference* between the signals is a more appropriate measure than the ratio (for oligonucleotide arrays, the signal difference has been shown to be quantitatively related to the change in mRNA abundance, see Color Figure 7.2b). In cases such as this, the ratio is also likely to be rather variable because it is difficult to know where to set "zero" and even if the transcript abundance is not strictly zero, the signal is not large relative to the background noise, and cannot be quantified with any confidence. In both of these cases, it is typical to have a minimum allowable value (often set by a measure of the background or the noise in the signals) for the denominator to avoid dividing by zero or an unreasonably small and overly noisy value. When at least one of the two values is too small, an approximate fold change can be given, but it must be remembered that it is an approximation that is completely dependent on the specifics of how the minimum denominator value was set, and that it is likely to be an underestimate of the true value.

7.2.8 SAMPLE AND DATA QUALITY CONTROL

To obtain results with the highest confidence, it is necessary to perform experiments in a consistent and careful fashion, and to perform quality control at several points during the experiments. It is very important to handle animals, tissue, and cells appropriately and to handle total RNA in ways that minimize degradation. To ensure that samples are of suitable quality before hybridizing them to arrays, the following procedures are employed:

1. Total RNA is run on gel to check the size distribution relative to rRNA bands and by spectrophotometer to ensure an OD 260/280 ratio of greater than or equal to 2.0.
2. Labeled, purified, and unfragmented cRNA is run on a gel to check for the correct size distribution relative to quality standards, and the amount of labeled product is quantitated using a measurement of the absorbance at 260 nm (based on a full absorption spectrum from 220 to 340 nm).
3. Following fragmentation, the labeled cRNA is run on a low molecular weight gel to check for a suitable distribution of fragment lengths (typically between 30 and 50 bases).

Following hybridization of a sample to an array, collection of an image, and basic image analysis, data "triage" is performed to make sure that the array data are of sufficient quality for further analysis and comparison with other data sets (see Table 7.1). The primary factors to monitor include background, noise, overall signal strength, the ability to detect spiked bacterial control RNA, the ratio of the 3′ and 5′ signals for actin and Gapdh mRNA (a measure of RNA length and quality — degraded RNA will result in high 3′/5′ ratios because only the region of the mRNA near the 3′ poly-A tail will be amplified and labeled), and the percentage of genes scored as "present." An example of some of the basic data quality measures that should be inspected for every sample and every array measurement, and an example of the consistency that can be obtained when experiments are performed well, is shown (see Table 7.1). Typically, we expect to see % Present values that are within 5% of each other, background and Q values (Q is a measure of the minimum background noise across the array image) within a factor of two of each other, scaling factors (SF) within a factor of two, and 3′/5′ ratios for both Actin and Gapdh of less than 2.5. When experiments meet these standards for sample and data quality, the false-positive rate can be expected to be acceptably low and the overall performance that can be expected is shown in Table 7.2.

7.2.9 THE IMPORTANCE OF WELL-CONTROLLED, REPLICATE MEASUREMENTS

For high-throughput, parallel measurements, data quality is of critical importance if one is attempting to identify with high confidence specific genes that are differentially expressed. The reason is that when monitoring, for example, ten thousand genes, even a low false-positive rate of 1% results in 100 incorrect difference calls, comparable to the number of true changes observed in many types of experiments (a false positive here is defined as an assignment of a gene as "differentially expressed" when in fact the mRNA abundance is not significantly changed). We find that when experiments are performed with sufficient care, the source of most of these false positives (which are in large part the result of setting the lowest possible thresholds in the interest of sensitivity) is random noise, small variations in sample preparation and other experimental steps, and the occasional array-specific physical defect. Because these various factors lead to largely random variations, observations made consistently in independent replicates can yield a false-positive rate closer to

TABLE 7.1
Expression Experiments Sample and Data Triage

File	Date	Sample	% Present	BG	Stdev (BG)	Q raw	SF	3'/5' Gapdh	3'/5' Actin
AFRSA99112401	11.24.99	129svEv Mb1	48%	117	3.5	3.8	1.18	1.2	1.0
AFRSA99112402	11.24.99	C57bl/6 Mb 1	48%	113	3.0	3.5	1.18	1.2	0.9
AFRSA99112404	11.24.99	129svEv Ag 1	49%	98	4.9	3.2	1.28	1.7	1.0
AFRSA99112403	11.24.99	C57bl/6 Ag 1	47%	127	3.7	3.7	1.12	1.4	0.9

Note: The minimal set of sample and data quality measures that should be recorded and checked for every array-based gene expression experiment. "%Present" is the percentage of probe sets that scored as present in a given data set (typically 25 to 50% for mammalian cells); BG is the overall, average background signal across the entire array; Stdev (BG) is the standard deviation of the background across different physical regions (a measure of background consistency); Qraw is a measure of the noise in the background signal and is used to set minimum thresholds in the data analysis algorithms; SF is the linear scaling factor that is applied to equalize the average signal (average PM-MM value set to a predefined level) for every array data set; and 3'/5' Gapdh and 3'/5' Actin are the ratios of the signals observed for probe sets derived from the 3' and 5' ends of these abundant transcripts (a value near 1.0 indicates that the two ends are approximately equally represented and that the original mRNA was not significantly degraded).

TABLE 7.2
Oligonucleotide Array Performance

	Routine Use	Current Limit
Starting material	1–5 µg total RNA	0.2 ng total RNA
Sample reuse	4–5 times	5–10 times
Detection specificity	1:100,000	1:2,000,000
Difference detection	1.8-fold	1.1-fold
False positives	<1%	<0.01%
Dynamic range	~1000-fold	>5000-fold
Probes per gene	32	4–8
Genes or EST per array	~12,000	>50,000

Note: Expected performance in array-based gene expression experiments with oligonucleotide arrays. The performance listed under "routine use" is what can be expected when data and sample quality meet the basic triage requirements (see Table 7.1). Samples can be rehybridized multiple times, but additional material may need to be added to keep the total volume high enough to correctly fill the hybridization chamber (small amounts of solution are lost on the walls of the hybridization chamber, and during filling and emptying). The false-positive value is given in terms of the number of genes called "differentially expressed" relative to the total number of genes or EST monitored. The 1% value is for a single comparison of two samples made independently and hybridized to different arrays (e.g., samples from two identical mice, with no replicates). The lower value under "current limit" is obtained by performing experiments at least in duplicate (i.e., at least two independent comparisons, involving four samples) and applying the multiparameter criteria described in the text.

0.01% (i.e., one percent of one percent), or only one false call of "different" ("increased" or "decreased") for every 10,000 genes monitored (see Figure 7.3).

7.2.10 ANALYSIS OF REPLICATE DATA

To obtain a low false-positive rate, it is important to use multiple criteria for assessing differences. One key to obtaining a low false-positive rate is good, consistent experimental technique while controlling as much as possible all sources of experimental variation (e.g., mouse handling, dissection protocols, tissue handling, RNA extractions, amplification and labeling reactions, hybridization and washing conditions, and array usage, see Figure 7.3). For example, in experiments done as independent duplicates using cell lines or different isogenic mice, we typically require that (1) a probe set score as "increased" or "decreased" in 2/2 comparisons, **and** (2) the fold change be at least 1.8-fold in 2/2 comparison, **and** (3) a probe set score as clearly "present" in at least one of the 4 (2×2) data sets, **and** (4) that the difference in the signal be at least 50 in 2/2 comparisons (in arbitrary units after scaling the overall intensity to a mean of 200 which corresponds to an RNA abundance in a mammalian cell of about 3–5 copies per cell — so a signal change of 50 corresponds to a change in mRNA abundance of roughly 1–2 copies per cell). These specific thresholds are somewhat arbitrary, but we have found that requiring ALL of the qualitative and quantitative criteria be met together makes it so each of the individual criteria can

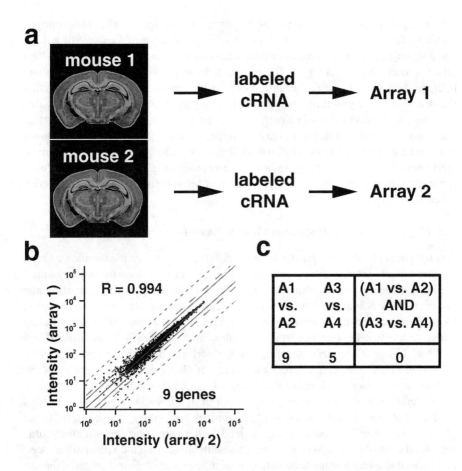

FIGURE 7.3 (a) General schematic of mouse brain gene expression experiments. Tissue is obtained by systematically dissecting the appropriate brain region(s) from at least two mice of the same age, sex, and genetic background, and that have been housed and handled identically. After obtaining tissue, total RNA is extracted, and from the total RNA is made labeled cRNA for array hybridization. Each sample is hybridized to a separate array. (b) A comparison of the quantitative results for independent replicates obtained for two different mice (C57BL/6). The correlation coefficient is very near 1.0 (0.994), and the number of genes that score as "differentially expressed" is small (9 of a possible 6584 in this example), indicating the high degree of reproducibility of the procedures, measurements, and analyses. (c) An analysis showing the low false-positive rate obtained when using stringent analysis criteria and independent replicates. Samples were prepared from dissected hippocampus from four different C57BL/6 mice. When the results for mouse 1 were compared to those for mouse 2, only nine genes scored as different. When mouse 3 was compared to mouse 4, only five genes were scored as different, and there were zero genes that scored as different in both of the independent comparisons. Multiple criteria were used in combination to assign a gene as differentially expressed (see text). When used in combination with carefully controlled, independent replicates, this approach produces an acceptably low false-positive rate of less than a few incorrectly assigned calls per 10,000 genes monitored.

be fairly permissive while the overall requirements are quite strict. For example, requiring only a signal change of 50 alone, or a quantitative fold change of at least 1.8 without the other requirements would lead to an increase in the false-positive rate by more than a factor of 10. Again, it is important to be cautious about interpreting a negative result because it is possible for some genes to miss passing the stringent set of criteria for being differentially expressed. It is always possible to reanalyze the data using more permissive criteria to pick up additional genes that may have changed, but these should be interpreted with greater caution than those that meet the stricter criteria. Also, it is straightforward to query the data to examine the behavior of any specific gene or any chosen set of genes in which one has a particular interest, apart from whether or not their behavior meets the global selection criteria.

7.2.11 NONISOGENIC MICE AND HUMAN SAMPLES

When there are likely to be greater intrinsic differences between the samples, a larger number of individual, independent measurements must be done to avoid misinterpreting expression variation that is due to the underlying genetic or experimental variation. This is important when using cell lines derived from nonisogenic mice, tissue from genetically different mice from an outbred strain, cells from different humans, or when the experimental procedures are intrinsically more difficult and variable. If larger numbers of replicates are performed, then the criteria can be relaxed to allow consistency across most, but not all, of the independent measurements. Another way to minimize the effects of animal-to-animal or cell line-to-cell line differences without increasing the number of hybridization reactions and arrays needed is to pool samples from multiple sources (e.g., the expression profile for pooled tissue from 5–10 outbred mice treated with a drug compared with the profile for a similar number of untreated mice). This helps smooth out individual differences, and gives the average expression behavior of a gene, but of course requires the use of more mice. This approach masks the underlying distribution of expression levels, but it does allow one to identify genes whose average level is different between two different sets of animals. If needed, the samples can be analyzed individually to determine the extent of variation within each of the sets.

We and others have also found that meaningful results can be obtained using human brain tissue from both fetal and postmortem adult brains. The mRNA in most regions of the brain appears to be surprisingly well preserved, even after postmortem intervals of 24 h or more. An essential requirement in these types of studies is that a sufficient number of experiments be performed across multiple individuals and multiple tissue samples to account for individual variation and possible tissue inhomogeneity.

7.2.12 ADDITIONAL ANALYSIS CONSIDERATIONS

The overall false-positive rate in array-based expression measurements can be made extremely low so that genes that meet the strict criteria for being called differentially expressed can be interpreted with high confidence (see Figure 7.3). However, it is difficult to assign a specific measure of confidence for any particular probe set without testing each one explicitly (which is difficult to do for tens of thousands of different probe sets). It is possible and often useful to apply standard statistical measures to the data, but these may not fully capture the uncertainty that results from nonrandom sources of variation (e.g., cross-hybridization to closely related sequences, probes that are based on incorrect sequence database information, systematic aberrant probe behavior). Even more difficult is the interpretation of a negative result (i.e., a call of "absent" or a call of "no change" in a comparison of different samples) and how to assign a measure of the probability of a false negative. As with any type of experiment, one should be very cautious about interpreting a negative, especially if the positive result has never been observed. For example, if a given gene is always scored as "absent," it may be because that gene is not expressed at a significant level in any of the samples tested, or it may mean that the probe set for that gene is somehow defective due to intrinsically poor hybridization behavior, or because the probes were designed based on incorrect sequence information. Instances of poorly performing probe sets for either experimental or informatics/database reasons appear to be rare, but they have been observed. They are likely to be rarer still as more complete and higher quality sequence information becomes available, as the annotation of sequence information improves, and as better rules are developed incorporating more direct observations for the selection of the best performing oligonucleotide probes.

There are some instances where replicate experiments may not be as important. The first is if one is only trying to determine the overall extent of the differences between two or more samples, and the general classes of genes that might be involved, rather than attempting to identify specific genes that are differentially expressed. In these cases, one is not trying to interpret the response of any given gene, so it is not as important that every individual gene be measured with such high confidence. The second is when there are multiple measurements that are part of a series, such as a time course or a dose–response, or just a collection of measurements across different conditions or even different cell types. If there are multiple measurements at different times or doses (or perhaps, gradations in a phenotype), then each point serves as a check on the others, and conclusions can be drawn based on multiple, related observations that are not simply identical replicates. The basic dictum guiding these recommendations is that if you want to be able to trust the data for any specific gene, don't believe anything you see once, but if it is seen independently more than once, it is likely to be real. Furthermore, confidence is increased as interpretations and conclusions are based not on single genes, but on the expression behavior of sets of genes.

7.2.13 ANALYSIS OF MULTIPLE RELATED DATA SETS AND CLUSTERING

An increasingly common approach to analyzing complex gene expression data involves the use of "clustering" methods. Clustering is simply grouping genes into sets that behave in a similar way across multiple experiments (e.g., going up, down, and remaining unchanged in a similar pattern), based solely on the data and without pre-defining expected patterns. It is often very useful to group genes that respond in the same way and to determine if the clustered genes have some functional relationships that may implicate certain multigene pathways, processes, and cellular mechanisms. Clustering and related methods for class discovery and class prediction are also useful for identifying "expression markers" that can be used as surrogates for a phenotype or cell state, as has been done to distinguish leukemias and tumor types.[11,12] There are a number of methods to cluster genes based on expression behavior, including manual examination of the data,[2] and statistical methods such as self-organizing maps,[13] K-tuple means clustering, or hierarchical clustering,[1,14,15] and a host of software packages are available. In terms of using clustering to help define the biological role of a protein, the basic underlying assumption is that genes with similar expression behavior are more likely to be functionally related. Genes without previous functional assignments or known biological roles can be given tentative assignments based on the functions of the other known genes in the same expression cluster.[16] Though not logically rigorous, the utility of the guilt-by-association idea has been demonstrated in yeast, since genes already known to be related do, in fact, tend to cluster together based on their experimentally determined expression patterns.[17,18] Assignments made in this way are certainly not definitive, but they can help indicate that a gene is involved with a particular cellular phenotype and that it may be related to a set of other genes and processes.

7.2.14 VERIFICATION AND FOLLOW-UP OF ARRAY-BASED OBSERVATIONS

It is important to emphasize that these new, parallel approaches do not replace conventional methods. Standard methods such as northerns, westerns, RT-PCR, immunohistochemistry, and *in situ* hybridization are extremely important because they provide independent confirmation of selected results and they can yield different types of information (e.g., concerning protein levels rather than mRNA abundance, or spatial and cell-type specificity, see Color Figure 7.4*). The more conventional, lower-throughput methods are simply used in a more targeted fashion to complement the broader measurements and to follow up on the genes, pathways, and mechanisms implicated by the array results. Because the false-positive rate can be made sufficiently low, it is not necessary to independently confirm every change for the results to be valid and trustworthy, especially if conclusions are based on the behavior of sets of genes rather than individual genes. More detailed follow up using complementary methods is recommended if a gene is being chosen, for example, as a drug target, as a candidate for population genetics studies, or as the target for the construction of a transgenic or knockout mouse.

---

* Color figures follow page 140.

7.2.15 AMPLIFICATION AND LABELING OF SMALL AMOUNTS OF mRNA

It is now routine to obtain high-quality gene expression data starting with 1 to 5 µg of total RNA from mammalian cells (the amount obtained from approximately a million cells). However, this amount of material is not always obtainable and inhomogeneous tissue may need to be examined in greater detail following microdissection, for example. Because the amount of RNA is sometimes limited, an important frontier in gene expression technology development involves reduction of the required amount of starting material. Efficient and reproducible mRNA amplification methods are required, and two approaches show significant promise. The first is a PCR-based approach that has been used to make single-cell cDNA libraries.[19-21] We have found that the amplification is efficient and reproducible, but that the relative abundance of the cDNA products is not well correlated with the original mRNA levels,[21a] although clever normalization and referencing strategies can be employed.[21b]

The second approach does not use PCR, but multiple rounds of linear amplification based on cDNA synthesis and a template-directed *in vitro* transcription (IVT) reaction (see Figure 7.5).[22-26] Each round of amplification (a "round" consists of a cDNA synthesis step, which yields no amplification, plus an IVT step to produce single-stranded RNA) yields an amplification of typically between 100 and 1000-fold. This basic method was developed specifically to characterize mRNA from single live neurons and even subcellular regions, and has been used more recently to amplify mRNA from 500 to 1000 cells from microdissected brain tissues for hybridization to spotted cDNA arrays.

In our hands, we find that the multiple-round cDNA/IVT amplification method produces sufficient quantities of labeled material (approximately 100 µg of labeled product obtained from 50 ng of total RNA, and approximately 25 µg of labeled product from 1 ng of total RNA), is reproducible (correlation coefficient of 0.975 between independent preparations starting with 50 ng of total RNA from the same source — this is nearly identical to the correlation found using the standard, established method starting with 5 µg of total RNA). A similar experiment starting with 20 ng of human heart total RNA (independent duplicates) yielded a correlation coefficient of 0.984, and a duplicate experiment using 1 ng of human lung total RNA showed a correlation of 0.978. These experiments indicate that the distribution of RNA in the final amplification product is very similar when using multiple rounds of amplification. The labeled and amplified products produce good signals on the arrays, the amplification appears to be highly reproducible (based on independent amplifications and array hybridization), and the number of mRNAs that do not amplify well appears to be small. The amplification can be performed directly from lysed cells in a tube, without any purification or extraction steps required prior to the first round of amplification. One caution is that the cRNA products tend to get shorter with each round, so the labeled products cover a shorter distance from the poly-A tail. This shortening is easily noticed as a significant increase in the 3'/5'

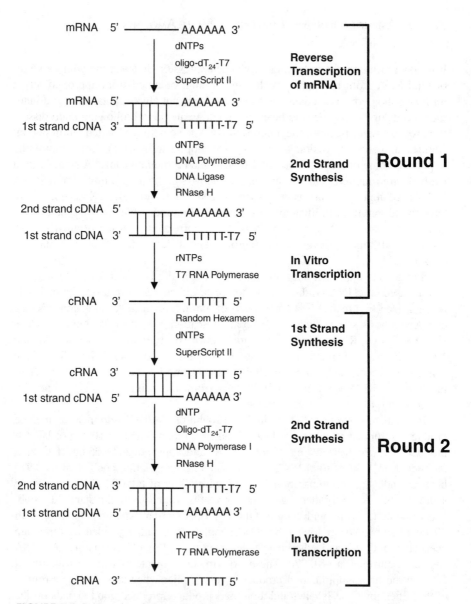

FIGURE 7.5 Schematic of the multiple-round, linear amplification strategy. Each round consists of a step in which RNA is converted to double-stranded DNA, which is then copied back into single-stranded RNA in an *in vitro* transcription (IVT) reaction driven off the T7 promotor. Each round produces an amplification of 100- to 1000-fold, and the amplification has been shown to be reproducible and to introduce a minimal amount of bias in the complex RNA population. It is possible to start with fewer than 10 cells, and 0.2 ng of total RNA, with final yields of more than 20 μg of labeled product, which is sufficient for an array hybridization reaction.

ratio for the Actin and Gapdh controls. Fortunately, as shown above, this shortening does not appear to significantly affect the results for the vast majority of genes, primarily because the probes for most genes are chosen from regions close to the poly-A tail (much closer than the 5′ Actin and Gapdh control probe sets). The use of this approach opens up the possibility of global, quantitative expression measurements starting with few cells from small nuclei, from laser capture microdissected cells, sorted cells, cells hand-picked based on location, morphology, or the expression of a marker or reporter, and potentially even single neurons. In preliminary studies, we have also found it possible to start directly with even fewer than 10 human or mouse cells (less than 0.2 ng of total RNA) for high quality, quantititative gene expression measurements.

7.2.16 GENE COVERAGE AND NEW ARRAY DESIGNS

While many current array designs (both commercially available and custom made) cover a very large number of genes, they do not cover all genes and all variant forms of the processed transcripts. Fortunately, EST and genome sequencing efforts continue to supply additional gene sequences which can be used for new array designs. In particular, there are ongoing efforts to sequence cDNA libraries made from mRNA from various regions of the mouse brain, as well as plans to sequence the entire mouse genome. The sequence information can then be used directly to design new arrays to increase the comprehensiveness of the gene coverage (one advantage of the oligonucleotide-based approach is that only the sequence information is needed for array design, with no requirements for physical intermediates such as clones, cDNA, or PCR products, etc.). This makes expansion of the data sets obtained using less comprehensive arrays rather straightforward because of an often unappreciated feature of the oligonucleotide array-based approach: samples can be hybridized repeatedly to multiple arrays without sacrificing data quality, and labeled samples can be stored for long periods of time without significant degradation or other changes that affect the results. We and others often hybridize the same sample as many as 8–10 times. This is possible because only a small fraction of the molecules in a sample remain bound to the oligonucleotide probes on the array surface — the vast majority of the molecules are recovered simply by removing the hybridization mixture (typically 200 to 300 µl) from the array cartridge after each hybridization reaction. Furthermore, after hybridization to a given set of arrays, samples can be stored (–80°C) for more than a year, thawed, and rehybridized to new arrays. This makes it possible to collect expression data using arrays that are available now, but then to add to these data very efficiently at a later date using the same samples without having to use additional animals or to perform additional tissue and RNA collections and sample preparations.

7.2.17 FURTHER IMPROVEMENTS

The power of these tools is impressive, but there is room for considerable improvement and technological development. The primary areas for further technical development include:

1. More complete gene coverage (essentially all genes and variant splice forms)
2. More genes per array without sacrificing data quality
3. Increased sensitivity and specificity to detect rare transcripts or low-abundance RNA in mixed-cell population (mRNA present at a relative abundance of less than one in 10^6)
4. Methods for routine experiments from small numbers of cells (1–100 cells)
5. More integrated, easier to use software tools for data analysis and visualization
6. Organized and more complete scientific "knowledge-bases" and software tools to accelerate and improve data interpretation

7.3 GENE EXPRESSION PROFILING IN NEUROBIOLOGY

We and others have used these techniques to study the brain (for a recent review see Lockhart and Barlow[27]). The next section provides an overview of some specific experiments, with an emphasis on appropriate experimental procedures and the potential uses of the technology for understanding brain function.

7.3.1 SEIZURE RESPONSE IN THE MOUSE HIPPOCAMPUS

As a simple experiment to test the feasibility of gene expression profiling to detect a biological response, we treated adult male mice with pentylenetetrazol (PTZ) to induce seizure and then determined the genes that were differentially expressed in the hippocampus (whole hippocampus) after 1 h.[28] Brain dissections were performed between 14.00–17.00 h on wet ice covered with paraffin. The hippocampus was removed after cutting the cortex sagitally. Hippocampi were prepared in duplicate from two different mice before treatment and two additional mice 1 h after seizure induction. Dissected tissue was frozen directly on dry ice and stored at –80°C until used for RNA preparation, cRNA labeling, and hybridization. From an analysis of the duplicate array results, we found that a total of 49 genes were induced and a total of 6 genes were repressed by a detectable amount (the transcriptional response of the two different animals was nearly identical). To test the validity of the results, we took advantage of the substantial literature on genes that are induced in response to seizure. As shown in Figure 7.6, the experiment successfully detected the induction of several known immediate-early genes, including members of the *fos* and *jun* family, growth factor inducible immediate early gene (*3CH134*), *cox-2*, and the transcription factors *KROX20* and *zif/268*. This study showed that array-based, parallel gene expression profiling across a large number of genes is a sensitive and accurate method for determining the transcriptional response to a biological stimulus.

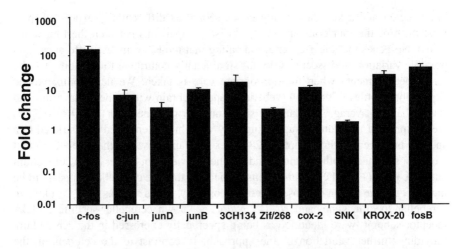

Immediate-early Gene

FIGURE 7.6 Gene expression changes in the hippocampus (C57BL/6) observed 1 h after seizure induction with PTZ. The difference in expression level (post-seizure vs. untreated mice) is shown for a set of known seizure-related immediate-early genes.

7.3.2 BRAIN-REGION-SPECIFIC GENE EXPRESSION MEASUREMENTS

Which genes are responsible for the unique structures and functions of specific brain regions? To begin to answer this question, we measured gene expression patterns in different brain regions. Again, all dissections were done between 14.00–17.00 h on wet ice covered with paraffin using 8-week-old male animals that had been singly housed for 1 week prior to sacrifice. Cortex (Cx), cerebellum (Cb), midbrain (Mb), and hippocampus (Hp) were prepared in duplicate from four different mice. In order to obtain sufficient tissue for RNA purification from amygdala (Ag) and entorhinal cortex (Ec), the microdissected regions of seven animals were pooled. Cortical dissections included the entire cortex except the olfactory bulbs. The cerebellum was dissected free of the brainstem, being cautious not to disrupt the paraflocculi. The "midbrain" consisted of the brain dissected free of cortex, pons, and medulla. The area lateral to and spanning from 3 mm ventral to the bifurcation of the external capsule was used as a landmark to define the borders of dissection for the entorhinal cortex. The area enclosed between the bifurcation of the external capsule was used to demarcate the dissection plane for the amygdala.

Of the 13,069 probe sets analyzed, 7,169 (55%) gave a hybridization signal consistent with a call of "present" in at least one brain region. This indicates that at least 55% of the genes covered on the murine arrays are detected in one or more areas of the adult male mouse brain.

We also identified genes that were uniquely expressed or highly enriched in one brain region relative to others. To determine the likelihood of error due to dissection inconsistency, we compared four independently obtained samples from the same brain region and asked, how many genes scored as differentially expressed in at least three of the four comparisons (pairwise comparisons between the four replicates). No genes met these criteria, indicating that mouse-to-mouse differences and possible variation in dissections do not significantly contribute to variability in the array measurements when the appropriate care is taken. We next compared the expression profiles of cortex, cerebellum, and midbrain within the same strain and found that, on average, a relatively small number of genes (70/13,069 or 0.54%) were expressed in a pattern suggesting they were highly enriched or restricted to a specific brain region. For example, 23 genes were expressed in the cerebellum that were not detected in other regions and another 28 were not expressed in cerebellum but were present in other brain regions, indicating that the cerebellum appears to be the most unique region of those tested. Importantly, genes such as *PCP-2*, a known cerebellar-specific gene, and *NMDA NR2C*, a known cerebellar-specific NMDA receptor subunit, were identified as being specifically expressed in the cerebellum, providing further validation of the approach. In contrast to the cerebellum, the structures of the medial temporal lobe (hippocampus, amygdala, and entorhinal cortex) showed extremely similar expression profiles. Only eight genes were unique to one of the three regions. Of the seven genes present in hippocampus but not amygdala or entorhinal cortex, six were also expressed outside of the medial temporal lobe. There was only one gene uniquely expressed in the amygdala and none in the entorhinal cortex. This suggests that forebrain structures, despite some functional differences, are highly similar at the molecular level. Finally, the midbrain was interesting in that, although there were ten genes uniquely expressed, no genes were exclusively "absent." In contrast to the very small number of differences between brain regions, 13.6% (1,780/13,069) of the monitored genes were found to be uniquely expressed between brain and fibroblasts, even though the two very different types of cells express a similar overall number of genes. This indicates, as might be expected, that various brain regions are considerably more similar to each other than to fibroblasts.

The high concordance between our data and published results for a set of known genes indicates that broad, array-based gene expression measurements are reliable for determining gene expression patterns in the brain. One important point, however, is that these studies compared large brain regions rather than subregions or specific cell types. It may be that differences in gene expression between various brain regions are much more pronounced in certain cell types, and that the high similarity in the expression patterns from different regions is due to averaging over all the cell types in the tissue. Recent data suggest that this may be the case. We have performed comparisons of the gene expression levels in whole mouse cortex and cerebellum and a similar analysis of more specifically dissected human cortex and cerebellum. In the case of the human tissue, only a small region of the cortex and cerebellum were used, whereas for the mouse the entire structures were included. The number of genes that differ, using the standard multiparameter criteria and a fold change of at least 1.8, between the two regions in mice is 138 (~2% of 6584 monitored), whereas in humans it is 399 (~6% of 7129 monitored). This experiment was performed on tissue from two different humans and in all cases the number of genes that were differentially expressed between cortex and cerebellum was similar.[28a] These data suggest that it is important to use microdissection of small regions or cell sorting in order to measure the precise gene expression patterns of specific subregions and cell types. As it becomes possible to use this technology for nuclei or even small cell populations in the CNS, higher resolution, region-specific, and cell-type specific information will be gained.

An important use of region-specific expression studies is to identify uniquely expressed genes and their promoters, which can be used to drive expression of a transgene in specific cell types or tissues in animal models. The paucity of site-specific tools in the mouse makes this an important use of the expression results. As has been demonstrated for yeast, when complete (or nearly complete) genomic sequence data are available, it is possible to use information about gene expression patterns to identify new cis-regulatory elements (genomic sequence motifs over-represented in the genomic DNA in the vicinity of similarly behaving genes) and "regulons" (sets of co-regulated genes).[1,2,17,18] In the mouse, the identification of genes that behave in similar ways across a range of experiments, along with the genome sequence, will allow the identification of new, previously unknown regulatory elements.

7.3.3 AUGMENTATION OF QTL ANALYSIS

Inbred mouse strains exhibit significant variation in many interesting CNS phenotypes. For example, inbred strains of mice vary greatly in their behavioral response to drugs of addiction, such as ethanol[29-30] and also show marked differences in some types of behavioral testing, such as prepulse inhibition.[31] We applied gene expression profiling of multiple brain regions in two commonly used inbred strains to help find genes that might account for the behavioral differences. Mouse embryonic stem cells are derived from 129 strains and this strain contributes to the genetic makeup of most mutants generated using homologous recombination. C57BL/6 is the strain most commonly used for outcrossing, the background strain of many spontaneous mutants, and is used in mapping of quantitative trait loci (QTL) and in many drug and neurobehavioral studies. In these studies we found that by housing animals individually for 1 week prior to sacrifice and by following careful dissection, RNA preparation, cRNA preparation, and hybridization protocols, the data were so consistent between mice that only two animals of each strain were required. We found that 0.6% of the monitored genes differed substantially between the two strains.[28] Based on what is already known about them, several of the differentially expressed genes are ideal candidates for modulating complex neurobehavioral phenotypes, and they also map to chromosomal regions where quantitative trait loci have been mapped. One example is the potassium channel *GIRK3* (more highly expressed in 129SvEv) which is located on chromosome 1 in a region that has been shown to contain one or more of the genes that contribute to strain differences for free-running period and locomotor activity, aspects of fear conditioned response (cued and contextual), open field emotionality, and acute pentobarbital-induced seizures. Although QTL analysis is powerful for mapping susceptibility loci to chromosome intervals, many genes reside in these large intervals, and in the standard approach, extensive additional work is required to identify the specific gene or genes involved. Our findings suggest that a more rapid expression-based strategy may be useful to identify candidate genes responsible for quantitative traits. Conventional QTL mapping approaches identify the gene(s) positionally through the genome. The alternative is to identify genes that contribute to a phenotype functionally based on gene expression behavior. Genes identified in the expression experiments are not necessarily genetically different between phenotypically distinct animals because the expression changes reflect the consequences of genetic differences, not necessarily the primary genetic causes. The two approaches are complementary in that the standard QTL approach identifies the genes or chromosomal regions that harbor genetic differences relevant to the phenotype, while the expression approach measures cellular differences that result from the genetic variation. It seems likely that by extending these studies to other phenotypically characterized strains and to nearly all genes in the entire genome, the set of interacting genes responsible for neurobehavioral variation in mice may be identified.

7.3.4 GENE EXPRESSION PROFILING IN NONISOGENIC MOUSE STRAINS

Based on our experience with different strains of mice, it is important to consider the implications of these results for studies that employ nonisogenic strains of mice. Many laboratories have generated mice with targeted mutations in genes (knockouts) or that over-expressed genes (transgenics), and reported novel behavioral phenotypes. The neurobehavioral phenotype of a particular mouse results not only from the specific alteration induced by a targeted mutation, the mis-expression of a particular gene or the administration of a particular drug, but also from the effects of modifiers, which may differ significantly based on genetic background. We have estimated that the 129SvEv gene expression profile is significantly different (~0.6% of monitored genes) from that of other strains commonly used in transgenic experiments, such as C57BL/6. The use of nonisogenic mouse strains is likely to lead to situations where expression differences may be identified, but it may be difficult to know whether they are due to the specific perturbation or are heavily influenced by other differences due to variation in genetic background even among genetically distinct littermates. By considering the results presented here, it may be possible to help exclude genes that tend to differ due to genetic background alone, but if possible, it is better to use genetically identical mice to minimize these additional sources of variation.

7.4 DISCUSSION

7.4.1 POTENTIAL SHORTCOMINGS OF THE METHOD

Knowing when and where a gene is expressed does not provide complete information about the cellular role of the gene product. Messenger RNA is only an intermediate on the way to production of the eventual protein products, and multiple methods have been developed for monitoring protein levels either directly or indirectly. Protein-based approaches are currently more difficult, less sensitive, and lower-throughput, in general, but it is clear that protein and RNA-based measurements are complementary.

It is of course possible that differences related to some interesting phenotypes may not be detectable by measuring mRNA abundance. Important events may involve, for example, changes in protein stability or posttranslational protein modifications, and the effects of these may be only indirectly reflected in the levels of mRNA. Even when cellular events are reflected in mRNA levels, some phenotypes may be the result of multiple, subtle changes in mRNA abundance that will be difficult to detect. It is possible to detect these smaller differences (e.g., changes of 10–50%), but to do so reliably requires performing a larger number of independent experiments so that sensitivity thresholds can be lowered and smaller differences can be trusted with reasonable confidence. It may be necessary to independently verify the smaller differences (using other methods such as northerns and quantitative RT-PCR) to make sure that they are real and reproducible, and to characterize the location and cell-type specificity using *in situ* hybridizations and microdissection. A final important point to remember is that gene expression measurements yield the consequences of a mutation or a treatment, and do not directly identify the causative genes. For example, a gene that is mutated may not be differentially expressed, but other genes may be as a consequence of the mutation. Similarly, a protein that is inhibited by a drug may not show up as differentially expressed, but what will be observed are the cellular effects of the inhibition.

7.4.2 EXPRESSION MARKERS

An intriguing application of expression experiments involves the identification of "expression markers," meaning genes whose expression levels are indicative of a phenotype or a physiological response. A related approach has been used to identify genes whose expression levels are indicative of cancer or tumor type, and this has sometimes been referred to as "molecular phenotyping." In order to find markers with predictive or diagnostic value, it is necessary to perform enough experiments across a large number of genes to find the 10 to 200 whose expression levels (or combinations of levels) are well correlated with a condition or trait. If such markers can be found, then their levels can serve as surrogates for more complicated, time-consuming or expensive physiological or behavioral tests, for example, and they can be used to monitor responses to treatments designed to modulate some characteristic or response. More global expression patterns as well as marker-based measurements of this type can also be used to characterized mutant mice that result from either site-directed or ENU-induced (random) mutagenesis.

7.4.3 TOO MUCH DATA?

It is sometimes thought that highly parallel genomics approaches of this type produce too much data, and that the large data sets are too unwieldy to be of immediate or significant scientific value. It is true that the amount of data that can be collected is impressively large, but not overwhelming given available computational power and storage capacity. It is also the case that the expression results covering thousands or even tens of thousands of genes and EST are only partially interpretable given the functional and biological information (and patience) available at the time they are initially generated. The biggest barrier, however, is not that there is too much data or that there are insufficiently sophisticated algorithms and software tools for querying and visualizing data on this scale. The greatest challenge instead may be the building of scientific "knowledge-bases" in which the data, facts, observations, and relationships that form the basis of our current scientific understanding are organized, and making this information available in ways that help scientists understand and interpret the complex observations that are becoming increasingly easy to make. Our ability to extract knowledge from global gene expression measurements tends to increase with time as new information becomes available, and collecting, organizing, annotating, curating, and linking scientific information is absolutely essential to take full advantage of the abilities of the new technologies and the rapidly increasing amount of sequence information. There is currently no substitute for the trained human brain when it comes to making biological sense of new results. But our brains need help because the data, scientific knowledge, and experimental observations are extensive and growing in volume every day. The greatest progress will come from bringing all the necessary computations, information, and relationships to scientists' fingertips so that the most insightful questions can be asked, and the most informed, complete, and meaningful biological interpretations can be made.

7.4.4 CONCLUSION

Genomic technologies, although already impressive, are still improving and becoming more powerful. The technical goal of being able to monitor essentially all genes from a mouse or human on an array or two starting with only 1–10 cells is now clearly in view. These highly parallel gene expression approaches allow one to look globally at the interactions of genes and modifiers and their effects, and will greatly enhance our ability to identify the genes that contribute to important phenotypes, and to define the role of developmental alterations, mutations, and compensatory mechanisms in causing or modifying particular behaviors. The studies described in this chapter demonstrate the feasibility and utility of expression profiling in the brain. The expression results serve as a framework to begin to understand, for example, the factors responsible for the variation in behavioral phenotypes, drug sensitivity, and neurotoxic-induced cell death. There is no doubt that the combination of gene targeting technology, robust behavioral analysis, genetics, biochemistry, and global gene expression measurements will provide new avenues for studying the brain and further our ability to understand the interplay between genes that give rise to unique brain functions and complex behaviors.

ACKNOWLEDGMENTS

We would like to thank Jo A. Del Rio, Dan Giang, Elizabeth Winzeler, Lisa Wodicka, Jane Gentry, and members of the Barlow and Lockhart groups for their ongoing patience and help.

REFERENCES

1. Spellman, P. T. et al., Comprehensive identification of cell cycle-regulated genes of the yeast *Saccharomyces cerevisiae* by microarray hybridization, *Mol. Biol. Cell,* 9, 3273–3297, 1998.

2. Cho, R. J. et al., A genome-wide transcriptional analysis of the mitotic cell cycle, *Mol. Cell,* 2, 65–73, 1998.

3. DeRisi, J. L., Iyer, V. R., and Brown, P. O., Exploring the metabolic and genetic control of gene expression on a genomic scale, *Science,* 278, 680–686, 1997.

4. Wodicka, L., Dong, H., Mittmann, M., Ho, M.-H., and Lockhart, D. J., Genome-wide expression monitoring in *Saccharomyces cerevisiae, Nat. Biotechnol.,* 15, 1359–1367, 1997.

5. Gingeras, T. R., et al., Simultaneous genotyping and species identification using hybridization pattern recognition analysis of generic mycobacterium DNA arrays, *Genome Res.,* 8, 435–448, 1998.

6. White, K. P., Rifkin, S. A., Hurban, P., and Hogness, D. S., Microarray analysis of Drosophila development during metamorphosis, *Science,* 286, 2179–2184, 1999.

7. Chambers, J. et al., DNA microarrays of the complex human cytomegalovirus genome: Profiling kinetic class with drug sensitivity of viral gene expression, *J. Virol.,* 73, 5757–5766, 1999.

8. *Nat. Genet. Suppl.,* 21, 3–50, 1999.

9. Lipshutz, R. J., Fodor, S. P., Gingeras, T. R., and Lockhart, D. J., High density synthetic oligonucleotide arrays, *Nat. Genet.,* 21, 20–24, 1999.

10. Lockhart, D. J. et al., Expression monitoring by hybridization to high-density oligonucleotide arrays, *Nat. Biotechnol.,* 14, 1675–1680, 1996.

11. Golub, T. R. et al., Molecular classification of cancer: Class discovery and class prediction by gene expression monitoring, *Science,* 286, 531–537, 1999.

12. Alizadeh, A. A. et al., Distinct types of diffuse large B-cell lymphoma identified by gene expression profiling, *Nature,* 403, 503–510, 2000.

13. Tamayo, P. et al., Interpreting patterns of gene expression with self-organizing maps: Methods and application to hematopoietic differentiation, *Proc. Natl. Acad. Sci. USA,* 96, 2907–2912, 1999.

14. Eisen, M. B., Spellman, P. T., Brown, P. O., and Botstein, D., Cluster analysis and display of genome-wide expression patterns, *Proc. Natl. Acad. Sci. USA,* 95, 14863–14868, 1998.

15. Wen, X. et al., Large-scale temporal gene expression mapping of central nervous system development, *Proc. Natl. Acad. Sci. USA,* 95, 334–339, 1998.

16. Chu, S. et al., The transcriptional program of sporulation in budding yeast, *Science,* 282, 699–705, 1998.

17. Roth, F. P., Hughes, J. D., Estep, P. W., and Church, G. M., Finding DNA regulatory motifs within unaligned noncoding sequences clustered by whole-genome mRNA quantitation, *Nat. Biotechnol.,* 16, 939–945, 1998.

18. Tavazoie, S., Hughes, J. D., Campbell, M. J., Cho, R. J., and Church, G. M., Systematic determination of genetic network architecture, *Nat. Genet.,* 22, 281–285, 1999.

19. Wang, A. M., Doyle, M. V., and Mark, D. F., Quantitation of mRNA by the polymerase chain reaction, *Proc. Natl. Acad. Sci. USA,* 86, 9717–9721, 1989.

20. Jena, P. K., Liu, A. H., Smith, D. S., and Wysocki, L. J., Amplification of genes, single transcripts and cDNA libraries from one cell and direct sequence analysis of amplified products derived from one molecule, *J. Immunol. Meth.,* 190, 199–213, 1996.

21. Dulac, C., Cloning of genes from single neurons, *Curr. Top Dev. Biol.,* 36, 245–258, 1998.

21a. Giang, D. and Lockhart, D.J., unpublished.

21b. de Graf, D. and Lander, E., personal communication.

22. Van Gelder, R. N. et al. Amplified RNA synthesized from limited quantities of heterogeneous cDNA, *Proc. Natl. Acad. Sci. USA,* 87, 1663–1667, 1990.

23. Eberwine, J. et al., Analysis of gene expression in single live neurons, *Proc. Natl. Acad. Sci. USA,* 89, 3010–3014, 1992.

24. Kacharmina, J. E., Crino, P. B., and Eberwine, J., Preparation of cDNA from single cells and subcellular regions, *Methods Enzymol.,* 303, 3–18, 1999.

25. Luo, L. et al., Gene expression profiles of laser-captured adjacent neuronal subtypes, *Nat. Med.,* 5, 117–122, 1999.

26. Wang, E., Miller, L. D., Ohnmacht, G. A., Liu, E. T., and Marincola, F. M., High-fidelity mRNA amplification for gene profiling, *Nat. Biotechnol.,* 18, 457–459, 2000.

27. Lockhart, D.J. and Barlow, C., Expressing what's on your mind: DNA arrays and the brain, *Nat. Rev. Neurosci.,* 2, 63–68, 2001.

28. Sandberg, R. et al., Regional and strain specific gene expression mapping in the adult mouse brain, *PNAS,* 97, 13209–13214, 2000.

28a. Del Rio, J.A. and Barlow, C., unpublished observations.

29. Crabbe, J. C., Gallaher, E. S., Phillips, T. J., and Belknap, J. K., Genetic determinants of sensitivity to ethanol in inbred mice, *Behav. Neurosci.,* 108, 186–195, 1994.

30. Bachmanov, A. A., Tordoff, M. G., and Beauchamp, G. K., Ethanol consumption and taste preferences in C57BL/6ByJ and 129/J mice, *Alcohol Clin. Exp. Res.,* 20, 201–206, 1996.

31. Crawley, J. N. et al., Behavioral phenotypes of inbred mouse strains: Implications and recommendations for molecular studies, *Psychopharmacology (Berl.),* 132, 107–124, 1997.

32. Lockhart, D. J. and Winzeler, E. A., Genomics, gene expression and DNA arrays, *Nature,* 405, 827–836, 2000.

8 DNA Microarrays and Human Brain Disorders

Károly Mirnics, David A. Lewis, and Pat Levitt

CONTENTS

8.1   INTRODUCTION

The development of novel sequence and expression analysis tools has revolutionized molecular biology, establishing the field of functional genomics virtually overnight.[1] Microarrays, with their capability to generate rapidly massive datasets, were at the forefront of this technical revolution.[2-4] Within the last several years, microarrays have been successfully used in comparative genomic hybridization,[5-8] on-chip sequencing,[9,10] and novel gene discovery.[11-13] However, the search for complex gene expression differences remains the dominant use of microarrays.[2-4,14-17] Knowledge of complex gene expression profiles or "transcriptomes" is essential to understanding the function of cells or tissue regions, and allows for the association of gene expression patterns with disease states or treatment efficacy. Unfortunately, but not surprisingly, microarray studies of the brain have been few to date.[13,16,18-26] The phenotypic diversity of nervous tissue, in which biologically meaningful changes often do not reach a 2-fold difference,[27-30] have resulted in serious reservations about the validity of microarray experiments involving brain material (for discusssion, see References 31 and 32). This skepticism has been even greater for experiments involving human postmortem tissue; well-founded questions were raised regarding the integrity of isolated RNA, human variation, the effects of lifestyle, clinical confounds (e.g., medication, drug abuse), and epigenetic factors on gene expression, and the problems associated with relatively small sample sizes that may lack statistical power.

The goal of most studies of complex brain disorders using gene microarrays is not simply to produce lists of altered genes, but rather to utilize the power of simultaneous expression profiling of thousands of genes that may yield complex molecular relationships that underlie the altered biology.[1,31] Thus, careful experimental design is essential for obtaining high-quality, meaningful data that puts what typically is extremely valuable material to its best use. Such factors include the right choice of starting material, use of well-founded neuroanatomical information for deciding on specific sample comparisons, preparative quality controls, the proper selection of microarrays and appropriate pairing paradigms across human subjects, careful collection and interpretation of data and well-controlled verification strategies (Figures 8.1 and 8.2). This chapter will focus on what we believe to be the most critical of these issues, which, when considered carefully, will facilitate clinical studies that use the latest molecular technologies.

8.2 CHOICE OF MATERIAL

The study of gene expression profiling in human brain tissue requires an understanding of both the advantages and limitations intrinsic to the use of postmortem material. In particular, the successful application of microarray technology to the study of diseases of the human nervous system depends upon the use of well-characterized tissue specimens, in which potential confounds are identified and addressed through critical experimental design, and the inclusion of appropriate control experiments. In this regard, the following three general issues are important to consider.

First, the relevant demographic and diagnostic information needs to be obtained in a standardized fashion for both cases and comparison subjects. Ideally, this information is acquired from both the review of available medical records and through structured interviews conducted with surviving relatives and other informants.[33] The latter typically provides data that are not available from the medical records, as well as an additional means to verify the information in the medical records through collateral sources. A substantial amount of information can be obtained through such interviews. These include demographic variables (e.g., age, gender, race), the clinical characterization of the illness (e.g., diagnostic signs and symptoms, age of onset, duration of illness, family history of illness), treatment of the illness (e.g., medications at the time of death, history of treatment with other medications, length and number of hospitalizations), and factors that may be comorbid with the primary diagnosis of interest (e.g., history of alcohol or other substance abuse, nicotine exposure). Knowledge of this information is essential for the selection of appropriate comparison subjects, including both normal controls and

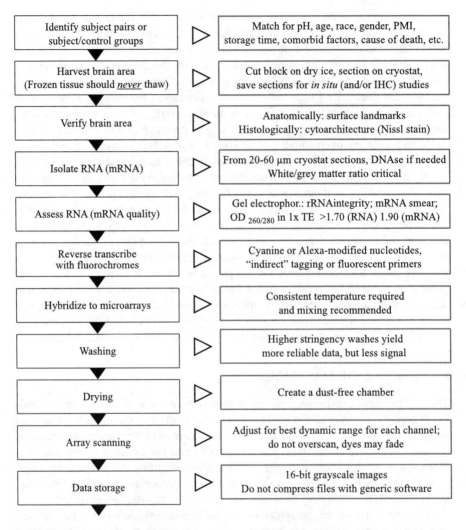

FIGURE 8.1 Anatomy of a cDNA microarray experiment: sample identification, isolation, labeling, hybridization, and image acquisition steps. (PMI, post-mortem interval; IHC, immunohistochemistry; S/B, signal/background; SOM, self-organizing maps.)

subjects with other disorders who share some features (e.g., medication history) with the cases of interest. In addition, the detailed clinical information suggests the types of parallel studies that may need to be conducted in animal model systems, where factors such as medication exposure can be assessed in a controlled fashion. For example, we have used macaque monkeys treated chronically with antipsychotic drugs in a manner that directly mimics their use in humans[34] as one way of assessing the potential contribution of these therapeutic agents to our microarray findings in subjects with schizophrenia.[16]

Second, in order to avoid the introduction of potential confounds secondary to geographic factors, differences in the handling of tissue specimens, or other variables,

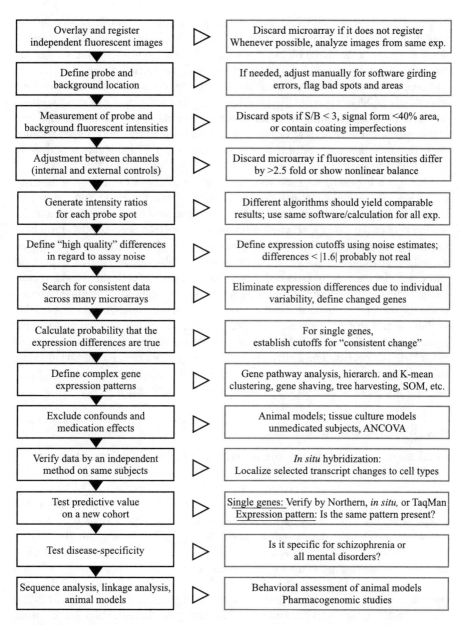

FIGURE 8.2 Anatomy of a cDNA microarray experiment: image analysis, data analysis, and data verification strategies.

all tissue specimens (cases and comparison subjects) should be obtained from the same source. Having the same group of individuals conduct all of the tissue processing also helps ensure that the same brain region of interest is analyzed across all subjects. In this regard, we cut brain specimens from all subjects into 1.0 cm thick blocks in a standardized fashion. This approach facilitates examination and sampling for neuropathological studies and permits a uniform dissection of the region(s) of interest at the time that a microarray study is initiated. Photographic documentation of the tissue blocks provides a means to confirm and record that the same macroscopic landmarks are present in the tissue blocks from all subjects. In addition, as described in more detail below, we also obtain slide-mounted sections from the tissue blocks that are used for the isolation of RNA. Some of these sections are used for Nissl staining in order to verify that the tissue block contains the characteristic cytoarchitecture of the brain region targeted for study. Other sections are saved for subsequent confirmation of array findings for individual transcripts by *in situ* hybridization.

Third, it is important to assess variables that may affect the expression or integrity of mRNA. These variables include agonal state events, the cause and manner of death, and the postmortem interval (PMI), the period between the time of death and the freezing of the tissue specimens. Although most mRNA species exhibit minimal to modest degradation following PMI of less than 24 h (comparable to freshly frozen animal brain tissue), premortem events can have substantial negative impact. In particular, antemortem hypoxia and/or acidosis, conditions frequently found in individuals who die in the hospital following medical complications or sustained periods of illness, appear to be associated with the loss of at least some mRNA species. Available studies suggest that postmortem brain pH provides an easily measured and reliable assessment of the severity of these factors.[35] Because the factors monitored by brain pH or PMI may not have linear effects on mRNA integrity, we prefer to match individual pairs of subjects as closely as possible on these variables, in addition to including pre- and postmortem conditions as covariates in statistical analyses.

8.3 RNA ISOLATION

In all microarray experiments, the integrity and purity of the starting material is critical.[36-38] We, as well as others, have successfully isolated intact mRNA up to 36 h after death.[16,39] Two isolation protocols have been used successfully in our laboratory.

In the first, RNA is isolated using RNAgents total RNA isolation kit (Promega Corporation, Madison, WI, USA), following manufacturer instructions for a double-extraction procedure.[16] However, we found that sample quality, yield, and consistency across starting material obtained from different human subjects were greatly improved by increasing the amount of all the solutions 3-fold over the manufacturer's recommendations. This procedure routinely provides ~0.5% yield of total RNA from postmortem human cortical tissue, with an optical density (OD) 260/280 >1.70. mRNA isolation is performed by a double-pass through OligoTex columns (Qiagen Inc., Valencia, USA). The final elution is performed in 25 µl at 75°C with $1 \times$ TE, pH = 7.0. To assure maximal mRNA recovery, a second elution, using the same parameters, is routinely performed. However, this second eluate typically does not contain significant levels of mRNA, and is not combined with the first eluate.

In a more recently adopted procedure, we have successfully used the Fast Track kit (Invitrogen, Carlsbad, CA, USA) for isolation of mRNA directly from tissue sections. The procedure is performed according to the manufacturer's instructions, although we have added an identical second-round mRNA re-isolation using the same procedure to obtain high-purity mRNA. Typical mRNA yields in the first round are 0.001–0.0015% of wet tissue weight, and about 40% of the material obtained in the first isolation is lost in the second round. The final mRNA yield is 0.5–1 μg/100 mg of wet weight tissue.

DNA contamination of the mRNA sample can affect labeling and distort the gene expression pattern obtained in microarray experiments. In our isolations, using either method noted above, DNA contamination generally does not carry over through the mRNA isolation. However, if DNA contamination persists, a standard DNase treatment is a viable option and can be incorporated into either protocol. The need for this step is readily assessed by real-time quantitative RT-PCR[40,41] using intron-spanning primers.

RNA yield using either isolation strategy does not appear to be a function of the postmortem interval.[39] Final concentration adjustments and buffer exchanges are done using YM-30 Microcon columns (Millipore Corp., Bedford, MA, USA) according to manufacturer's instructions. We prefer this method to ethanol or isopropanol precipitation because of ease of use and minimal sample loss (usually <10%).

8.4 SAMPLE QUALITY CONTROL

An aliquot of the isolated total RNA (mRNA) is routinely run on a 1.5% agarose gel. A sample is usually of sufficient quality for the microarray experiment if the 23S ribosomal band is more intense than the 18S band and if ~50% of the mRNA smear is >2 kb (Figure 8.3). Furthermore, measured in 1xTE, the 260/280 OD for total RNA must exceed 1.70 (>1.90 for mRNA).

For Affymetrix GeneChips®,[4,38] the use of test arrays to verify the quality of the labeled cRNA is strongly recommended — the T7 *in vitro* transcription amplification procedure, while usually reliable, is prone to sample degradation if nuclease-free conditions are compromised. Using the recommended T7-aRNA technology, applied to human cortical samples, we were able to achieve 3′ vs. 5′ ratios of ~3:1 to actin and GAPDH. In our experiments, this integrity is clearly of sufficient quality to obtain measurable gene expression above background.

Real time quantitative RT-PCR[40,41] also can provide a relatively precise assessment of RNA/mRNA integrity after tissue extraction, though it is rarely used for such quality control issues. Following a quantitative reverse transcription with Superscript II (Life Technologies, Gaithersburg, MD, USA) primers and probes are designed against the 3′ and 5′ ends of selected genes. The threshold cycles of the two amplifications are compared, precisely assessing abundance differences between the 3′ and 5′ ends of the same transcript.

8.5 SAMPLE MATCHING PARADIGMS

Affymetrix GeneChip® technology currently requires hybridizing each sample to a single oligonucleotide array.[4,38,42] This is followed by comparison of transcriptomes across multiple arrays using proprietary software from Affymetrix. In this process, the software adjusts mathematically for interarray variability using spiked-in standards and

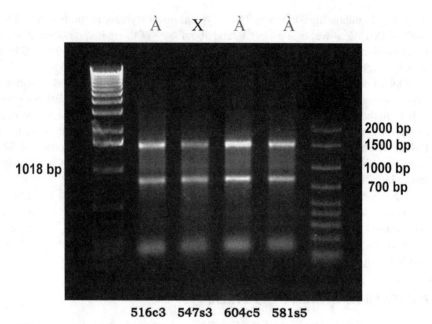

FIGURE 8.3 mRNA quality is essential in postmortem brain experiments: total RNA from four PFC Area 9 human samples was subjected to electrophoresis on a 1% nondenaturing agarose gel. DNA size markers are located in outside lanes, numbered lanes represent individual samples. Due to the signal inversion of rRNA bands and reduced mRNA smear, sample 547s3 was not used for microarray analyses.

overall array signal intensity. If there are qualitative differences between individual arrays within the dataset, data mining could be affected adversely, leading to higher false-negative or false-positive calls. The vast improvements in the production of affordable oligonucleotide arrays, and the potential of new technology on the horizon to allow for individual laboratory custom oligonucleotide-based array platforms,[43,44] will facilitate the expanded use of these arrays applied to complex human brain disorders. The use of arrays that do not require pairing allows sample comparisons across the entire group of analyzed subjects, providing an additional level of data mining opportunities and complexity.

Typically, cDNA arrays are used in dual-fluorescent hybridizations, reducing interarray variability in a two-sample comparison. To take full advantage of this same-array hybridization, pairing strategies become an essential part of the experimental design. There are several possible strategies to consider (Color Figure 8.4).*

8.5.1 PAIRWISE COMPARISONS

In this design, each experimental (disease) sample is hybridized against a different sample from the identical anatomical area of a control brain. The experimental-control

---

* Color figures follow page 140.

pairs, whenever possible, are matched for as many parameters as possible, including race, postmortem interval (PMI), gender, and agonal state. The pairwise experimental design has the distinct advantage of preserving the individuality of each sample,[16,45] which may be critical in understanding polygenic disorders that may produce a finite number of difference patterns. Multiple factors may contribute to the expression differences obtained from each microarray: (1) genetic variability between two humans; (2) disease-related effects; (3) differences in lifestyle and environment; (4) clinical treatment of the disease; (5) false-positive results (assay noise = labeling noise + microarray noise). In each pairwise comparison, >95% of the observed differences will be due to individual human variability, but these individual differences rapidly filter out as inconsistent (nonrepeating) noise over multiple pairwise comparisons.

8.5.2 POOLED CONTROL VS. INDIVIDUAL SAMPLES

In the first variety of this paradigm, control samples from many individual brains are pooled into a common reference sample. The control samples should still closely match the experimental group population for as many parameters as possible, including agonal state, age, sex, PMI, race, and co-morbid factors (cf. substance abuse). Each experimental sample is hybridized on a single microarray against this pooled control. This comparison paradigm greatly reduces observed gene expression changes due to variation within the control subjects, creating an internal reference profile used in every microarray hybridization. This is useful for comparing experiments across microarrays and helps to evaluate differences in labeling efficiency and microarray performance. However, the individuality of the control samples is lost, and the transcriptomes of individual control subjects remain unknown. Gene expression profiles in a pooled control sample, particularly with limited human material, may be somewhat artificial — expression levels of genes in a brain region may change physiologically as a function of age, and these dynamic differences are reduced into an artificial baseline. Furthermore, if the pooled control sample doesn't include material from many individuals, expression levels for individual genes will become vulnerable to extreme outliers, shifting control expression averages into a misleading range.

The creation of a common reference sample from many different control brains and brain regions (or even other tissues) is another variation of this type of comparison.[15] This reference is used in every hybridization, paired with either a control or experimental sample. The resulting experimental data is normalized across microarrays using the reference, and control-experimental expression values are *indirectly* compared across arrays. While this design may be very powerful when >3-fold expression differences are expected, we generally do not recommend use of this paradigm in experiments involving human brain tissue. This approach eliminates the most valuable feature of cDNA arrays — direct comparison of samples on the *same* microarray. Indirect comparisons are likely to be much more noisy than direct comparisons, and introduction of any additional noise (e.g., microarray variation) will reduce the quality of the obtained data.

8.5.3 Pooled Control vs. Pooled Sample

This experimental design is the least desirable. This strategy yields data that are very difficult to interpret, due to outliers and loss of sample individuality. These data will not reveal the number of experimental samples that contributed to the observed change. However, this approach may be informative when used in conjunction with the pairwise design.

8.6 CHOICE OF PREMANUFACTURED MICROARRAYS

The use of all DNA microarrays is based on the principle of complementary hybridization between genetic material bound to a solid support and a labeled sample. With regard to terminology in discussing array experiments, we follow the most commonly used cDNA microarray convention, defining the "probe" as the anchored nucleic acid with known identity, and "target" as the free nucleic acid sample whose identity/abundance is being measured. However, it should be noted that some manufacturers (e.g., Incyte Genomics, Inc.) uses the inverse terminology, where targets are the immobilized oligonucleotides and probes refer to the labeled samples.

While many different microarray technologies show great promise, two technologies have become widely used within the research community over the last several years. Affymetrix manufactures oligonucleotide microarrays (GeneChips®) by a procedure called photolithographic synthesis, a process where short oligonucleotides are built directly onto a glass surface (for an overview of this technology, see References 4, 17, and 42). GeneChips® experiments require small amounts of starting material (≥50 ng mRNA); they are easy to process, provide quantitative results, and have good nominal sensitivity. There also are several drawbacks to this technology, including the use of amplified starting material that may add to experimental noise, the limited choice of microarrays, the need to use a proprietary analysis system, and the prohibitive cost (>$100,000) for producing custom microarrays that contain specific gene sets, which may be desirable for projects involving complex brain samples.

The other widely used type of microarray, the cDNA microarray, utilizes deposited DNA fragments onto a solid support. Virtually any type of DNA can be printed: chromosomes, BAC, YAC, PCR products, plasmids, or oligonucleotides. Using relatively simple robotic equipment, thousands of different probes, each with a unique sequence, can be planted onto several square inches of a DNA-binding coated glass surface.[14,46,47] These immobilized DNA probes, under the appropriate conditions, will capture complementary sequences from different samples,[48] which allows the comparison of two or more samples on the *same* microarray (Color Figure 8.5).* For gene expression studies, an experimental and a control sample are usually labeled in a standard reverse transcription, either by incorporating fluorescent nucleotides[2,36,49] into the cDNA strand or by incorporating substances to which fluorescent labels may be attached in a secondary reaction.[50,51] The incorporated fluorochromes have distinct excitation/emission wavelengths (currently Cy3 and Cy5

---

* Color figures follow page 140.

typically are used). The labeled cDNA are combined into a single reaction mix, and hybridized onto the same microarray. A series of washes removes noncomplementary binding, and the remaining target-probe-fluorochrome is quantified by measuring the fluorescence for each of the genes across the microarray. To compensate for a number of variables,[52] including different absolute intensities of fluorochromes, different incorporation rates during reverse transcription, and scanning differences, the two channels are mathematically equalized to external and internal standards. The adjusted fluorescent intensities of the experimental and control samples are compared and reported as relative expression differences (usually in folds of change).

cDNA microarray technology, pioneered by Stanford University and NIH investigators,[2,3,49] is not proprietary, and microarray production and scanning tools are now affordable, even for individual laboratories. Furthermore, many excellent data analysis tools are freely available on the World Wide Web. The drawbacks of cDNA arrays include instability of the fluorescent dyes, poor or uneven efficiency of dye incorporation during the reverse transcription, and lower nominal sensitivity than the Affymetrix GeneChips®.

cDNA arrays produced on nitrocellulose membranes are the least expensive, viable alternative to microarray approaches.[53] Although they will not be separately discussed, most of the cDNA microarray limitations also apply to the membrane arrays. Their attractive price, ease of manufacturing and processing, reliability, and minimal requirement of starting material makes them an ideal choice for low throughput experiments where major expression differences are expected (Color Figure 8.6).* Indeed, membrane-based arrays with radioactively labeled samples will continue to be used with great success in the analysis of complex mental disorders.[20,22]

As a simple rule, the choice of the specific type of microarray depends primarily on the representation of the probes on the microarray and the biological problem to be investigated (cf. complexity and/or degree of expected differences). Probe lists are downloadable for all commercially available microarrays and should be extensively studied *before* the initiation of the experiments.

8.7 DESIGN OF CUSTOM MICROARRAYS

Currently, there is no commercially available microarray that contains all, or even most probes that a neuroscientist may desire. Many investigators use a two-step experimental approach. In the initial studies, the transcriptome differences are compared using high-density, commercially available microarrays for >10,000 genes. This is followed by exhaustive data analysis, yielding a more definitive understanding of the biological components that will be studied in greater depth. Custom-made microarrays are ideal for these hypothesis-driven, follow-up experiments. The arrays should include a more complete probe representation related to a defined biological function (cf. all the genes encoding proteins involved in glutamate signaling) and greater, same-probe redundancy than afforded by the commercially available, high-density microarrays. To learn more about the available clones, visit SOURCE at Stanford University (http://genome-www4.

---

* Color figures follow page 140.

stanford.edu/cgi-bin/SMD/source/sourceSearch). Several critical considerations in the design of custom microarrays are noted below.

8.7.1 CLONE SELECTION AND PROBE GENERATION

3′cDNA library-derived clone sets with >10,000 genes are available from both the NIH and commercial entities (e.g., Research Genetics, Clontech, Incyte Genomics). These clone sets are large, convenient to use, and are usually purchased and managed by an institutional core facility, rather than an individual laboratory. To minimize cross-contamination and human error, these clones should be kept in redundant clone sets and handled by robotic equipment. For custom arrays, a subset of these clones is usually re-amplified using universal vector primers (e.g., M13) in a standard PCR reaction. The reaction products should always be checked by gel electrophoresis to ensure that the PCR in all cases generated a single-class product of the expected length. To avoid annotation errors (which are quite common), critical clones should be sequence-verified. The reaction product is purified.[36] However, some investigators reported successful microarray production using unpurified reaction products.

An alternative approach to clone amplification is direct PCR amplification of the gene from a known sample. Gene-specific primers (unmodified or amino-modified) are designed against the desired region of each molecule, and the resulting amplicons are sequence-verified, purified, and printed onto the microarrays. Cross-contamination concerns are greatly reduced due to gene-specific product amplification. This approach is less suited for high-scale array generation, but is reasonable for low-density microarray production.

The simplest way to generate microarray probes is by direct synthesis. In this process longer oligonucleotides (50–100) are synthesized in microgram quantities, and printed as probes onto the microarrays.[54]

The advantage of the latter two approaches is that the investigator controls the length, region and G/C content of the future probe. This creates a uniform collection of probes on the microarrays with similar optimal hybridization properties. Furthermore, the selection of the correct location with respect to the mRNA position often enables creation of probes that are selective for splice variants of genes, something that usually is not achieved when 3′-derived clones are used.

8.7.2 SELECTION OF INTERNAL AND EXTERNAL CONTROLS

Control probes must be included on each custom microarray. As internal controls, one should designate a set of genes with consistently unchanged gene expression between experimental and control conditions. These internal control genes should be expressed in the labeled sample at different absolute levels and should make up ~15% of the total microarray probes. Nonmammalian gene sequences (e.g., yeast gene fragments) with no known homology to mammalian genes usually serve as external controls. They are polyadenylated and added into each sample before reverse transcription at varying ratios between the two samples. Pre-labeled external controls may also be added after the reverse transcription is completed. In the first paradigm, the spiked-in control is reverse transcribed with the tissue sample, providing feedback about the linearity and overall quality of the labeling reaction. Adding pre-labeled controls after the reverse transcriptions will provide feedback regarding the quality of the hybridization against the microarray.

8.7.3 SELECTION OF ARRAYING SURFACE

Many different arraying surfaces have been used with success. Poly-L-lysine-coated slides are commonly used, typically because they are inexpensive and are produced readily in-house. Aminosilane-coated slides also are available from a number of manufacturers (e.g., TeleChem., Cell Associates, Corning, Sigma, AP Biotech), and in our system exhibit more intense fluorescent signals, better spot consistency, and lower background than the poly-L-lysine slides. Nitrocellulose-coated glass slides[55] retain the deposited DNA very well, but their use has not been optimized for many of the currently available hybridization procedures. Unfortunately, none of the currently available arraying surfaces are ideal, and they show great variability from batch to batch, making optimization of array printing a process that requires continued monitoring.

8.7.4 ARRAY PRINTING

Commonly used microarray printers deposit DNA onto the solid surface utilizing a pin-and-ring system, pins (quill, split, or solid), capillary transport, or dispensing (piezoelectric or inkjet style) technologies.[56,57] Each has distinct advantages and drawbacks. Pin-and-ring arrayers produce consistent spot shapes and do not need dedicated environmental controls, but require a long printing time and relatively high volumes of the DNA probes to be arrayed. Quill and split pin printers usually have higher throughput capacity, require smaller volumes, and print better on nitrocellulose-coated slides, but may be prone to pin clogging and often produce a less consistent shape of the probe spots. However, if maintained and used properly, most printing robots will produce microarrays of good quality.

There are several key variables to test and control for the production of high-quality arrays, including printing chamber humidity and temperature, printing solution, and DNA concentration. The most commonly used arraying solutions are 3X SSC and 50% DMSO, but even nuclease-free water has produced acceptable results in our laboratory. Concentrations of 0.1–1 µg/µl of DNA in the arraying solution seem to work well for production of most microarrays.

Array probes corresponding to the same gene should be redundant whenever possible. A four to eight spot redundancy is recommended, and these repeated probe spots must be scattered over different areas of the microarray to control for potential regional differences in array hybridization. To assess the quality of the printed microarrays, probes may be visualized using SYBR green II staining.[58]

8.8 AMPLIFICATION OF STARTING MATERIAL

cDNA microarrays may require >1 µg of high-quality transcript for a single experiment. When this amount of mRNA is prohibitive, starting material can be amplified.

PCR-based or T7-RNA polymerase based amplifications are two of the most commonly used approaches. PCR-based amplifications are easy to perform and are developed commercially by Clontech (SMART® amplification),[59,60] however, this method has not yet been extensively validated in microarray experiments. The T7-aRNA protocol[61] involves a linear amplification without a significant bias in mRNA species representation and increases the amount of the starting material in a single reaction by >100-fold. Several laboratories have used this method with great success even for single cell transcriptome amplification using two rounds of T7-aRNA amplification.[21,23,24] However, these sequential reactions, separated by a new RT step, are likely to show some preferential amplification of certain mRNA species. Hence, these experiments provide an extremely valuable, but often a binary (present/absent) snapshot of cellular transcriptome; less robust expression differences between individual cells may represent amplification artifacts rather than true transcriptome differences.

The GeneChip® technology is based on one round of T7-aRNA amplification, and the sample amplification is usually performed by RT-aRNA amplification kits developed by Affymetrix. This technology has been extensively validated over the last 5 years.[17,38,42]

8.9 SAMPLE LABELING

In a laboratory setting, the most common approach of sample labeling incorporates fluorescently modified nucleotides (usually Cyanine 3 — Cy3 and Cyanine 5 — Cy5) directly into the cDNA strand of the targets.[2,14,36] This method has been used in the vast majority of studies using cDNA microarrays, producing outstanding data.[11,15,60,62,63] However, this procedure is very sensitive to environmental influences (e.g., ambient temperature), impurities (e.g., DNA or protein contamination), and ionic concentrations.[37] Furthermore, due to the bulkiness of the Cy dyes and their partial inhibition of some reverse transcriptases, the RT reaction is less efficient than incorporating radionucleotides, and high dye incorporation can interfere with the hybridization reaction.[64] The incorporation is also sequence-dependent, and may result in more efficient labeling of certain mRNA classes. As alternatives to the bulky Cy dyes, Alexa fluorochrome-labeled[65] nucleotides or indirect labeling via amino-allyl-dNTP may be used. In the latter method, aa-dNTP is incorporated into the RT reaction. This step is followed by a linkage reaction to couple the appropriate Cy molecules (or other fluorochromes) to the aa-dNTP. Several companies (Clontech, Molecular Probes, Stratagene) have developed kits utilizing this approach. To circumvent the often prohibitive amount of starting material required, and Cy incorporation problems, NEN Life Sciences offers an alternative indirect labeling protocol that reduces material requirements by over 20-fold.[51] Briefly, in separate reverse transcription reactions biotinylated nucleotides are incorporated into cDNA of one sample, fluorescein-coupled nucleotides into cDNA of the other sample, the two are combined and hybridized to the same microarray. Hybridization is followed by microarray incubation with anti-Fluorescein-HRP and deposition of Cyanine-3-Tyramide. Finally, to visualize the biotin incorporation, the microarray is incubated with Streptavidin-HRP, and Cyanine-5-Tyramide is deposited.

A novel, innovative and promising approach to primer labeling is a two-step labeling procedure pioneered by Genisphere, Inc.[50] A capture sequence is bound to the oligodT primer and a standard reverse transcription is performed. Following the RT, the cDNA with the attached capture sequence is mixed with a fluorescently labeled "3DNA"(dendrimere). This Cy3 or Cy5 pre-labeled 3DNA has a pre-attached sequence recognition site that will bind the capture sequence incorporated into the cDNA, forming a cDNA-3DNA-flourochrome complex. The 3DNA binds to ~250 fluorochromes simultaneously,[50] resulting in very bright signal from each of the cDNA molecules, thus decreasing sample requirements to as little as 0.25 g of total RNA.

Incyte Genomics, Inc. (Fremont, CA, USA) labels the targets using prelabeled random 9-mers (Operon Technologies, Inc., Alameda, CA, USA) and unmodified nucleotides in an mRNA-based reverse transcription reaction.[16,66] Purity of mRNA is essential — if present, ribosomal RNA will also be reverse transcribed in this reaction. Hybridization with such targets generally leads to diminished detection of expression differences between experimental and control samples, yielding the false impression of little variability between the two samples.

Several laboratories are currently using total RNA (rather than mRNA) as starting material in oligo-dT primed labeling reactions.[36,37,50] This seems to decrease the starting material requirement without compromising signal intensity or increasing microarray background.

There are several excellent web sites that have posted detailed protocols for probe design, nucleic acid purification, labeling, custom array design strategies, and microarray printing, including http://www.microarrays.org, http://cmgm.stanford.edu/pbrown, http://www.nhgri.nih.gov/DIR/LCG/15K/HTML/, and http://www.gene-chips.com.

8.10 HYBRIDIZATION

Several web-published protocols are used successfully for labeled probe hybridization on microarrays (for Internet links, see Reference 36). However, even with high-quality probes, this procedure is very sensitive to temperature changes. The newly developed microarray hybridization stations (AP Biotech, Genomic Solutions, Ventana Medical), due to their strict ambient control, seem to yield more consistent hybridization results than manual procedures.

8.11 IMAGE ACQUISITION

Dedicated microarray scanners perform readout of fluorescently hybridized microarrays. GeneChips® are scanned using a specialized scanner (not to be confused with a universal slide scanner also made by Affymetrix), while the commonly used cDNA microarray scanners can analyze any standard 24×75 mm glass slide. Microarrays are scanned with spectrally nonoverlapping lasers, exciting the appropriate probe-bound fluorochromes (peak absorbance ~550 nm for Cy3 and ~650 nm for Cy5), which will in response emit fluorescent light of longer wavelengths (peak emission ~570 nm for Cy3 and ~670 nm for Cy5). Microarray scanners either simultaneously scan array slides at two wavelengths using a dual-beam laser scanning system (Axon Instruments) or perform independent scans for each fluorochrome, much like in confocal microscopy (GMS-Affymetrix, GSI Lumonics). In the scanning process, images are obtained at 5–10 μm/pixel resolution and stored separately for each fluorescent channel as 16-bit grayscale images. Other well-established manufacturers producing high-quality scanners, using somewhat different technologies, include Virtek, AP Biotech/Molecular Dynamics, and Genomic Solutions.

We strongly recommend using the acquisition software that accompanies the specific microarray scanner that will be utilized. These recommendations typically ensure that the optical data is well within linear range and ready for the process of image analysis. Because Cy5 is more susceptible to photobleaching than Cy3, the former channel should be scanned first.

8.12   IMAGE ANALYSIS

Image analysis is often incorrectly considered "data analysis" — these two terms should not be confused. In the image analysis process, data readout is obtained from the fluorescent scans, whereas the data analysis process involves an evaluation of the relationship between data points, often across many samples.

The 65,536 shades of gray (2^{16}) are translated into numerical signal intensities across the appropriate array regions. Briefly, software is used to define microarray probe position, make adjustments to the "grid" layout, and determine those probe areas with usable signal for subsequent analysis. Global microarray and local target backgrounds are established, and compared to probe signal intensity, generating a signal-to-background ratio (S/B). For each target, the signal intensity and background analysis is performed independently for the two fluorescent channels. Based on user-defined criteria, probes that do not meet S/B cutoffs or probe area requirements are flagged as unreliable data. Using operator-defined internal and external controls, the two channels are mathematically adjusted for comparable intensity. This adjustment[67] compensates for different amounts of input material, quality of fluorochromes, different labeling efficiency between the two samples, and different signal detection settings between channels. This is most often achieved by multiplying all probe intensities of the Cy3 or Cy5 channel with a single "adjustment factor." The adjustment factor is defined by calculating overall Cy3 and Cy5 signal intensities from all probes across the microarray, and defining a multiplying factor that "equalizes" channel intensities. Finally, the "adjusted" individual probe spots are compared for signal intensity, and the Cy3/Cy5 (or other fluorochrome) signal intensities are reported as a ratio, usually as fold increase or decrease between the two samples.

Image analysis software is available from many manufacturers and they often are bundled with the hardware purchase. In contrast to using the scanner, you should be able to analyze the obtained images with the software of your choice. Before committing to image analysis software, we strongly recommend that you evaluate it for ease of use, reliability, integration into existing databases, platform, and desired features (e.g., radioactive membrane analysis capabilities or automatic spot finding). There are a number of available software packages used successfully by the arraying community, including ScanAlyze (free from Stanford University/Eisen lab for academic investigators), QuantArray (GSI), GenePix Pro (Axon Instruments), Pathways (Research Genetics), ImaGene (BioDiscovery), and MicroArray Suite (Scanalytics).

8.13 ESTABLISH RELIABILITY IN THE INITIAL EXPERIMENTS

8.13.1 ASSESS THE NOISE OF THE SYSTEM

For cDNA microarrays, labeling two aliquots of a single, control sample with Cy3 and Cy5 and hybridizing these against each other on the same microarray addresses fundamental assay error, including the combined noise generated by reverse transcription/label incorporation, hybridization, preferential binding, microarray quality, and scanning bias. Whenever resources permit, this experiment should be repeated several times, starting with a new sample pair and a new labeling reaction. It is noteworthy that labeling and hybridizing different tissues may generate errors of different magnitude on the same type of microarray.

For GeneChips®, analyzing repeated hybridizations of the same sample to different oligonucleotide arrays also is essential. In these experiments, aliquots of the same sample should be used to generate new probes in the IVT reaction, thus including the potential noise from variations in the T7 amplification step.

8.13.2 ASSESS THE RELIABILITY OF A SINGLE HYBRIDIZATION

For cDNA arrays, repeated hybridizations should be done with reversed Cy3 and Cy5 labels. In repeated experiments, values that exhibit reciprocal changes should be excluded from further analysis. Data reliability increases the more times the hybridization is repeated, with statistical analyses capable of uncovering valid gene expression differences on the order of 15–50%.

8.13.3 ASSESS THE RELIABILITY OF THE BIOLOGICAL EFFECT

To assess the biological differences between two states, repeated hybridizations of the same sample or sample pair is encouraged,[68] but not required. Comparing many different sample pairs, where the same biological effect is expected, has greater statistical and biological power than repeating same-sample hybridizations many times. For example, in comparing normal and diseased states, repeating five matched sample comparisons four separate times will establish the precise gene expression differences for each of the sample pairs, but will be less informative about the disease process itself than comparing 20 disease-control subject pairs — even if there is a real possibility that some observations were false positive or false negative. Appropriate statistical analysis will evaluate consistency and significance.

8.13.4 DETECTABLE BIOLOGICAL CHANGES

The gene expression cutoffs and the true expression difference cutoffs will depend primarily on the efficiency of target labeling, the quality of the microarrays, the number of redundant probes on the microarrays, and the number of repeated experiments. Differences in experimental designs and the variability of parameters noted above make it difficult to recommend a single set of guidelines. However, several issues should be considered:

1. Performing more sample comparisons will facilitate the discovery of real biological differences. The critical question is *how consistent* are the gene expression changes across the different experimental-control pairs. This can be assessed by nonparametric statistical measurements.
2. Greater probe redundancy will result in a more reliable microarray measurement. The scattering of redundant probes randomly over the entire microarray, rather than clustered in one area, eliminates potential sided-bias during hybridization or scanning. Outlier measurements should be discarded and the intensity over the remaining spots should be averaged, and treated in further analysis as a single measurement. We found empirically that for a gene to be reliably expressed in the sample, the probes should report a S/B ratio greater than 3. This cutoff will vary across experimental conditions, array types, and designs.
3. It is more likely that small expression differences will be detected when the same sample hybridization is repeated across many microarrays. For both the GeneChips® and cDNA microarrays, repeated measures of individual samples will facilitate the detection of real biological changes in the 1.2- to 1.5-fold range. In a well-controlled single experiment, expression differences are likely to be real if they report a >1.6-fold change.
4. Expression differences between samples are more likely to be false if the gene is expressed at a low level. The most reliable expression differences usually are seen for genes expressed at moderate to high levels, although for GeneChips®, features of genes expressed at the highest level may saturate, potentially masking real expression differences in the comparisons.

8.14 BIOLOGICAL CHANGES LOST IN THE NOISE

The fundamental, heterogeneous nature of brain tissue at the cellular level will result in the failure of the microarray to detect some genes that are present in the samples.[32,45] This is especially common for receptors and all transcripts that are present at low levels in a small subpopulation of the cells in the tissue sample. At the level of each sample, microarrays theoretically can detect 1:250,000 mRNA molecules. If *all* the cells in the tissue expressed the same transcript at similar levels, this sensitivity would allow the detection of only few transcript copies per cell.[69] Unfortunately, in brain, it is typical that only subpopulations of cells express specific transcripts, and these mRNA species oftentimes may be diluted to the limits of microarray resolution. For example, if a marker transcript only is expressed in neurons, glial transcripts will effectively dilute the marker by 50%, halving the sensitivity. Moreover, if the specific mRNA is expressed in only interneurons, there will be an approximate 3-fold dilution with projection neuronal transcripts (normal ratio of interneuron to projection neurons in primate cortex is ~1:3[70,71]). If only 50% of the interneurons express the marker, there is an additional 2-fold dilution by the nonexpressing interneuronal transcripts. The result is technically overwhelming — the nominal microarray sensitivity is reduced by 36-fold. Consequently, in complex samples, many low and rare abundance transcripts are not detectable by microarrays, especially if the corresponding probes are not optimized (see above).

Because of the nature of the tissue samples and limitations in sensitivity, many differences in gene expression between samples will be masked.[32,45] This occurs particularly for transcripts that are present in the majority of the cells in the samples, but expression changes occur only in a subpopulation. For example, if 10% of the interneurons in the cortex up-regulate a gene by 10-fold, and if 30% of interneurons and 50% of projection neurons express normal levels of the same gene, the measurable mRNA in the comparisons will be altered by ~20%. To uncover expression changes of this magnitude by microarrays, the same-sample hybridizations should be repeated many times.

Difficulties in data interpretation also arise for diseases associated with general down-regulation of mRNA production. As an accepted practice, we hybridize a constant amount of labeled target to the microarrays. It is important to keep in mind, however, that the same amount of mRNA may originate from a different number of cells in the two brains. Furthermore, because the amount of starting material is generally not controlled carefully, a modest generic cell loss may not be detectable. Thus, a somewhat greater amount of tissue in which there is general cell loss could still yield the same amount of mRNA isolated from both samples.

8.15 COMMON TECHNICAL PROBLEMS WITH MICROARRAYS

Technical problems are typical of microarray experiments. Recognition of these will save wasted effort and resources.

8.15.1 No to Low Variability

If the variability between two samples is less than the variation observed in the self-hybridization control experiment (see above), the data are unreliable and should not be analyzed. Absence of gene expression differences between samples is often due to rRNA or DNA contamination.

8.15.2 Low Signal Intensity

When microarrays do not detect spiked-in controls and transcripts more abundant than 1:25,000 copies, data should be considered unreliable. Reasons for such false-negatives often include poor quality RT, fluorochromes, hybridizations, and/or microarrays.

8.15.3 Disproportionate Labeling

Absolute Cy3 and Cy5 intensities should be comparable for most of the genes on the microarray. If the balancing coefficient for signal adjustment between the two channels >2.5, the data are unreliable. The limited labeling of one channel could be due to many variables, including contamination of the RNA sample, poor fluorochrome, or a difference in the concentration of samples that are hybridized. If the big variation is due to improper image acquisition, the microarray should be rescanned.

8.15.4 High and Uneven Background

This is a common technical problem. Microarrays with exceedingly high background should be discarded, and if data is interpreted on subareas with low background, only local background measurements should be used. There are numerous reasons for obtaining high background, and it often is due to inappropriate blocking, drying out in processing, or undesired temperature fluctuations during the hybridization.

8.15.5 Dust Particles, Deposition Defects, and Coating Imperfections

Discard target spots from further analysis if they: (1) have significantly different shapes viewed under the two excitation wavelengths; (2) have small, high-intensity spots that give signal in only one channel when the overall target signal intensities are comparable; or (3) produce signal from less than 40% of the measured area.

8.15.6 Uneven Hybridization Across Microarray Regions

First establish that this is not due to problems with the scanner. Probe spot intensity should not change if the microarray is rotated by 180° and rescanned. If uneven hybridization across regions is minimal and symmetric for both channels, the data may be usable, but should be interpreted cautiously.

8.15.7 Nonlinear Signal Balance

This occurs when a single numerical factor is not sufficient for intensity adjustment between the red and green channels. When the adjustment factor for low intensity signals is different from the high-intensity adjustment by >25%, data should be discarded, as it may reflect nonlinear RT in one of the samples.

8.15.8 Internal/External Controls Balance Differently Than Overall Average

Many investigators, especially when using high-density microarrays, determine the red/green balancing factor[67] by assuming that overall average gene expression is similar between the two samples (see above). Indeed, this usually is an accurate assumption for high-density microarrays, and the adjustment factor calculated by using global, internal, or external controls is almost identical. However, if these adjustment coefficients are different by >25%, this may reflect biased selection of internal controls or pipetting errors for the external controls. Fortunately, if no other microarray problems are detected, the hybridization data are not likely to be compromised, and with careful interpretation, valuable information can be extracted.

8.15.9 SCANNING PROBLEMS

Scans should *never* be saturated, because such data lead to flawed comparisons and biased interpretations. Although usually not required, if desired, two scans can be performed — one at a low-intensity laser setting for assessment of high signal intensity targets, followed by a high-intensity scan for assessment of targets with weak fluorescent intensity. In general, we do not recommend this strategy, because high-intensity laser scans often enhance both signal and background equivalently, thus failing to yield better S/B than the lower intensity scans.

Whenever differential intensity is reported across microarray regions, the microarrays should be rotated by 180° and rescanned at the same settings. Differential intensities often are due to an improperly leveled microarray stage or misaligned optics.

8.15.10 ARRAY PRINTING PROBLEMS

Printing problems are numerous, typically being dependent upon platform type, sample, type of surface, environment in which the printing is being performed, and even the individual printer. Practice and experience makes a master. Systematically vary the parameters until you obtain consistently high-quality printing. It is critical to properly maintain the pins and keep track of individual pin performance within and across arrays.

8.16 VERIFICATION OF OBTAINED MICROARRAY DATA

Data obtained with a single genomics approach are prone to difficulties in interpretation, and perhaps even lead to the questioning of data reliability in the absence of independent validation of the findings. With newer methods like microarrays, data verification remains an important issue, but is it realistic to investigate thousands of data points with additional methods? Regardless of the technical strategies employed, one still needs to determine *what* and *when* to verify. We believe that the particular answers will vary, depending upon the experimental situation. Moreover, a strategy should be decided upon in a biological context. For example, if a study reports expression changes of only a few genes, verification of data with an independent method is rather simple to incorporate into the experimental design. The verification of unique gene expression pattern, involving hundreds of different transcripts, is more complex. We were recently faced with this problem. While it was possible to identify select, critical data points to be verified, a full reproduction of the study by an independent method was unreasonable.

Complex transcriptome changes, where simultaneous increases and decreases coupled with the presence or absence of other gene products define the disease, may be verified in a three-step procedure: (1) based on the data obtained in the initial set of experiments, a disease-specific gene expression pattern is defined; (2) on a new group of subjects, the predicted gene expression pattern is validated; (3) and selected individual changes, across the two datasets, are verified by an independent method.

Brain samples, especially in gyrencephalic species, where cortical surface is individually variable, are vulnerable to sampling errors that will affect both microarrays and other molecular techniques. Most notably, tissue blocks may have different gray/white matter ratio, and the extracted mRNA from two samples could contain different relative amounts of neuronal or glial transcripts. If such a control and experimental sample are compared on the same microarray, glial markers will be reported increased and neuronal markers (often receptors) decreased in the experimental sample. This opens the possibility of false interpretation of data. Unfortunately, real-time quantitative RT-PCR and Northern analysis of brain samples are equally vulnerable to gray/white matter bias (especially if the same sample is used for verification), and due to this, sample bias may report the same false-positive data as the microarray analysis.

Furthermore, the vast majority of the biologically important expression changes, at the level of a brain region sample, may not reach 2-fold.[27-30,72] Although real-time RT-PCR technology has been extensively validated for detection of expression changes >2-fold, its value in verification of smaller changes surprisingly has not been established. For this reason and at this time, when expression changes are <2-fold, Northern hybridization is the preferred verification method over real-time quantitative RT-PCR.

It should not be forgotten that the effect of a gene expression change on neural networks will depend on its precise anatomical localization. Contrary to the methods that assess gene expression in a combined tissue sample, *in situ* hybridization offers anatomical information about transcript distribution, and this information is essential to put the changes into the biological context of the disease. Because differences can be detected across even a small subpopulation of cells in a sample area, *in situ* hybridization will reliably detect changes that are prone to dilution in tissue samples, thus exceeding the sensitivity of the above methods. *In situ* hybridization also offers the advantage of requiring substantially less starting material for analysis, and it is not affected by white/grey matter bias. Furthermore, one can collect tissue sections from the *same* block of tissue that was used for RNA isolation (Mirnics, 2001 #534; Mirnics, 2000 #240; Mirnics, 2001 #454). Data verification for individual gene products can be extended further using immunocytochemistry, although this method is not amenable to quantitative analysis of the gene product. Rather, major qualitative differences in expression can be discerned. If one is interested in such quantitative information, Western blot analysis of tissue homogenates is a reliable alternative.

8.17 DATA ANALYSIS CONSIDERATIONS

Current strategies for reporting gene expression differences between microarrays can be divided into three classes (Figure 8.7).

QUESTION ASKED   DATA ANALYSIS METHOD

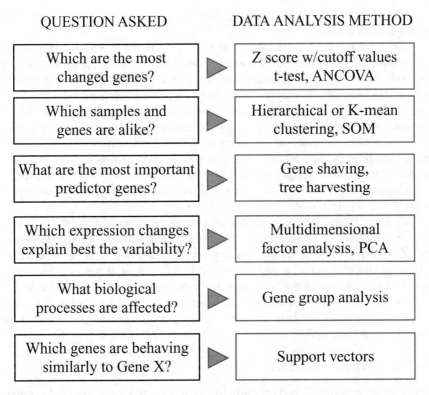

Which are the most changed genes?	▷	Z score w/cutoff values t-test, ANCOVA
Which samples and genes are alike?	▷	Hierarchical or K-mean clustering, SOM
What are the most important predictor genes?	▷	Gene shaving, tree harvesting
Which expression changes explain best the variability?	▷	Multidimensional factor analysis, PCA
What biological processes are affected?	▷	Gene group analysis
Which genes are behaving similarly to Gene X?	▷	Support vectors

FIGURE 8.7 Each analytical method is paired with the question that provides a "best" answer. By "predictor genes," we mean gene changes that best explain individual factors related to the disease, such as patient survival or therapeutic responsiveness.

8.17.1 REPORTING THE MOST CHANGED GENES

Although microarrays provide a unique opportunity to perform complex data analyses, investigators typically ask as a first step, *"What are the most consistently changed genes across different comparisons?"* This is most often determined by identifying those genes that consistently exceed previously established cutoff values across the pairwise comparisons. Such differences in expression are analyzed for statistical probability across the pairwise comparisons.

One can report a predefined fraction (1 or 5% confidence interval "cutoffs") of the most changed observations, irrespective of the quality of the individual microarray. However, this type of data manipulation may be misleading. Across different experiments — comparing the same sample to itself, two different brain samples or a cancer tissue to normal tissue, the number of differentially expressed genes will vary (Color Figure 8.8).* It is biologically incorrect to assume that a fixed percentage of genes will be changed between different pairwise comparisons — the similarities

---

* Color figures follow page 140.

or differences between transcriptomes will depend upon biological factors, not predetermined ratios.

cDNA microarrays do not always report precisely the magnitude of increases or decreases. A 2-fold increase for gene X and 3-fold increase for gene Y suggest that both genes are expressed more abundantly in the experimental than in the control sample. However, the assumption that gene X was relatively more increased than gene Y may not be valid — the reported increase on *different* targets will depend on many different parameters, including the quality and sequence of the spotted probe, as well as the labeling procedure used. In contrast, different magnitudes of change reported by the *same target* across microarrays are usually real: if gene V is increased by 2-fold in comparison A, and the same gene V is increased 3-fold in pairwise comparison B, it is likely true that the difference for gene V was greater in comparison B than in A.

Similarly, if gene Z is expressed at an absolute intensity of 500 units, and gene W is expressed at 1000 units in the same sample comparison, one cannot assume that Z/W is 1:2. Absolute target intensity levels, within the same sample, among others, will depend on the sequence and length of the target. Whereas cDNA arrays exhibit these caveats, GeneChips®, due to strictly controlled oligonucleotide production of the bound probe and sample fragmentation, are quantitative.[38] As such, GeneChips® provide valuable data concerning the relative abundance of mRNA species in the samples.

8.17.2 SORTING BY SIMILARITY

This method is accomplished typically by clustering samples according to similarities of gene expression patterns and by sorting genes with similarly changing patterns of expression.

Hierarchical clustering is, to date, the most popular form of data sorting. The method identifies common patterns between samples and gene expression profiles.[73] The approach, however, is influenced considerably by expression outliers, and will report clustering even in data that is not inherently hierarchic (e.g., in two aliquots of control sample). *Gene shaving* is a more advanced combination of sorting and principal component analysis, employing a sequential discarding of data that is least informative about the imposed question.[74] *K means clustering* allows the investigator to factor into the data analysis prior knowledge regarding the number of clusters, whereas using *self-organizing maps* (SOM),[75,76] one can impose partial structure on the data during hypothesis testing. *Principal components analysis* (PCA) and *factor analysis* are used for determining the key variables in a multidimensional data set that explain the differences in the observations.[77] *Supervised three harvesting*[78] models the outcome variable of the hierarchical clustering, identifying both genes with effects of their own and genes that act in unison. *Support vector machines* use supervised computer learning and knowledge of prior gene function to identify unknown genes of similar function.[79] While powerful, most of these analytical approaches require large datasets, often not available in studies of brain disorders that utilize postmortem material.

8.17.3 Assessing Potentially Affected Biological Functions

At the core of this method are two concepts: (A) the altered function of individual genes, whose products cooperate to perform a defined biological function, may give rise to similar deficits, and (B) when a discrete biological function is affected, multiple members of the functional pathway will show adaptive gene expression changes.

In a *group analysis*,[16] genes may be clustered together by structure (e.g., transmembrane proteins), chemical function (e.g., kinases), or biological function (e.g., presynaptic release) (Color Figure 8.9).* Statistical parameters can be used to evaluate central tendency measurements, with the distribution of these groups compared to overall gene expression distribution in the comparison and the mean value of all the expressed genes. This method is very powerful in detecting putative function-related expression changes, but negative data should be interpreted with caution. The strategy of functional clustering is complicated, because gene group effects may be abolished if functionally redundant isoforms of a single gene exist within a cascade. Furthermore, to date our detailed knowledge of the interacting molecular elements in specific cellular functional cascades is limited. Such subjective clustering reaches an additional level of complexity when one considers that individual gene products typically perform multiple, seemingly unrelated biological functions. Thus, the most difficult questions to answer are *"To which functional pathway does gene X belong?"* and *"Where does my functional cascade begin and end?"*

It is clear that data analysis methods require substantial improvements. In a short time, however, it is likely that the integration of microarray and bioinformatics approaches will result in the development of extensive and interactive databases of molecular cascades,[80,81] which will be able to assess the entire transcriptome comprehensively, linking expression changes to disease mechanisms and phenotypic adaptations.

8.18 SCHIZOPHRENIA: AN EXAMPLE OF HOW AND WHAT WE LEARNED FROM THE MICROARRAY APPROACH[16,19,45]

The following is an abbreviated version of our experimental and data analysis strategies used to investigate altered gene expression in subjects with schizophrenia. Data acquisition and analysis have led to a newly proposed model that incorporates altered biological function into a neurodevelopmental concept of the disease.

8.18.1 Harvesting and Storage

Brain specimens were obtained at autopsy, cut into 1.0 cm-thick coronal blocks, frozen, and stored at –80°C.

---

* Color figures follow page 140.

8.18.2 SUBJECTS

Two groups of schizophrenic subjects, consisting of six and five pairs of subjects with schizophrenia and matched controls. Diagnoses for all subjects were made using information obtained from structured interviews with surviving relatives and medical records, followed by a consensus diagnosis conference of experienced clinical researchers.

8.18.3 HISTOLOGICAL VERIFICATION OF BRAIN REGIONS

Toulidine blue, 20 μm sections, standard staining procedure.

8.18.4 PAIRING

Individual pairing design, each subject with schizophrenia matched to the best available control. Matching parameters included race, gender, PMI, age, and brain pH.

8.18.5 NUCLEIC ACID ISOLATION

From 50 μm cryostat sections, total RNA was isolated using RNAgents (Promega), mRNA with a double-pass through Oligotex columns (Qiagen).

8.18.6 MICROARRAYS

UniGEM-V and UniGEM-V2 (Incyte Genomics, Inc) cDNA microarrays, with >7,000 and >10,000 genes and EST, respectively.

8.18.7 LABELING

At Incyte Genomics, with Cy3- and Cy5-labeled random 9-mers.

8.18.8 HYBRIDIZATION

Modified Brown protocol.

8.18.9 IMAGE ACQUISITION

GenePix scanner and software (Axon Instruments). Operators and investigators were blind to the specific category to which each sample belonged.

8.18.10 ARRAY EXCLUSION CRITERIA

(A) Non-linear signal balance, (B) detection levels <1:50,000 mRNA copies according to external spikes, (C) Cy3/Cy5 balance factor >| 2.5|, (D) compressed arrays with less variability than the control experiment.

8.18.11 PROBES EXCLUDED FROM DATA ANALYSIS

(A) Spiked-in controls, (B) S/B ratio <5-fold for either the Cy3 or Cy5 channel, (C) signal from <40% of the probe area, (D) target gel electrophoresis reported no amplification or multiple bands, and (E) those showing evidence of major coating imperfections or target impurities.

8.18.12   CONTROL EXPERIMENTS

(A) The same aliquot of a control sample was hybridized against itself, yielding essentially identical assay noise measurements as Incyte's control experiments (Color Figure 8.10A).* (B) One sample pair hybridization was repeated three times on different UniGEM-V microarrays, essentially reproducing the gene expression differences on all microarrays.

8.18.13   CHANNEL INTENSITY ADJUSTMENT

For each microarray, we used overall gene expression average between the two hybridized samples.

8.18.14   EXPRESSION CUTOFFS

Based on our control data and Incyte's control experiments, we considered signal intensity $>|1.6|$ as probably true and $>|1.9|$ as true differences between the samples (Color Figure 8.10B).

8.18.15   DATA ANALYSIS

Using the established cutoff criteria, we identified several individual genes that were consistently changed across multiple comparisons. Statistical analysis included calculations of the probability of each individual gene to be changed by chance.

Using the gene group approach, we identified four biological functions/cascades that were consistently impaired with all subjects with schizophrenia. Statistical analysis of data included t-test and Chi-square test.

Using *in situ* hybridization, selected gene expression changes were verified on the tissue blocks from the same subjects. The microarray-reported changes were also assessed in a new, independent cohort of subjects. Changes in gene expression patterns (using gene groups) were also verified on the second cohort of subjects. Effects of antipsychotic treatments on gene expression were assessed using a chronic neuroleptic-treated primate model, both by microarray and *in situ* hybridization. By analyzing subjects with major depressive disorder without psychosis, diagnosis-specificity of selected gene expression changes were assessed by *in situ* hybridization. A multifactorial ANCOVA analysis did not uncover confounds.

8.18.16   SPECIFIC FINDINGS

N-ethylmaleimide sensitive factor (NSF), Synapsin 2 (SYN2), regulator of G-protein signaling 4 (RGS4) were all consistently decreased across the subjects with schizophrenia, but not in haloperidol-treated monkeys or subjects with major depression.

---

* Color figures follow page 140.

Changes in the expression of these genes have not been associated with schizophrenia previously (Color Figure 8.10C). Expression changes for AMPA2 receptor[82,83] and glutamic acid decarboxylase 1,[29] already described in schizophrenia with conventional methods, were also confirmed.

Functional data mining across all subjects with schizophrenia revealed consistent gene expression decreases in genes whose products are responsible for presynaptic release (PSYN).

The individual pattern of specific PSYN gene decrease was different across schizophrenic subjects, arguing for the existence of different molecular signatures across subjects with schizophrenia. However, all these signatures were related to a deficit in a presynaptic secretory function.

8.18.17   GENERATED HYPOTHESIS FROM MICROARRAY DATA

In our model,[16,45] impaired synaptic transmission is present from factors operating early in development,[84,85] but does not manifest in clinical symptoms because exuberant synaptic connections compensate for the deficits of individual synapses. When synaptic pruning ends in late adolescence,[86,87] the decreased number of synapses is not able to compensate for impaired synapse function, and the clinical manifestations of the disease emerge. This may be followed by postsynaptic changes, including (but not limited to) down-regulation of RGS4.[19] This may result in prolonged signaling through G-protein coupled receptors.[88,89] However, as a result of impaired synaptic release (and consequent lack of normal synaptic drive), overpruning of the presynaptic neuropil may occur.[90]

8.19   FUTURE DIRECTIONS

Microarray sensitivity, complexity, and availability will increase. Eventually, a whole genome will be represented on a single microarray. The obtained data will be quantitative, and it will be able to detect even the most diluted transcripts. However, the exact technology that this future microarray will utilize is unclear at the present time.

Profiling of single cells and neuronal phenotypes will be a major focus of microarray research. Characterization of individual neurons will define molecular phenotypes that will be associated with cellular function. Molecular phenotype changes will be associated with diseases and altered physiological states.[21,24]

Data analysis tools will become more sophisticated; databases will allow data sharing and data linking across platforms. Data analysis tools will uncover disease-specific gene expression patterns, where a disease (or its symptom) may be defined by a specific combination of increase, decrease, presence *and* absence of individual gene expressions. Genomic, gene expression, behavioral, and clinical information databases[91-93] will become transparent, and data will be analyzed across different platforms and methods. Data mining, using combined knowledge from multiple sources, will assess individual subjects for functional deficits.

Protein arrays[94-97] and tissue arrays[98,99] will further develop. Within the next several years, high-throughput protein assays will become reliable and affordable. Ultimately, samples will be analyzed routinely for gene expression, informative DNA sequence variations, protein expression, and the critical data will be further evaluated on tissue microarrays.

Sequence-based microarray analysis approaches will become routinely used in clinical diagnostics. As the SNP map will become dense, sequence-based microarrays will correlate expression changes with informative polymorphisms. Genomic material of premorbid subjects will be assessed for individual risk of developing a long list of diseases.

Pharmacogenomics approaches will enable targeted supplementation or repression of gene functions. Ultimately, the combined basic science and clinical data will define the ideal drug target genes.[100,101] Gene-specific genetic and xenobiotic interventions will become feasible, and individual treatment strategies will be developed even before the onset of the disease.

The most exciting discoveries are yet to come.

ACKNOWLEDGMENTS

We are grateful to the many members of the Levitt, Lewis, and Mirnics laboratories for their encouragement and helpful discussions throughout our microarray studies. We add a specific note of thanks to Dr. Daniel H. Geschwind, Dr. Frank A. Middleton, and Ms. Deborah Hollingshead for useful suggestions regarding the organization and content of this chapter. Our research summarized in this chapter was supported by Projects 1 (DAL) and 2 (PL, KM) of NIMH Center Grant MH45156 (DAL).

REFERENCES

1. Kozian, D.H. and Kirschbaum, B.J., Comparative gene-expression analysis, *Trends Biotechnol.,* 17(2), 73–78, 1999.
2. Schena, M. et al., Quantitative monitoring of gene expression patterns with a complementary DNA microarray, *Science,* 270(5235), 467–470, 1995.
3. DeRisi, J. et al., Use of a cDNA microarray to analyse gene expression patterns in human cancer, *Nat. Genet.,* 14(4), 457–460, 1996.
4. Lockhart, D.J. et al., Expression monitoring by hybridization to high-density oligonucleotide arrays, *Nat. Biotechnol.,* 14(13), 1675–1680, 1996.
5. Geschwind, D.H. et al., Klinefelter's syndrome as a model of anomalous cerebral laterality: Testing gene dosage in the X chromosome pseudoautosomal region using a DNA microarray, *Dev. Genet.,* 23(3), 215–229, 1998.
6. Drmanac, S. et al., Accurate sequencing by hybridization for DNA diagnostics and individual genomics, *Nat. Biotechnol.,* 16(1), 54–58, 1998.
7. Cheung, V.G. et al., Linkage-disequilibrium mapping without genotyping, *Nat. Genet.,* 18(3), 225–230, 1998.
8. Hacia, J.G. et al., Detection of heterozygous mutations in BRCA1 using high density oligonucleotide arrays and two-colour fluorescence analysis, *Nat. Genet.,* 14(4), 441–447, 1996.
9. Gunthard, H.F. et al., Comparative performance of high-density oligonucleotide sequencing and dideoxynucleotide sequencing of HIV type 1 pol from clinical samples, *AIDS Res. Hum. Retroviruses,* 14(10), 869–876, 1998.

10. Kozal, M.J. et al., Extensive polymorphisms observed in HIV-1 clade B protease gene using high-density oligonucleotide arrays, *Nat. Med.*, 2(7), 753–759, 1996.

11. Schena, M. et al., Parallel human genome analysis: Microarray-based expression monitoring of 1000 genes, *Proc. Natl. Acad. Sci. USA*, 93(20), 10614–10619, 1996.

12. Tanaka, T.S. et al., Genome-wide expression profiling of mid-gestation placenta and embryo using a 15,000 mouse developmental cDNA microarray, *Proc. Natl. Acad. Sci. USA*, 97(16), 9127–9132, 2000.

13. Geschwind, D. et al., A genetic analysis of neural progenitor differentiation, *Neuron*, in press.

14. Duggan, D.J. et al., Expression profiling using cDNA microarrays, *Nat. Genet.*, 21(1 Suppl), 10–14, 1999.

15. Alizadeh, A.A. et al., Distinct types of diffuse large B-cell lymphoma identified by gene expression profiling, *Nature*, 403(6769), 503–511, 2000.

16. Mirnics, K. et al., Molecular characterization of schizophrenia viewed by microarray analysis of gene expression in prefrontal cortex, *Neuron*, 28(1), 53–67, 2000.

17. Lockhart, D.J. and Winzeler, E.A., Genomics, gene expression and DNA arrays, *Nature*, 405(6788), 827–836, 2000.

18. Sandberg, R. et al., From the cover: Regional and strain-specific gene expression mapping in the adult mouse brain, *Proc. Natl. Acad. Sci. USA*, 97(20), 11038–11043, 2000.

19. Mirnics, K. et al., Disease-specific changes in regulator of G-protein signaling 4 (RGS4) expression in schizophrenia, *Mol. Psychiatry*, in press.

20. Whitney, L.W. et al., Analysis of gene expression in mutiple sclerosis lesions using cDNA microarrays, *Ann. Neurol.*, 46(3), 425–428, 1999.

21. Luo, L. et al., Gene-expression profiles of laser-captured adjacent neuronal sub-types, *Nat. Med.*, 5, 117, 1999.

22. Vawter, M. et al., Application of cDNA microarrays to examine gene expression differences in schizophrenia, *Brain Res. Bull.*, in press.

23. Ginsberg, S.D. et al., Predominance of neuronal mRNA in individual Alzheimer's disease senile plaques, *Ann. Neurol.*, 45(2), 174–181, 1999.

24. Ginsberg, S.D. et al., Expression profile of transcripts in Alzheimer's disease tangle-bearing CA1 neurons, *Ann. Neurol.*, 48(1), 77–87, 2000.

25. Livesey, F.J. et al., Microarray analysis of the transcriptional network controlled by the photoreceptor homeobox gene Crx, *Curr. Biol.*, 10(6), 301–310, 2000.

26. Niculescu III, A.B. et al., Identifying a series of candidate genes for mania and psychosis: A convergent functional genomics approach, *Physiol. Genomics*, 4(1), 83–91, 2000.

27. Collinge, J. and Curtis, D., Decreased hippocampal expression of a glutamate receptor gene in schizophrenia, *Brit. J. Psychiatry*, 159, 857–859, 1991.

28. Akbarian, S. et al., Gene expression for glutamic acid decarboxylase is reduced without loss of neurons in prefrontal cortex of schizophrenics, *Arch. Gen. Psychiatry*, 52(4), 258–266; discussion 267–278, 1995.

29. Volk, D.W. et al., Decreased glutamic acid decarboxylase67 messenger RNA expression in a subset of prefrontal cortical gamma-aminobutyric acid neurons in subjects with schizophrenia, *Arch. Gen. Psychiatry*, 57(3), 237–245, 2000.

30. Woo, T.U. et al., A subclass of prefrontal gamma-aminobutyric acid axon terminals are selectively altered in schizophrenia, *Proc. Natl. Acad. Sci. USA*, 95(9), 5341–5346, 1998.

31. Watson, S.J. et al., The "chip" as a specific genetic tool, *Biol. Psychiatry*, 48(12), 1147–1156, 2000.

32. Geschwind, D.H., Mice, microarrays, and the genetic diversity of the brain, *Proc. Natl. Acad. Sci. USA,* 97(20), 10676–10678, 2000.

33. Glantz, L.A. and Lewis, D.A., Reduction of synaptophysin immunoreactivity in the prefrontal cortex of subjects with schizophrenia. Regional and diagnostic specificity, *Arch. Gen. Psychiatry,* 54(10), 943–952, 1997.

34. Pierri, J.N. et al., Alterations in chandelier neuron axon terminals in the prefrontal cortex of schizophrenic subjects, *Am. J. Psychiatry,* 156(11), 1709–1719, 1999.

35. Harrison, P.J. et al., The relative importance of premortem acidosis and postmortem interval for human brain gene expression studies: Selective mRNA vulnerability and comparison with their encoded proteins, *Neurosci. Lett.,* 200(3), 151–154, 1995.

36. Hegde, P. et al., A concise guide to cDNA microarray analysis, *Biotechniques,* 29(3), 548, 2000.

37. Wildsmith, S.E. et al., Maximization of signal derived from cDNA microarrays, *Biotechniques,* 30(1), 202–208, 2001.

38. See Chapter 7.

39. Barton, A.J. et al., Pre- and postmortem influences on brain RNA, *J. Neurochem,* 61(1), 1–11, 1993.

40. Freeman, W.M. et al., Quantitative RT-PCR: Pitfalls and potential, *Biotechniques,* 26(1), 112–122, 124–125, 1999.

41. Heid, C.A. et al., Real time quantitative PCR, *Genome Res.,* 6(10), 986–994, 1996.

42. Lipshutz, R.J. et al., High density synthetic oligonucleotide arrays, *Nat. Genet.,* 21(1 Suppl), 20–24, 1999.

43. LeProust, E. et al., Digital light-directed synthesis. A microarray platform that permits rapid reaction optimization on a combinatorial basis, *J. Comb. Chem.,* 2(4), 349–354, 2000.

44. Singh-Gasson, S. et al., Maskless fabrication of light-directed oligonucleotide microarrays using a digital micromirror array, *Nat. Biotechnol.,* 17(10), 974–978, 1999.

45. Mirnics, K. et al., Analysis of complex brain disorders with microarrays: Schizophrenia as a disease of the synapse, *Trends Neurosci,* in press.

46. Cheung, V.G. et al., Making and reading microarrays, *Nat. Genet.,* 21(1 Suppl), 15–19, 1999.

47. Bowtell, D.D., Options available — from start to finish — for obtaining expression data by microarray, *Nat. Genet.,* 21(1 Suppl), 25–32, 1999.

48. Southern, E. et al., Molecular interactions on microarrays, *Nat. Genet.,* 21(1 Suppl), 5–9, 1999.

49. Shalon, D. et al., A DNA microarray system for analyzing complex DNA samples using two- color fluorescent probe hybridization, *Genome Res.,* 6(7), 639–645, 1996.

50. Stears, R.L. et al., A novel, sensitive detection system for high-density microarrays using dendrimer technology, *Physiol. Genomics,* 3(2), 93–99, 2000.

51. Adler, K. et al., MICROMAX: A Highly Sensitive System for Differential Gene Expression on Microarrays, in *Microarray Biochip Technology,* Schena, M., Ed., Eaton Publishing, Natick, MA, 221–230, 2000.

52. Schuchhardt, J. et al., Normalization strategies for cDNA microarrays, *Nucleic Acids Res.,* 28(10), E47, 2000.

53. Bertucci, F. et al., Sensitivity issues in DNA array-based expression measurements and performance of nylon microarrays for small samples, *Hum. Mol. Genet.,* 8(9), 1715–1722, 1999.

54. Chambers, J. et al., DNA microarrays of the complex human cytomegalovirus genome: Profiling kinetic class with drug sensitivity of viral gene expression, *J. Virol.,* 73(7), 5757–5766, 1999.

55. Stillman, B.A. and Tonkinson, J.L., FAST slides: A novel surface for microarrays, *Biotechniques,* 29(3), 630–635, 2000.

56. Mace, M.L. et al., Novel microarray printing and detection technologies, in *Microarray Biochip Technology,* Schena, M., Ed., Eaton Publishing, Natick, MA, 39–64, 2000.

57. Rose, D., Microfluidic technologies and instrumentation for printing DNA microarrays, in *Microarray Biochip Technology,* Schena, M., Ed., Eaton Publishing, Natick, MA, 19–38, 2000.

58. Battaglia, C. et al., Analysis of DNA microarrays by non-destructive fluorescent staining using SYBR green II, *Biotechniques,* 29(1), 78–81, 2000.

59. Zhumabayeva, B. et al., Use of SMART-generated cDNA for gene expression studies in multiple human tumors, *Biotechniques,* 30(1), 158–162, 2001.

60. Heller, R.A. et al., Discovery and analysis of inflammatory disease-related genes using cDNA microarrays, *Proc. Natl. Acad. Sci. USA,* 94(6), 2150–2155, 1997.

61. Eberwine, J. et al., Analysis of gene expression in single live neurons, *Proc. Natl. Acad. Sci. USA,* 89(7), 3010–3014, 1992.

62. Golub, T.R. et al., Molecular classification of cancer: Class discovery and class prediction by gene expression monitoring, *Science,* 286(5439), 531–537, 1999.

63. Chen, J.J. et al., Profiling expression patterns and isolating differentially expressed genes by cDNA microarray system with colorimetry detection, *Genomics,* 51(3), 313–324, 1998.

64. Worley, J. et al., A systems approach to fabricating and analyzing DNA microarrays, in *Microarray Biochip Technology,* Schena, M., Ed., Eaton Publishing, 2000, 65–86.

65. Panchuk-Voloshina, N. et al., Alexa dyes, a series of new fluorescent dyes that yield exceptionally bright, photostable conjugates, *J. Histochem. Cytochem.,* 47(9), 1179–1188, 1999.

66. Lewohl, J.M. et al., Gene expression in human alcoholism: Microarray analysis of frontal cortex, *Alcohol Clin. Exp. Res.,* 24(12), 1873–1882, 2000.

67. Chen, Y. et al., Ratio-based decisions and the quantitative analysis of cDNA microarray images, *J. Biomed. Optics,* 24, 364–374, 1997.

68. Lee, M.L. et al., Importance of replication in microarray gene expression studies: Statistical methods and evidence from repetitive cDNA hybridizations, *Proc. Natl. Acad. Sci. USA,* 97(18), 9834–9839, 2000.

69. Kane, M.D. et al., Assessment of the sensitivity and specificity of oligonucleotide (50 mer) microarrays, *Nucleic Acids Res.,* 28(22), 4552–4557, 2000.

70. Hendry, S.H. et al., Numbers and proportions of GABA-immunoreactive neurons in different areas of monkey cerebral cortex, *J. Neurosci.,* 7(5), 1503–1519, 1987.

71. Winfield, D.A. et al., An electron microscopic study of the types and proportions of neurons in the cortex of the motor and visual areas of the cat and rat, *Brain,* 103(2), 245–258, 1980.

72. Akbarian, S. et al., Selective alterations in gene expression for NMDA receptor subunits in prefrontal cortex of schizophrenics, *J. Neurosci.,* 16(1), 19–30, 1996.

73. Eisen, M.B. et al., Cluster analysis and display of genome-wide expression patterns, *Proc. Natl. Acad. Sci. USA,* 95(25), 14863–14868, 1998.

74. Hastie, T. et al., "Gene shaving" as a method for identifying distinct sets of genes with similar expression patterns, *Genome Biol.,* 1(2), 0003.0001–0003.0021, 2000.

75. Toronen, P. et al., Analysis of gene expression data using self-organizing maps, *FEBS Lett.,* 451(2), 142–146, 1999.

76. Tamayo, P. et al., Interpreting patterns of gene expression with self-organizing maps: Methods and application to hematopoietic differentiation, *Proc. Natl. Acad. Sci. USA,* 96(6), 2907–2912, 1999.

77. Raychaudhuri, S. et al., Principal components analysis to summarize microarray experiments: Application to sporulation time series, *Pac. Symp. Biocomput.*, 455–466, 2000.

78. Hastie, T. et al., Supervised harvesting of expression trees, *Genome Biol.*, 2(1), 0003.0001–0003.0012, 2001.

79. Brown, M.P. et al., Knowledge-based analysis of microarray gene expression data by using support vector machines, *Proc. Natl. Acad. Sci. USA*, 97(1), 262–267, 2000.

80. Ermolaeva, O. et al., Data management and analysis for gene expression arrays, *Nat. Genet.*, 20(1), 19–23, 1998.

81. Kanehisa, M. and Goto, S., KEGG: Kyoto encyclopedia of genes and genomes, *Nucleic Acids Res.*, 28(1), 27–30, 2000.

82. Eastwood, S.L. et al., Immunoautoradiographic evidence for a loss of alpha-amino-3-hydroxy-5-methyl-4-isoxazole propionate-preferring non-N-methyl-D-aspartate glutamate receptors within the medial temporal lobe in schizophrenia, *Biol. Psychiatry*, 41(6), 636–643, 1997.

83. Eastwood, S.L. et al., Decreased expression of mRNA encoding non-NMDA glutamate receptors GluR1 and GluR2 in medial temporal lobe neurons in schizophrenia, *Brain Res. Mol. Brain Res.*, 29(2), 211–223, 1995.

84. Carpenter, W.T. and Buchanan, R.W., Schizophrenia, *New Eng. J. Med.*, 330, 681–690, 1994.

85. Lewis, D.A. and Lieberman, J.A., Catching up on schizophrenia: Natural history and neurobiology, *Neuron*, 28(2), 325–334, 2000.

86. Huttenlocher, P., Synaptic density in human frontal cortex — Developmental changes and effects of aging, *Brain Res.*, 163, 195–205, 1979.

87. Bourgeois, J.P. et al., Formation, elimination and stabilization of synapses in the primate cerebral cortex, in *The Cognitive Neurosciences*, Gazzaniga, M., Ed., MIT Press, Cambridge, MA, in press.

88. Berman, D.M. et al., The GTPase-activating protein RGS4 stabilizes the transition state for nucleotide hydrolysis, *J. Biol. Chem.*, 271(44), 27209–27212, 1996.

89. De Vries, L. et al., The regulator of G protein signaling family, *Annu. Rev. Pharmacol. Toxicol.*, 40, 235–271, 2000.

90. Selemon, L.D. and Goldman-Rakic, P.S., The reduced neuropil hypothesis: A circuit based model of schizophrenia, *Biol. Psychiatry*, 45(1), 17–25, 1999.

91. Liao, B. et al., MAD: A suite of tools for microarray data management and processing, *Bioinformatics*, 16(10), 946–947, 2000.

92. Sherlock, G. et al., The Stanford microarray database, *Nucleic Acids Res.*, 29(1), 152–155, 2001.

93. Wheeler, D.L. et al., Database resources of the National Center for Biotechnology Information, *Nucleic Acids Res.*, 29(1), 11–16, 2001.

94. Lueking, A. et al., Protein microarrays for gene expression and antibody screening, *Anal. Biochem.*, 270(1), 103–111, 1999.

95. Luo, L.Y. and Diamandis, E.P., Preliminary examination of time-resolved fluorometry for protein array applications, *Luminescence*, 15(6), 409–413, 2000.

96. Borrebaeck, C.A., Antibodies in diagnostics — from immunoassays to protein chips, *Immunol. Today*, 21(8), 379–382, 2000.

97. Dutt, M.J. and Lee, K.H., Proteomic analysis, *Curr. Opin. Biotechnol.*, 11(2), 176–179, 2000.

98. Kononen, J. et al., Tissue microarrays for high-throughput molecular profiling of tumor specimens, *Nat. Med.*, 4(7), 844–847, 1998.

99. Schraml, P. et al., Tissue microarrays for gene amplification surveys in many different tumor types, *Clin. Cancer Res.,* 5(8), 1966–1975, 1999.

100. Kawanishi, Y. et al., Pharmacogenomics and schizophrenia, *Eur. J. Pharmacol.,* 410(2–3), 227–241, 2000.

101. Destenaves, B. and Thomas, F., New advances in pharmacogenomics, *Curr. Opin. Chem. Biol.,* 4(4), 440–444, 2000.

102. Korade-Mirnics, Z. et al., DNA microarray analysis of *g-csf* dependent, *lyn*-dependent genes, *Exp. Hematol.,* 28(12), 1497, 2000.

Section 4

9 Full-Length cDNA Libraries: Reagents for Functional Studies of the Nervous System

Munetomo Hida, Yutaka Suzuki, and Sumio Sugano

CONTENTS

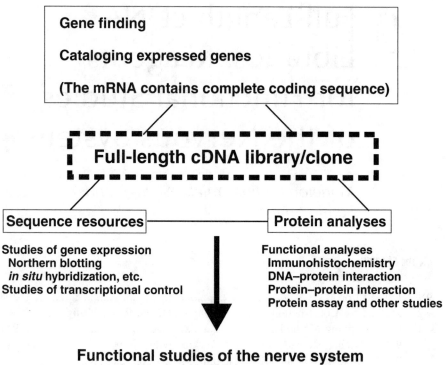

FIGURE 9.1 Analyses using full-length cDNA libraries and clones.

9.1 INTRODUCTION

The genomic revolution manifested by completion of human and other mammalian genome sequences and the ready public access to complementary DNA (cDNA) libraries, sequences, and clones will impact greatly on conceptual approaches and experimental strategies for our understanding of nervous system function and complex behaviors. While the development of systematic cDNA resources utilizing the expressed sequence tag (EST) strategy has led to discovery of EST clones and generation of nonredundant collection of genes found in the nervous system, cDNA libraries enriched in full-length cDNA clones are often required for assessing gene functions in their full biological context. A "full-length cDNA library" contains a significantly higher percentage of full-length DNA clones than a cDNA library constructed by standard methods.[1] The full-length cDNA serves as effective material for functional analysis of a gene (Figure 9.1), e.g., recombinant protein synthesis,[2,3] protein characterization[4–7] of newly isolated genes, generation of knockout mice,[8,9] and expression profiling by microarray analysis.[10]

In many cases, full-length cDNA is not available because the largest part of the cDNA library constructed by the standard methods represents non-full-length cDNA that usually lacks the 5′-end of the messenger RNA (mRNA). At the time of the early planning stages of the Human Genome project in the late 1980s, the value of cDNA libraries, EST sequences, and clones was clear; however, it was also clear that the development of an annotated and complete catalog of full-length cDNA required advances in methodology and strategy, as well as improved reagents. Substantial advances in cDNA technology during the past few years have occurred. This chapter discusses an efficient approach by which representative full-length coding sequences and clones for longer mammalian gene transcripts may be obtained.

There are several critical factors in the process of synthesis and cloning of full-length cDNA. The starting mRNA needs to be intact and free of any contaminating genomic DNA and heterogeneous nuclear RNA, since mRNA degraded in the isolation and preparation steps yields truncated first strand cDNA. Alternatively, they are derived from incompletely synthesized cDNA during the first strand synthesis mainly caused by secondary structures in the mRNA. In the first strand synthesis, RNA reverse transcriptase tends to stop during the synthesis and fall short of the 5′-end. The contamination of non-full-length cDNA in cDNA libraries is an inevitable result of the fragile nature of mRNA and the enzymatic properties of RNA reverse transcriptase. Another important contributing factor is the differential cloning efficiency of full-length cDNA as compared with the truncated cDNA. Thus, full-length cDNA molecules generated during cDNA synthesis are not often faithfully represented after the cloning step.

Several methods have been developed for constructing a full-length cDNA library and selecting for the completely synthesized cDNA from a large amount of the truncated ones (see References 15–19). These methods take advantage of the unique features found in both the 3′- and the 5′-ends of mRNA. As for the 3′-end, mRNA has a poly(A)+ stretch.[11] By using oligo dT primer for the first strand cDNA synthesis, 3′-end complete mRNA must be selectively synthesized. Furthermore, all cellular eukaryotic mRNA, with the exception of organelle mRNA, possess a 5′ terminal m7GpppN (where N is any nucleotide) structure, called the CAP.[12] The CAP structure is added in the nucleus during early transcriptional step by RNA polymerase II, and is critical for mRNA biosynthesis and subsequent processing. Although a number of CAP binding proteins have been identified in the cytoplasm[13] and in the nucleus,[14] the CAP structure, unlike the poly(A)+ tail, cannot be used to select 5′-end complete mRNA directly unless the CAP structure is first replaced with a sequence tag.

We have developed a novel method to enzymatically replace the CAP structure of mRNA with a synthetic RNA oligonucleotide (5′-oligo), which we named "oligo capping."[15] Taking the mRNA tagged with 5′-oligo as starting material, a new procedure to construct a full-length cDNA library was developed. Based on the scheme shown in Figure 9.2, the cDNA containing both the 3′-end (polyA) and 5′-end (5′-oligo) sequence tag is selectively cloned.

Other groups have also presented several methods to construct a full-length cDNA library. Kato et al.[16] combined the oligo capping[15] and Okayama-Berg[17] methods using DNA–RNA chimeric oligos for the CAP replacement. Edery et al.[18] used the CAP binding protein to select full-length cDNA (CAP Retention Procedure). Carninci et al.[19] chemically modified and biotinylated the CAP structure ("CAP Trapper"). Both of the latter two methods make use of the CAP-dependent retention of the full-length cDNA to the solid supports.

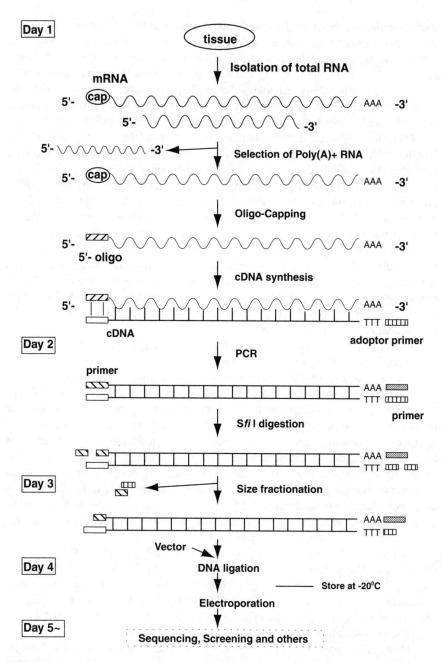

FIGURE 9.2 Scheme to construct full length-enriched cDNA libraries using the oligo capping method.

We now describe the detailed protocol for constructing full-length cDNA libraries, and demonstrate the utility of our full-length cDNA libraries for the analysis of genes expressed in various regions of *Macaca* monkey brain.

9.2 METHOD FOR CONSTRUCTION OF A FULL-LENGTH cDNA LIBRARY USING THE OLIGO CAPPING METHOD

9.2.1 SPECIAL CONSIDERATIONS FOR WORKING WITH RNA

When handling tissue and equipment in the isolation of high-purity intact RNA for full-length cDNA library construction, it is extremely critical to observe laboratory practices that minimize or eliminate contamination by ribonucleases (RNases). Unlike deoxyribonucleases (DNases), which require metal ions for activity and can therefore easily be inactivated with agents such as EDTA, RNases are single-strand specific endoribonucleases that are active over a wide pH range, resistant to metal chelating agents, and can survive prolonged boiling or autoclaving.[20]

The following guidelines will help to prevent accidental contamination of samples with RNase, allowing the isolation of intact mRNA. A separate work area needs to be maintained with a set of materials used exclusively for mRNA isolation, i.e., pipettes, pipette tips, Eppendorf tubes, buffer solutions, glassware, and reagents. Gloves are to be worn at all times and precautions taken to avoid RNase contamination from skin and airborne particles when opening and closing reagent tubes. Metal tools like spatulas should be inactivated by holding in a burner flame for 1 min prior to use, and all glassware should be baked at 180°C or higher for 3 or more hr. RNase-free solutions and buffers not containing primary amino group (e.g., Tris) can often be prepared using H_2O treated with a 0.1% solution of diethyl pyrocarbonate (DEPC). RNase A-type enzymes rely on histidine residues within the active site for catalytic activity and can be inactivated by alkylating DEPC.[21] DEPC is a potential carcinogen and should be handled in a chemical fume hood. Although it can be decomposed into CO_2 and ethanol by autoclaving, a minute trace amount of residual DEPC in the solutions may inhibit enzyme reactions. We have found DEPC treatment is not necessary for our RNA work; therefore, we prepare all reagents with sterile MilliQ water.

9.2.2 OVERVIEW OF THE OLIGO CAPPING METHOD

The scheme to construct a full-length cDNA library using the oligo capping method is shown in Figure 9.3. First, total RNA is isolated from tissue (or cultured cells) and poly(A)$^+$ RNA is purified from total RNA. Using isolated poly(A)$^+$ RNA, the

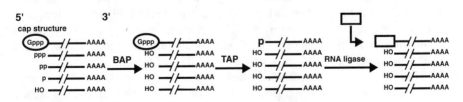

FIGURE 9.3 Oligo capping procedure: RNA molecules are represented as solid lines and 5′-oligo as boxes. Poly(A)$^+$ RNA consists of RNA molecules with various types of 5′-ends as shown at the left margin. Gppp: cap structure, p: phosphate, OH: hydroxyl.

CAP (7 - methylated GTP)

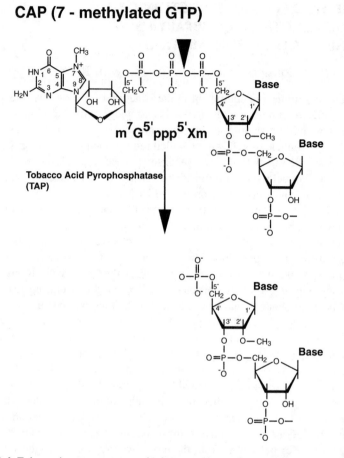

FIGURE 9.4 Eukaryotic cap structure and tobacco acid pyrophosphatase (TAP) activity. TAP hydrolyzes the eukaryotic cap structure at the position suggested by the arrowhead.

CAP structure with the synthetic 5′-oligo is replaced after three steps of enzyme reactions.

Bacterial alkaline phosphatase (BAP) hydrolyzes the phosphate from the 5′-ends of truncated mRNA, which are non-capped. The CAP structure itself remains intact during this reaction. Then, tobacco acid pyrophosphatase (TAP) cleaves the CAP structure itself at the position indicated by the arrowhead in Figure 9.4, leaving a phosphate at the 5′-end.[22] Finally, T4 RNA ligase selectively joins the synthetic oligonucleotide to the phosphate at the 5′-end. As a result, the oligonucleotide would be introduced only to the 5′-ends, which originally had the CAP structure.

With the oligo-capped mRNA as starting material, first strand cDNA is synthesized with the reverse transcriptase enzyme and a dT adapter primer. After first strand cDNA synthesis, the template mRNA is alkaline-degraded and the synthesized first strand cDNA is PCR-amplified. The PCR 5′-primer is the cap-replaced 5′-oligo sequence, and the 3′-primer has a part of the dT adapter primer sequence. The amplified cDNA fragments are digested with a restriction enzyme, size fractionated, and cloned into a plasmid vector.

9.2.3 DETAILED PROTOCOLS FOR THE OLIGO CAPPING METHOD

The total time required for construction of a cDNA library following this procedure is 5 or 6 days. Although the procedure can be paused at any time, we do not stop the procedure until the first strand cDNA synthesis is completed. Because the oligo capping method consists of multistep enzyme reactions with long reaction times, the uttermost care must be taken that all the reagents are prepared in an RNase-free condition. The pH of each reagent should also be accurately adjusted. The list of reagents and buffers are specified in Table 9.1.

9.2.3.1 Experimental Procedures on Day One
        (approximately 15 Hours)

9.2.3.1.1 Extraction of Total RNA from Brain Tissue

RNA isolation is one of the most important steps in the construction of a full-length cDNA library, and the starting RNA material must be of the highest quality obtainable. The protocol for total RNA extraction from brain tissue is presented in Table 9.2. Whenever possible, fresh tissue should be used. Due to a high lipid content in the brain tissue, it is difficult to obtain a clean separation of RNA from other contaminating subcellular fractions; therefore, extra care should be given to the RNA extraction process in order to isolate intact RNA from the brain.

Our method of total RNA isolation combines the use of the two commercially available RNA preparation reagents — Trizol (Life Technologies) and RNeasy (Qiagen). The quality of RNA isolated using Trizol alone and a combination of Trizol and RNeasy is assessed by gel electrophoresis in Figures 9.5a and b, respectively. While it is a convenient RNA isolation method for a wide variety of tissues, the AGPC (acid guanidium-phenol-chloroform) method[23] of isolation (commercially available as Trizol and Isogen) yields total RNA containing fragmented RNA, genomic DNA, and the lipids. Use of columns supplied in the RNeasy kit removes such unfavorable unwanted fractions. The NP-40 method[21] is another approach to RNA isolation that may be utilized by employing 1×10^7 to 5×10^7 cultured cells as an RNA source instead of tissue samples. The NP-40 method[21] will only give cytoplasmic RNA.

9.2.3.1.2 Selection of poly(A)+ RNA

A typical mammalian cell contains 10–30 pg of RNA, most of which is localized in the cytoplasm. More than 90% of the RNA is rRNA, tRNA, and small nuclear RNA. Usually 1–5% of the total RNA represents mRNA, which is heterogeneous in both size and sequence.

Table 9.3 presents a protocol by which affinity chromatography on oligo(dT)-cellulose columns is used to separate poly(A)+ RNA from the bulk of cellular RNA. When working with a small amount of the tissue, oligo(dT)-cellulose is preferable to other reagents such as Oligo-Tex (Nippon-Roche, Tokyo, Japan). Many commercial kits for poly(A)+ RNA selection procedure use latex or magnetic beads for the

TABLE 9.1
Materials and Reagents for Full-Length cDNA
Library Construction

Materials

1. Total RNA extraction kit: Trizol (Life Technologies, Rockville, MD) and RNeasy (Qiagen)
2. Oligo(dT)-cellulose (Collaborative, Bedford, MA) for poly(A)$^+$ RNA isolation.
3. Ethachinmate (WAKO); a carrier for the ethanol precipitation.
4. 3M Sodium acetate, pH adjusted to 5.5 with CH_3COOH.
5. RNasin (40 U/µl Promega, Madison, WI); an RNase inhibitor.
6. Bacterial alkaline phosphatase [BAP] (0.25 U/µl; TaKaRa).
7. Tobacco acid pyrophosphatase [TAP] (20 U/µl; TAP can be purified from tobacco cells, BY-2, as described[22] or obtained commercially (Nippon Gene Inc., Tokyo, Japan).
8. T4 RNA ligase (25 U/µl; TaKaRa).
9. 24 mM ATP.
10. 50 mM $MgCl_2$.
11. 50% (w/v) PEG 8000 (Sigma). Add ddH_2O to PEG 8000 so that the w/v is 50%. Dissolve PEG 8000 at 65°C. and sterilize the solution by filtration through the Millipore membranē (ø = 0.20 µm).
12. DNase I (RNase-free).
13. Spin Column S-400HR (Pharmacia Biotech).
14. Superscript II (200 U/µl; Life Technologies).
15. 5X First-Strand Synthesis Buffer (Life Technologies).
16. 1 M DTT (Life Technologies).
17. 5 mM dNTP.
18. 2.5N NaOH.
19. 7.5 M NH_4OAc (Ammonium Acetate).
20. Gene Amp PCR kit from Perkin-Elmer.
21. *Sfi* I (20 U/µl; New England Biolabs).
22. *Dra* III (6.0 U/µl; WAKO, Tokyo, Japan).
23. Gene Clean II (Bio-101).
24. DNA Ligation Kit Ver.1 (TaKaRa).
25. 5′-oligoribonucleotide A: 5′ — AGC AUC GAG UCG GCC CUU GUU GGC CUA CUG G — 3′ (100 ng/µl).
26. Oligo(dT) adapter primer B: 5′ — GCG GCT GAA GAC GGC CTA TGT GGC CTT TTT TTT TTT TTT TTT — 3′ (5 pmol/µl).
27. PCR 3′-primer C: 5′ — AGC ATC GAG TCG GCC TTG TTG — 3′ (10 pmol/µl).
28. PCR 5′-primer D: 5′ — AAC ACC AGC AGC AAC AAT CAG — 3′ (10 pmol/µl).
29. pME18S-FL3; plasmid vector.

TABLE 9.1 (continued)
Materials and Reagents for Full-Length cDNA Library Construction

Reaction Buffers

2× Loading Buffer		(Final conc.)
SDS (Sodium dodecylsulfate)	0.8 g	(2%)
NaCl	2.34 g	(1 M)
1 M Tris-HCl (pH7.0)	1.6 ml	(4 mM)
0.5 M EDTA (pH8.0)	160 µl	(0.2 mM)
ddH$_2$O	25 + X ml	
Total	40 ml	

5× BAP Buffer		(Final conc.)
1 M Tris-HCl (pH7.0)	250.0 µl	(500 mM)
14 M 2-mercaptoethanol	1.8 µl	(50 mM)
ddH$_2$O	248.2 µl	
Total	500.0 µl	

5× TAP Buffer		(Final conc.)
3 M Sodium acetate (pH5.5)	41.7 µl	(250 mM)
14 M 2-mercaptoethanol	1.8 µl	(50 mM)
0.5 M EDTA (pH8.0)	5.0 µl	(5 mM)
ddH$_2$O	451.5 µl	
Total	500.0 µl	

10× Ligation Buffer		(Final conc.)
1 M Tris-HCl (pH7.0)	250.0 µl	(500 mM)
14 M 2-mercaptoethanol	1.8 µl	(50 mM)
ddH$_2$O	264.6 µl	
Total	500.0 µl	

10× STE		(Final conc.)
1 M Tris-HCl (pH7.0)	100.0 µl	(100 mM)
5 M NaCl	200.0 µl	(1 M)
0.5 M EDTA (pH8.0)	5.0 µl	(5 mM)
ddH$_2$O	680.0 µl	
Total	1000.0 µl	

oligo-dT supports; however, purification of high-quality poly(A)$^+$ RNA is often difficult. We first hydrate and equilibrate oligo(dT)-cellulose powder, and then proceed to pack a separation column. This procedure affords us the flexibility to adjust the column bed volume and the washing conditions according to the quality and

TABLE 9.2
Protocol 1 — Extraction of Total RNA from Brain Tissue

Trizol Step

1. Put 20 ml of Trizol reagent in a sterile 50-ml centrifuge tube.
2. Add 2–3 g of tissues and homogenize the sample in Trizol reagent using the Polytron which has been washed with 6 times with ddH$_2$O and EtOH.
3. Centrifuge 12,000 × g, 20 minutes, 4°C.
4. Decant upper aqueous phase into a new tube.
   Note: After centrifugation, white flocculent materials will make up most of the volume of the aqueous phase. These white materials likely contain lipids and do not form a tight interface. The contamination of lipids disturbs the isolation of clean and intact RNA. Be careful not to transfer the white material.
5. Add 4 ml of chloroform, shake vigorously for 15 s, and let it stand for 3 min at room temperature (RT).
6. Centrifuge 30 min, 12,000 × g, 4°C.
7. Transfer the supernatant to new tube.
8. Add 10 ml of isopropanol, mix well, and let it stand for 10 min at RT.
   Note: Store at –80°C as EtOH precipitate unless you do the next step immediately.
9. Centrifuge 12,000 × g, 20 minutes, 4°C and save the RNA pellet and discard the supernatant.

RNeasy Step

1. Dissolve the RNA pellet in 2 ml of ddH$_2$O on ice.
2. Immediately add 15 ml RLT buffer (RNeasy kit) and add 15 ml of 70% EtOH into the tube and mix vigorously for 30 s.
3. Apply a half of the solution to a RNeasy column in a sterile 50-ml tube.
4. Close tube, and centrifuge for 3 min at 3000–5000 × g at RT.
5. After centrifugation, dump the flow-through.
6. Apply the remaining solution to the RNeasy column in the tube.
7. Close tube, and centrifuge for 5 min at 3000–5000 × g at RT.
8. After centrifugation, discard the flow-through.
9. Add 15 ml RWI buffer (RNeasy kit) to the column, close tube, centrifuge for 5 min at 3000–5000 × g at RT, and discard the flow-through.
10. Add 10 ml RPE buffer (RNeasy kit) to the column, close tube, centrifuge for 2 min at 3000–5000 × g at RT, and discard the flow-through fraction.
11. Repeat Step 10 by adding another 10 ml RPE buffer to the column, close tube, and centrifuge for 10 min at 3000–5000 × g at RT, and discard the flow-through fraction.
12. To elute, add 1.2–1.5 ml of ddH$_2$O to the column in new 50-ml tube.
13. Close tube, and let it stand for 1 min at RT.
    Note: At this step, start preparing the selection of poly(A)$^+$ RNA (Steps 1 to 5).
14. Centrifuge for 3 min at 3000–5000 × g at RT, and collect RNA solution in the tube.
    Note: The electrophoresis of the isolated total RNA solution (5 μl) and the measurement of OD$_{260}$ of the RNA solution diluted 250-fold are recommended for quantification and assessing the quality of the RNA (Figure 9.5b).

quantity of the total RNA. Figure 9.5c shows RNA isolated after separation of poly(A)$^+$ RNA by oligo(dT)-cellulose column chromotography.

9.2.3.1.3   Oligo Capping Procedure[15]

The oligo capping procedure described in Table 9.4 consists of multistep enzymatic reactions. Each step should be carried out on ice to avoid RNA degradation. In order

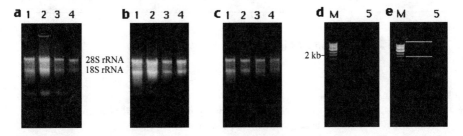

FIGURE 9.5 Example of electrophoresis of isolated RNA and PCR products. In (a), (b), and (c), the bands representing 28s and 18s ribosomal RNA are indicated. Lanes contain RNA from the following tissues: 1, human small intestine; 2, monkey liver; 3, monkey heart; 4, dog brain (neocortex). In (d) and (e), lane 5 contains RNA from monkey brain (brain stem), and lane M contains molecular markers. Total RNA was isolated using Trizol (a) and Trizol-RNeasy (b); (c) Selection of poly(A)⁺ RNA by oligo(dT)-cellulose is decreased; (d) after PCR for amplifying the first strand cDNA, the PCR products were observed as a smear; (e) the fraction of PCR products longer than 2 kb were recovered in the step of size fractionation.

to maximize the yield and fidelity of the oligo capping, complete recovery of the aqueous layer after phenol:chloroform extraction should not be attempted; proteins, lipids and carbohydrates often located in the intermediate layer as floating particles will inhibit the subsequent enzymatic reactions. After ethanol precipitation, dissolve the RNA pellet completely since undissolved RNA remaining in the solution will lead to cloning of short cDNA.

Although BAP exhibits the highest enzymatic activity at 65°C, at this temperature the effect of RNase inhibitor diminishes and the activity of RNase increases. Thus we recommend that the BAP reaction be carried out at 37°C in the presence of an excess amount of BAP. The reaction buffer accompanying the enzyme should not be used, because RNA is easily hydrolyzed at high pH (the buffer supplied by TaKaRa has pH 9).

TAP is commercially available from several suppliers (Nippon Gene; No. 313-04021, Wako; No. 201-12981, Epicenter; No. T19050), but we find some lots of commercially available TAP are highly contaminated with endonucleases and RNases. The endonuclease hydrolyzes RNA, leaving the phosphate at the 5′-end; this serves as an acceptor of the 5′-oligo at the RNA ligation step. The CAP structure of the mRNA at the 5′-end is replaced with the 5′-oligo, as described in Figure 9.4.

9.2.3.1.4   First Strand cDNA Synthesis

We and other laboratories[24,25] have made modifications in the protocol to improve the efficiency of the reverse transcriptase reaction, including the first strand cDNA synthesis at 70°C using the Thermoscript (Life Technologies) enyzme. The protocol is shown in Table 9.5. However, we have not observed a significant difference between the reverse transcriptases Super Script II and Thermoscript. We use 10 pmols of the oligo-dT primer in this protocol, since a higher concentration of the primer increases internally primed cDNA products, e.g.,. the cDNA lacking the 3′-end of the mRNA. Figure 9.5d shows the results of first strand cDNA synthesis, as visualized by gel electrophoresis.

TABLE 9.3
Protocol 2 — Selection of Poly(A)$^+$RNA

Reagents

Cool 100% EtOH at –20°C or lower.

0.1 N NaOH: Dilute 0.4 ml of 2.5 N NaOH into 9.6 ml of ddH$_2$O in a sterile plastic tube.

1 × loading buffer: Dilute 2 × loading buffer with ddH$_2$O in a sterile tube.

Procedure

1. Suspend 0.5–1.0 g of oligo(dT)-cellulose in 10 ml ddH$_2$O.
2. Let it stand for 1–5 min at RT, remove the supernatant, and resuspend the oligo(dT)-cellulose in 1 ml ddH$_2$O.
3. Pour a column of oligo(dT)-cellulose (0.4–0.6 ml packed volume) in a sterile Dispocolumn (Bio-Rad).
4. Add 5 ml (1 ml × 5 times) of 0.1 N NaOH to the column.
5. Wash and neutralize the column with 5 ml (1 ml × 5 times) of sterile 1 × loading buffer. Equilibrate the oligo(dT)-cellulose column by washing 1 × loading buffer until the pH of the flow-through is lower than 8.0 to avoid RNA degradation at an alkaline pH.
6. Add equal amount of 2 × loading buffer into the isolated total RNA solution. After dissolving the RNA, you may heat the solution to 65°C for 2 min and subsequently cool the solution. Heating the RNA solution disrupts regions of secondary structure that might involve the poly(A)$^+$ tail.
7. Apply the RNA solution to the column and collect the elute in a sterile tube. Proceed quickly from Step 7 to Step 11 (within 20 min) to minimize RNA degradation during these steps.
8. When all of the solution has eluted, reapply the elute to the column. Again collect the flow-through fractions. Before reapplying, heat the solution to 65°C for 2 min.
9. Repeat Step 8 twice.
10. Wash the column with 5 ml (1 ml × 5 times) of 1 × loading buffer.
11. Elute the poly(A)$^+$ RNA from the oligo(dT)-cellulose with 3 ml (1 ml × 3 times) of sterile elution buffer. Collect the flow-through fraction in a sterile plastic tube containing 8 ml of pre-chilled EtOH and 330 μl of 3 M NaOAc.
12. After all of the solution has eluted, mix well, ethanol precipitate the RNA by centrifuging the tube and wash the pellet with 80% EtOH.
13. Resuspend the RNA pellet in 67.3–80 μl of ddH$_2$O on ice.
14. The electrophoresis and the determination of the OD260 of the isolated RNA solution are done for the quantitative estimation and the quality assessment of the RNA (Figure 9.5c).
15. Oligo(dT)-cellulose can be recycled. Prior to use it is rinsed in 0.1 M NaOH once, then in binding buffer twice. After use it is rinsed in 0.1 M NaOH once, then in elution buffer twice, then stored in elution buffer at 4°C.

9.2.3.2 Experimental Procedures on Day Two (approximately 5 Hours)

9.2.3.2.1 Alkaline Degradation of the Temperate mRNA

The main purpose of this procedure is to degrade the RNA template that may interfere with the subsequent PCR reaction step. The full protocol is shown in Table 9.6.

TABLE 9.4
Protocol 3 — the Oligo Capping Method

a. BAP Treatment to Remove the Phosphate at the 5′-End of the mRNA

1. Set up the following reaction with 100 to 200 µg of purified poly(A)$^+$ RNA in a total volume of 100.0 µl: 67.3 µl poly(A)$^+$ RNA/ddH$_2$O, 20.0 µl 5× BAP buffer, 2.7 µl RNasin (40 U/µl), and 10.0 µl BAP(0.25 U/µl)
2. Incubate the reaction mixture at 37°C for 60 min.
3. Extract the solution with phenol:chloroform (1:1).
   Note: The phenol:chloroform extraction is performed to remove proteins (e.g., enzyme). Don't be tempted to completely remove the aqueous layer, to transfer any of the flocculent material at the interface.
4. Repeat the phenol:chloroform (1:1) extraction one time to remove residual BAP.
5. Precipitate the RNA with ethanol.
   a. Add 2.5 volume of ethanol in the presence of 0.3 M NaOAc (pH 5.5) and 1 µl of Ethachinmate as a carrier and mix well.
   b. Centrifuge the sample at 4°C for 10 min.
   c. Remove the supernatant and rinse the pellet with 150 µl of 80% ethanol. Drying up the pellet is not necessary as the less dry the pellet, the easier it is to solubilize RNA.
   Note: The wash with 80% ethanol is performed to remove any residual salt from the pellet. The salts but not RNA/DNA will be soluble in 80% ethanol.
6. Dissolve the RNA pellet in 75.3 µl of ddH$_2$O on ice.

b. TAP Treatment to Cleave the CAP Strucuture

1. Set up the reaction with BAP-treated poly(A)$^+$ RNA as follows:
   75.3 µl RNA solution (from the Step *a*-6), 20.0 µl 5× BAP buffer, 2.7 µl RNasin (40 U/µl), and 2.0 µl TAP (0.25 U/µl) in a total volume of 100.0 µl
2. Incubate at 37°C for 60 min.
3. Extract the solution with phenol:chloroform (1:1).
4. Ethanol precipitate the RNA.
   After washing with 80% ethanol, remove securely the supernatant with a pipette, since the volume of the following solution for resolving the RNA is small. Don't dry the pellet in a vacuum centrifuge (e.g., speed vac).
5. Dissolve the RNA pellet in 11.0 µl of ddH$_2$O on ice.

c. Ligation of the BAP/TAP-Treated Poly(A)$^+$ RNA with 5′-Oligo

1. Set up the reaction with BAP/TAP-treated poly(A)$^+$ RNA as follows:
   11.0 µl RNA solution, 4.0 µl 5′-oligoribonucleotide (100 ng/µl), 10.0 µl 10× ligation buffer, 10.0 µl 50 mM MgCl$_2$, 2.5 µl 24 mM ATP, 2.5 µl RNasin (40 U/µl), 10.0 µl T4 RNA ligase (25 U/µl), and 50.0 µl PEG 8000 (50% w/v) in a total volume of 100.0 µl
2. Incubate at 20°C for 3 h.
3. Add 200.0 µl of ddH$_2$O.
4. Extract the solution with phenol:chloroform (1:1).
5. Ethanol precipitate the RNA.
6. Dissolve the RNA pellet in 77.3 µl of ddH$_2$O on ice.

TABLE 9.4 (continued)
Protocol 3 — the Oligo Capping Method

d. DNase I T(RNase-Free) Treatment of the "Oligo-Capped" mRNA to Degrade the Remnant of Genomic DNA

1. Set up the reaction in a total volume of 100.0 µl as follows:
   77.3 µl the "Oligo-Capped" mRNA solution, 16.0 µl 50 mM $MgCl_2$, 4.0 µl 1 M Tris-HCl (pH 7.0), 5.0 µl 0.1 M DTT, 2.7 µl RNasin (40 U/µl), and 2.0 µl DNase I (5.0 U/µl).
2. Incubate at 37°C for 10 min.
3. Extract the solution with phenol:chloroform (1:1).
4. Ethanol precipitate the RNA.

e. Spin-Column Purification

1. Dissolve the "Oligo-Capped" mRNA in 45 µl of ddH_2O and add 5 µl of 10× STE.
2. Pass the RNA solution through the spin column (S400-HR) to remove the excess 5'-oligo.
3. Ethanol precipitate the RNA.
4. Dissolve the RNA pellet in 21.0 µl of ddH_2O on ice.

TABLE 9.5
Protocol 4 — cDNA Synthesis with RNase H-Free Reverse Transcriptase (Super Script II)

1. Set up the reaction mixture in a total volume of 50.0 µl as follows:
   21.0 µl the "Oligo-Capped" mRNA, 10.0 µl 10× First Strand buffer, 8.0 µl 5 mM dNTP, 6.0 µl 0.1 M DTT, 2.0 µl oligo-dT adapter primer (5 pmol/µl), 1.0 µl RNasin (40 U/µl), and 2.0 µl RTase I (Super Script II).
   Use of random primers in this step will give 5'-enriched cDNA libraries.[24]
2. Incubate at 42°C for more than 3 h. In order to avoid the misannealing of dT primers, do not incubate at lower temperature.
   Setting the extension time longer than usual will ensure that reverse transcription reaction goes to completion.

Remaining 5'-oligos and the temperate mRNA serve as a PCR primer in the PCR reaction. Some DNA–RNA heteroduplexes consisting of the first strand cDNA and the template RNA may not be denatured sufficiently to single strand molecules at 94°C in the PCR, because DNA–RNA heteroduplexes are more stable than DNA–DNA duplexes. Thus it is critical to remove any RNA complementary to the cDNA molecules before PCR amplification.

9.2.3.2.2 PCR Amplification, Size Fractionation, and Cloning of the cDNA Fragments

The synthesized first strand cDNA are amplified by PCR; however, PCR mutations can be introduced during the PCR reaction. We estimate the rate of the deletion/insertion

TABLE 9.6
Protocol 5 — Alkaline Degradation of the Temperate mRNA

Procedure

1. Extract the reaction mixture from Day One with phenol:chloroform (1:1).
2. Add 2 μl of 0.5 M EDTA (pH 8.0) to stop the reaction thoroughly.
3. In order to degrade the template RNA, add 15 μl of NaOH (0.1 M) and heat the solution at 65°C for 60 min.
4. Add Tris-HCl (1 M, pH 7.0) to neutralize the solution.
5. Precipitate the first strand cDNA with ethanol. For this step use ammonium acetate to remove the fragmented nucleic acid.
   a. Add 2.5 volume of ethanol in the presence of 2.5 M NH_4OAc and 1 μl of ethachinmate as a carrier.
   b. Centrifuge the sample at 4°C for 10 min.
   c. Remove the supernatant and rinse the pellet with 150 μl of 80% ethanol. Drying up the pellet is not necessary.
      Note: Do not use the ammonium ion until RNA ligation is finished because ammonium ion interferes the enzymatic activity of RNA ligase.
6. Dissolve the cDNA pellet in 50.0 μl of ddH_2O.
   Note: This cDNA solution is useful to detect the 5′-end of an mRNA by PCR with the primer D and a gene-specific primer.

mutation and that of the substitution mutation at 1/8000 bp and 1/1000 bp, respectively. The full protocol is described in Table 9.7.

9.2.3.3 Experimental Procedures on Day Three (approximately 3 Hours)

As shown in Figure 9.5e, cDNA fractions containing cDNA longer than 2 Kb were excised from the gel and extracted using a Gene Clean II kit (Bio101). The detailed protocol for DNA size fractionation is described in Table 9.8.

9.2.3.4 Experimental Procedures on Day Four (approximately 4 Hours)

Size-fractionated cDNA is inserted into the *Dra* III-digested pME18S-FL3 plasmid vector (Genbank Acc. No. AB009864) following the protocol presented in Table 9.9. This vector has the following advantages: (1) pME18S-Fl3 contains a eukaryotic promoter, SR alpha, upstream to the cloning site; therefore, a cloned cDNA can be expressed when transfected into the cultured eukaryotic cells; and (2) the cDNA can be directionally cloned in an orientation defined manner, since the *Dra* III sites located at the 5′-end and the 3′-end cloning site have different sequences.

The two main drawbacks of this method for constructing a full-length cDNA library are as follows:

1. PCR, which is used for the amplification of the first strand cDNA, sometimes introduces a mutation into the cDNA and frequently causes a strong

TABLE 9.7
Protocol 6 — PCR Amplification of the First Strand cDNA and Digestion of the PCR Amplified Products with *Sfi* I Restriction Endonuclease

1. Use ½ to ⅓ of the first strand cDNA solution for the PCR amplification.

   Set PCR reactions up as follows: 52.4 μl first strand cDNA solution and ddH$_2$O, 30.0 μl 3.3× Reaction buffer II, 8.0 μl 2.5 mM dNTP, 4.4 μl 25 mM Mg(OAc)$_2$, 1.6 μl Primer C, and 1.6 μl Primer D in a total volume of 98.0 μl.

   Store with the rest of the first strand cDNA solution at –20°C.

2. Add DNA taq polymerase rTth (2 μl) when the solution temperature becomes 94°C.

3. Perform PCR as follows:

   a. Initial denaturation at 94°C for 5 min.

   b. Amplification for 12-15 cycles. 94°C 1 min, 52°C 1 min, 72°C 10 min.

   c. Final extension at 72°C for 10 min.

4. Extract the solution with phenol:chloroform (1:1).

5. Ethanol precipitate the PCR products.

6. Dissolve the pellet in 87.0 μl of ddH$_2$O.

7. Digest the PCR products with *Sfi* I by incubating the reaction mixture at 50°C overnight as follows: 87.0 μl PCR products and ddH$_2$O, 10.0 μl 10 × NEB buffer 2, 1.0 μl 100× BSA, and 2.0 μl *Sfi* I in a total volume of 100.0 μl.

TABLE 9.8
Protocol 7 — Size Fractionation

Procedure

1. Extract the *Sfi* I-digested DNA solution with phenol:chloroform (1:1).

2. Ethanol precipitate the DNA.

3. Separate the *Sfi* I-digested PCR products by electrophoresis in an 1% TAE agarose gel. (See Figure 9.5d.)

4. Recover the fraction longer than 2 kb. (See Figure 9.5e.)

   Note: Use the long 365-nm ultraviolet (UV) wavelength of a transilluminator as a short-wavelength such as 312 nm causes DNA damages.

5. Crush the gel in 1 ml of NaI solution and incubate at 42°C for 15 min.

6. Add 20 μl of grass powder and Spin down for 30 s.

7. Wash the pellet with 0.8 ml of wash buffer.

8. Repeat the wash step once more.

9. After removing the debris thoroughly, resuspend the pellet with 100 μl of ddH$_2$O.

10. Incubate at 37°C more than 5 h. A long elution time will ensure a complete recovery of the cDNA fragments.

bias in the cDNA expression. The latter reflects differences in PCR efficiency among cDNA.

2. The cDNA was occasionally cut in the *Sfi* I digestion step, thereby leading to a loss of some cDNA in the library.

TABLE 9.9
Protocol 8 — Cloning of cDNA into a Plasmid Vector pME18S-FL3

Advance Preparation of the Cloning Vector
1. Digest the 10 µg of pME18S-FL3 in 80 µl of ddH$_2$O and 10 µl of 10× H buffer and 10 µl of 10× BSA and 2 µl of *Dra* III (6.0 U/µl) at 37°C for 6 or more hours. After the phenol:chloroform (1:1) extraction and the ethanol precipitation, repeat the reaction once more in order to reduce the remnant of the uncut vector.
2. Perform electrophoresis of the *Dra* III-digested DNA on an 1% agarose gel and purify the 3.0 kb-plasmid vector using GeneClean kit. Save the stuffer band (0.4 kb) to use as a control insert.
   Note: After incubating, load a portion of the sample with uncut plasmid vector DNA to check out the completeness of the digestion reaction If the reaction is incomplete, repeat the digestion.
3. Ligate the 10–50 ng of the cloning vector with the equal amount of the mock insert. Put the insert (–) control to estimate the background level of vector. Transform the *E. coli* with each ligate and compare the number of formed colonies between the insert (–) and (+). Until the ratio insert (+) to (–) becomes greater than 100:1, repeat the digestion and the purification of the vector. Store the vector solution at –20°C.

Cloning Procedure
1. Ethanol precipitate the eluted cDNA fragments.
2. Dissolve the pellet with 8 µl of ddH$_2$O.
3. Add 1 µl of *Dra* III-digested pME 18S-FL3.
4. Add 80 µl of solution A and 10 µl of solution B.
5. Incubate at 16°C for 3 h.
6. Extract the solution with phenol:chloroform (1:1).
7. Ethanol precipitate the DNA.
8. Dissolve the pellet with 30 µl of ddH$_2$O.
9. To evaluate the size of the library, transform the host *E. coli* with the cDNA and plate out a certain volume after 45 min of shaking at 37°C.
   Note: Use 1 µl per transformation of the cDNA solution. Usually the library size is 10^5–10^6 recombinant colonies for 200–300 µg of the poly(A)$^+$ RNA.

9.3 ANALYSIS OF FULL-LENGTH cDNA LIBRARIES

We have constructed 9 full-length cDNA libraries from different regions of the brains of cynomolgus monkeys (*Macaca fascicularis*) with the oligo capping method (Table 9.10), albeit with different procedures for RNA isolation and poly(A)$^+$ RNA selection. The average length of cDNA inserts was approx. 2.0 kb. Many clones were randomly isolated from these libraries and 13,162 expressed sequence tags (EST) from the 5′-end were obtained by one-pass DNA sequencing.

A homology analysis was performed using a BLAST search[25–27] against the NCBI GenBank.[28] About 60% of these EST corresponded to the known mRNA (mostly that of human), 19% matched with only the sequences in dbEST, and 14% matched with the sequences of others (e.g., pseudogenes) (Table 9.10). These EST were classified as "known," "EST," and "others," respectively. The rest of the EST (7%) that had no matches to the sequences in the database were classified as

TABLE 9.10

Representative Full-Length cDNA Clones in Monkey Brain Libraries Made by the Oligo Capping Method

Organ	Library Name #	Known #	Full/Near-full #	Not Full #	Full %	EST #	Others #	Unknown #	Total #
Brain (temporal lobe)	QntC	354	(153)	(201)	43.22	91	166	119	730
Brain (cerebellum cortex)	QccA	58	(25)	(33)	43.10	26	55	97	236
Brain (mesencephalon)	QmbA	17	(7)	(10)	41.18	7	16	19	59
Brain (hypophysis)	QphA	18	(8)	(10)	44.44	0	12	8	38
Brain (cerebral cortex)	QncA	54	(25)	(29)	46.30	7	15	3	79
Brain (parietal lobe)	QnpA	3834	(2552)	(1282)	66.56	911	993	486	6224
Brain (cerebellum cortex)	QccE	3481	(2262)	(1219)	64.98	1408	591	269	5749
Brain (temporal lobe)	QntD	16	(7)	(9)	43.75	5	4	9	34
Brain (whole brain)	QwbB	3	(2)	(1)	66.67	0	6	4	13
	7835				64.34	2455	1858	1014	13,162

Note: Sequencing from the QnpA and QccE libraries was performed by Osada, N., Kusuda, J., Tanuma, R., Iseki, K. and Hashimoto, K. at Division of Genetic Resources, National Institute of Infectious Diseases (http://www.nih.go.jp/yoken/genebank/gbjcrb.html). Column headings are explained in the text. Numbers in columns denoted by "#" represent frequencies; those denoted by "%" represent percentages.

"unknown." In the EST classified as "unknown," some clones had an open reading frame (http://www.nih.go.jp/yoken/genebank/QccE/QccE-FL.html#list). The three most abundant "known" EST are alpha-tubulin isotype M-alpha-4, elongation factor 1 alpha subunit (EF-1 alpha), and myelin basic protein (MBP).

Using the macaque EST data classified as known, we have estimated "fullness" of the EST data. We scored EST as full length if it had the same or the longer 5'-end than any other previously reported cDNA sequences. We scored the EST as near-full length, if it had shorter 5'-end but still contained the ATG initiation codon. If EST did not cover the initiator ATG, they scored as not full length. The percentage of full-length cDNA clones was about 64% (Table 9.10). The majority of full-length cDNA were less than 3,500 bp long, whereas some cDNA represented full-length messages of 11 kb.

Several groups also have reported novel methods to construct a full-length cDNA library.[16,18,19] Using mouse cDNA libraries constructed by "Cap Trapper,"[19] Hayashizaki's group in RIKEN (http://genome.rtc.riken.go.jp/) has collected approximately 20,000 novel mouse genes. We have not yet compared the advantage or disadvantage of our methods with this approach.

The QnpA, QccE, and QwbB libraries in Table 9.10 were constructed according to the method described in this chapter. In the construction of the remaining six libraries in Table 9.10, poly(A)[+] RNA extracted only with Trizol (without RNeasy) and purified with Oligo-Tex was used as starting material. The average proportion of full or near-full EST was higher in the three libraries constructed with the methods described in this chapter (66%) than in the remaining six libraries (44%).

9.4 DISCUSSION

Brain libraries constructed according to the methods described in this chapter (Table 9.10: QnpA, QccE, and QwbB) contained more full and near-full clones than those constructed using our previous protocols. The quality of the library, as assessed by the presence of nonredundant full-length clones, seems to be highly dependent on the quality of the isolated mRNA prior to application of our oligo capping method. While it worked well for constructing conventional cDNA libraries, the poly(A)[+] RNA isolated by these earlier protocols did not work well in our oligo capping protocol for full-length cDNA library construction. At present, only the combined use of Trizol and RNeasy results in isolation of poly(A)[+] RNA from the brain (at least Macaque brain) has worked well in our efforts to construct high-quality full-length cDNA libraries. We do not know the reason for this difference.

We usually recover cDNA fractions of greater than 2,000 bp. Following this procedure, size fractionation is possible to obtain cDNA of up to 3–4 kb long. Because it is difficult to clone the full-length cDNA of longer mRNA species all at once, we think the 5'-end enriched cDNA library is useful for this population.[26]

In the known class of EST, which corresponded to known mRNA, 64% of the EST in our brain cDNA libraries were scored as full or near-full length. This percentage is similar to that observed in other full-length cDNA libraries we prepared by our oligo capping method.[26] The insert size distribution of full clones in the libraries was also similar (data not shown). In addition, the brain cDNA libraries we constructed using the methods and procedures outlined in this chapter have a much higher content of full-length cDNA clones than the cDNA libraries constructed by standard methods. The 5' EST sequences that we generated are good resources for the analysis of upstream regions of the genes in the nervous system. The identification of transcription start site(s) is in progress (manuscript in preparation), and will provide valuable information on the structural features within the promoter region of these genes once the genomic sequence becomes available.

The preparation and analysis of full-length cDNA libraries provide us with a repertoire of the entire mRNA sequences and the corresponding full-length cDNA clones. These are useful reagents to accelerate our understanding of the underlying biology of the mammalian nervous system. For example, full-length cDNA and clones may be used to map genes to specific chromosome using florescent *in situ* hybridization, as well as analyze quantitative and qualitative alterations in genes expressed during development and during pathophysiologic states.[19-21,29-31] Full-length cDNA clones will constitute an extremely valuable reagent set for use in the massively parallel analysis of protein expression. Global proteomics in turn will be a prerequisite for studying structural motifs important for protein–protein interactions in cellular signal transduction pathways that are involved in nervous system function and complex behavior, and for elucidating molecular targets for discovery of therapeutic agents for diseases of the nervous system.

REFERENCES

1. Gubler, U. and Hoffman, B.J., A simple and very efficient method for generating cDNA libraries, *Gene,* 25, 263, 1983.
2. Coligan, J.E., et al., *Current Protocol in Protein Science,* John Wiley & Sons, New York, 1999.
3. Harlow, E. and Lane, D., *Using Antibodies: A Laboratory Manual,* Cold Spring Harbor Laboratory Press, Cold Spring Harbor, NY, 1999.
4. Sefton, B. and Hunter, T., *Selected Methods in Enzymology: Protein Phosphorylation,* Academic Press, 1998.
5. No authors listed, Protein phosphorylation. Part B. Analysis of protein phosphorylation, protein kinase inhibitors, and protein phosphatases, *Methods Enzymol.,* 201, 1, 1991.
6. Nishimune, A. et al. Detection of protein-protein interactions in the nervous system using the two-hybrid system, *Trends Neurosci.,* 19, 261, 1996.
7. Fire, A. et al., Potent and specific genetic interference by double-stranded RNA in *Caenorhabditis elegans, Nature,* 39, 806, 1998.
8. Hogan, B. et al., *Manipulating the Mouse Embryo: A Laboratory Manual,* 2nd ed., Cold Spring Harbor Laboratory Press, Cold Spring Harbor, NY, 1994.
9. Tsien, J.Z. et al., Subregion- and cell type-restricted gene knockout in mouse brain, *Cell,* 87, 1317, 1996.
10. Kurian, K.M., Watson, C.J., and Wyllie, A.H., DNA chip technology, *J. Pathol.,* 187, 267, 1999.
11. Jackson, R.J. and Standart, N., Do the poly(A) tail and 3' untranslated region control mRNA translation? *Cell,* 62, 15, 1990.
12. Furuichi, Y. and Miura, K., A blocked structure at the 5' terminus of mRNA from cytoplasmic polyhedrosis virus, *Nature,* 253, 374, 1975.

13. Izaurralde, E. et al., A nuclear cap binding protein complex involved in pre-mRNA splicing, *Cell,* 78, 657, 1994.
14. Sonnenberg, N., Cap-binding proteins of eukarypotic messenger RNA: Functions in initiation and control of translation, *Prog. Nucl. Acid Res. Mol. Biol.,* 35, 173, 1988.
15. Maruyama, K. and Sugano, S., Oligo capping: A simple method to replace the cap structure of eucaryotic mRNA with oligoribonucleotides, *Gene,* 138, 171, 1994.
16. Kato, S. et al., Construction of a human full-length cDNA bank, *Gene,* 150, 243, 1994.
17. Okayama, H. and Berg, P., High-efficiency cloning of full-length cDNA, *Mol. Cell. Biol.,* 2, 161, 1982.
18. Edery, I. et al., An efficient strategy to isolate full-length cDNA based on an mRNA cap retention procedure (CAPture), *Mol. Cell. Biol.,* 15, 3363, 1995.
19. Carninci, P. et al., High-efficiency full-length cDNA cloning by biotinylated CAP trapper, *Genomics,* 37, 327, 1996.
20. Lewin, B. et al., *Genes,* 7th ed, Oxford University Press, U.K., 2000.
21. Sambrook, J. et al., *Molecular Cloning,* 2nd ed., Cold Spring Harbor Laboratory Press, Cold Spring Harbor, NY, 1989.
22. Shinshi, H. et al., A novel phosphodiesterase from cultured tobacco cells, *Biochemistry,* 15, 2185, 1976.
23. Chomczynski, P. and Sacchi, N., Single-step method of RNA isolation by acid guanidinium thiocyanate-phenol-chloroform extraction, *Anal. Biochem.,* 162, 156, 1987.
24. Schwabe, W., Lee, J.E., Nathan, M., Xu, R.H., Sitaraman, K., Smith, M., Potter, R.J., Rosenthal, K., Rashtchian, A., and Gerard, G.F., ThermoScript RT, a new avian reverse transcriptase for high-temperature cDNA synthesis to improve RT-PCR, *Focus,* 20, 30, 1998.
25. Mizuno, Y., Carninci, P., Okazaki, Y., Tateno, M., Kawai, J., Amanuma, H., Muramatsu, M., and Hayashizaki, Y., Increased specificity of reverse transcription priming by trehalose and oligo-blockers allows high-efficiency window separation of mRNA display, *Genome Res.,* 10, 1345, 2000.
26. Suzuki, Y. et al., Construction and characterization of a full length-enriched and a 5'-end-enriched cDNA library, *Gene,* 200, 149, 1997.
27. Altschul, S.F. and Gish, W., Local alignment statistics, *Methods Enzymol.,* 266, 460, 1996.
28. Dennis, A.B. et al., GenBank, *Nucleic Acids Res.* 28, 15, 2000.
29. Larrick, J.W. and Siebert, P.D., *Reverse Transcriptase PCR,* Ellis Horwood, London, 1995.
30. Wikinson, D.G., *In situ Hybridization,* IRL Press, Oxford, 1992.
31. Andy Choo, K.H., *In Situ Hybridization Protocols, Method in Molecular Biology,* Vol. 33, Humana Press, Boston, 1994.

10 Methods for Gene Delivery to Neural Tissue

Jürgen A. Hampl, Alice B. Brown,
Nikolai G. Rainov, and Xandra O. Breakefield

CONTENTS

10.1 GENE DELIVERY TO THE NERVOUS SYSTEM

10.1.1 OVERVIEW

Gene delivery to the brain presents several unique challenges based on limited and risky access through the skull/vertebrae; sensitivity to volumetric changes; critical nuclei controlling life-functions; and a highly specialized blood-brain-barrier designed to keep particles and viruses out. The architectural heterogeneity of the neurocircuitry in the central nervous system (CNS) provides the opportunity to target focal areas of disease pathogenesis, but further complicates delivery for more global CNS diseases, such as brain tumors. This chapter will focus on modes of gene delivery to the nervous system, including vectors, cells, and routes of administration.

Transgene delivery and expression involves a number of steps at the cellular level: binding to and entrance into cells; cytoplasmic transport to the nucleus; entrance into the nucleus; fate in the nucleus; and transcriptional regulation of the transgene. In general, given the small volumes of inoculum that can be introduced into the brain, virus vectors, rather than synthetic vectors, have proven most efficient at gene delivery to date. Virus vectors currently used for delivery into the CNS of rodents include those derived from herpes simplex virus (HSV), adenovirus (Ad), adenoassociated virus (AAV), and retroviruses (derived both from Moloney murine leukemia virus [MoMLV or RV] and human immunodeficiency virus [HIV]). In addition to direct vector injection into the CNS, extensive studies have implanted cells, which have been previously genetically modified in culture, to deliver secretable transgene products to the brain. Retrovirus vectors are typically used to genetically modify such cells, since they can efficiently integrate transgenes into the genome of dividing cells.

A number of modes of delivery have been developed to tackle both focal and global delivery using various vector and cell vehicle designs.[1-3] Most gene delivery to the brain has involved direct stereotactic injection of replication-defective vectors into the parenchyma or into experimental tumors in rodents. By this route, the vector or other forms of DNA are taken up by cells only in the immediate vicinity of the injection site, since diffusion is very limited, with slower injection rates allowing somewhat wider dispersion. Newer modalities of delivery have included on-site generation of vectors by injection of packaging cells to produce retrovirus vectors, which can infect residually dividing cells in the brain, including glia and neuroprogenitor cells, as well as tumor cells,[4-6] and limited propagation of replication-conditional HSV or adenovirus vectors, used mostly in the context of tumors where toxicity is a component of the therapy.[7-9] On-site vector generation can potentially be combined with the use of migratory cells, such as neuroprogenitor cells[10-12] or endothelial cells,[13] so that the range of vector distribution is extended over a larger region. Other routes of delivery to cells in the CNS have included: (1) injection into fluid spaces, such as the vitreous humor in the eye[14] and the cerebral spinal fluid (CSF) through intrathecal or intraventricular routes for delivery to the choroid plexus, ependymal/meningeal layers, and onto the neighboring parenchyma through processes extending into these layers;[15,16] and (2) infusion through the blood-brain or blood-tumor barriers by intraarterial injection combined with temporary osmotic[17,18] or pharmacologic[19-21] disruption of the vasculature. Brain tumors can be preferentially targeted by virture of the relatively high permeability of the tumor neovasculature, as compared to the blood-brain barrier, with the latter having tight junctions between endothelial cells and being surrounded by an astrocytic sheath.[22]

Studies are ongoing to try to restrict the cell types that express transgenes within the zone of delivery. This can be done in three basic ways (Figure 10.1): (1) by taking advantage of intrinsic cell properties, e.g., only some glia, neuroprogenitor cells, and tumor cells divide in the adult brain and thus only they will integrate and express transgenes delivered by MoMLV retrovirus vectors; and only neurons have retrograde and anterograde viral transport mechanisms, so that at sites distant from HSV vector injections, only projecting neurons will be transduced; (2) by modifying the vector coat or virion such that it binds to or enters only specific cell types, e.g.,

TARGETING

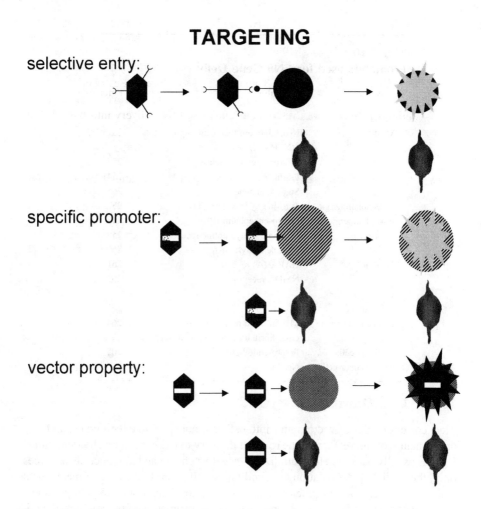

FIGURE 10.1 Modes of targeting specific cells with vectors. Vectors are typically targeted by one of three means, as shown here for tumor cells (round) vs. normal neural cells (with processes). Top: Ligands on the vector surface bind selectively to receptors present on the tumor cell and not on the normal cell, allowing selective entry and expression (star) in the tumor cell. Middle: The vector enters both cell types, but the transgene is placed under the control of a promoter active only in the tumor cell, so the transgene is expressed selectively in that cell. Bottom: Properties inherent in the vector determine its ability to transduce the tumor cell and not the normal cell, e.g., retrovirus vectors enter both cells but the transgene can only enter the nucleus and thereby be expressed in dividing cells.

adenovirus virions expressing ligands for or antibodies to the EGF receptor preferentially infect tumor cells in the brain which overexpress this receptor on their cell surface;[23] and (3) by using cell-specific promoters, e.g., the tyrosine hydroxylase promoter to confine transgene expression to neurons, which use dopamine or norepinephrine as transmitters,[24] and the glial fibrillary acidic protein (GFAP) promoter for astrocytes and some gliomas[25] (Table 10.1).

TABLE 10.1
Promoters used for CNS Gene Delivery

Cell Type	Promoter	Reference
Neurons:		
Most	Neurofilament heavy and light chains (NFH & NFL)	251, 252
	Neuron-specific enolase	253
	Prion	254
	Sodium channels	255, 256
Catecholaminergic	Tyrosine hydroxylase (TH)	257, 24
Enkephalinergic	Prepro-enkephalin (PE)	258, 139
Cholinergic	Nicotinic acetylcholine receptor	259
	Choline acetyltransferase	260
GABergic	GABA (A) receptor	261
	NMDA receptor	262
Glia:		
Schwann cells	PO	263
Oligodendrocytes	Myelin basic protein (MBP)	264
Astrocytes	Glial fibrillary acidic protein (GFAP)	25
Endothelial cells:	Prepro-endothelin-1	265
Neuropresursor cells:	Nestin	266

10.1.2 GENE DELIVERY

Most current modes of vector entry into cells are not specific to cell types, and rely on the nature of the vector and the route of delivery to target certain cell populations. DNA is usually packaged to neutralize its negative charge and to reduce shear forces on it, thus avoiding degradation and facilitating diffusion. In the case of mechanical delivery, DNA can be precipitated with calcium phosphate; bound to gold particles; encapsulated in artificial membranes, termed liposomes; and/or complexed with proteins, termed molecular conjugates or polymers. Particles can be delivered by electroporation or bombardment, in which case they pass directly through the cell membrane causing temporary disruption of it. Alternatively, liposomes can fuse to the membrane, thus releasing their contents directly into the cytoplasm, or conjugates can be taken up by endocytosis or pinocytosis. In the latter case they end up within membrane vesicles in the cytoplasm and must exit from those vesicles to avoid the degradative endosomal–lysosomal pathway. Mechanical means of gene delivery can be very inefficient with a low level of cellular uptake, degradation within cells, and poor access to the nucleus. Synthetic vectors frequently incorporate specific viral proteins or DNA sequences, sometimes from several different viruses, to promote cell delivery, DNA stability, and transgene expression.

In the case of virus vectors, the transgene is contiguous with viral sequences and is packaged with viral DNA in virus capsids, thus allowing entrance into cells by the mode characteristic of that particular virus. For most virus vectors in current use, the cellular recognition molecules are present on most cells and thus entry is not cell specific, although there may be differences in the relative efficiency of infection among cell types and species. For example, for viruses commonly used for gene delivery: HSV and retrovirus enter by direct fusion of the viral envelope to the cell membrane, such that the capsid is released into the cytoplasm; while adenovirus and AAV vectors are taken up by receptor-mediated endocytosis. Viral proteins involved in cell uptake, e.g., the Sendai virus membrane fusion protein and the vesicular stomatitis virus glycoprotein (VSVG), have been incorporated into synthetic vectors to enhance this process.

Most transgenes are designed to be expressed within the cell nucleus, but the means by which vectors move across the cytoplasm into the nucleus is not well understood. Transgenes introduced by most vectors can access the nucleus through nuclear pores even in nonmitotic cells. In general, however, breakdown of the nuclear membrane facilitates nuclear entry, and host cell DNA replication promotes integration of the foreign DNA and expression of transgenes. Some virus vectors, like retrovirus, can only access the nuclear space when the nuclear membrane breaks down during mitosis; others like AAV and HIV can enter the nucleus and integrate into the genome of dividing, as well as nondividing cells. "Free" vector DNA in the nucleus can be transcriptionally active for some time, but is eventually degraded or lost during mitosis. Some viruses, e.g., HSV and Epstein Barr virus (EBV), can maintain their genome as a stable extra-chromosomal (episomal) element in the host cell nucleus. Latent HSV establishes a nucleosomal configuration in some postmitotic cells, e.g., sensory neurons; while EBV maintains itself as a circular DNA element that replicates in-phase with host lymphoid cells.

10.1.3 POTENTIAL THERAPEUTIC USES

Although improved means of gene delivery to neurons has many immediate applications in basic neuroscience, the technology is not well enough developed for widespread applications in human disease. Problems that remain to be resolved are toxicity of the vectors, the inability to sustain and control levels of transgene expression, and the difficulty in delivering genes to neural cells *in vivo*. Still therapeutic studies are being considered for severely debilitating or life-threatening diseases, especially ones where it may be sufficient to deliver the gene to a subset of cells or for which some toxicity can be tolerated. For example, some beneficial gene products can provide functions to the transduced as well as to surrounding cells. Examples include growth factors or neurotransmitters, which are released from cells, and intracellular enzymes that can degrade toxic products released by endogenous cells and taken up by transduced cells. This strategy has proven effective in experimental models of gene replacement in lysosomal storage diseases, neuroprotection via growth factors for neurodegenerative diseases, and synthesis of neurotransmitters, like dopamine and acetylcholine for functional enhancement of movement and memory. Still in the context of the need for global therapy in the brain, e.g., for lysosomal storage disorders, in which case transduced cells can release the missing enzyme and it is taken up and used by nontransduced cells, it is not clear that the therapy will be sufficient to "cure" the disease and may only serve to prolong the debilitated state of the patients. Possibly the focus on intervention in this early stage of development of therapeutic gene delivery to neural cells should be for diseases affecting a focal set of neurons using nontoxic vectors where the stable expression of transgenes may not be critical, e.g., pain and epilepsy,[25a] or for brain tumors where therapy can tolerate or utilize vector toxicity and where transgene expression can be effective even when transitory.[26] Still there is great pressure to extend human trials to other neurologic disease, such as neurodegenerative diseases by delivery of growth factors, antioxidant or antiapoptotic molecules, and if properly designed these interventions can provide critical knowledge about the dynamics of gene delivery to the human brain.

To date very few gene therapy phase I trials have involved delivery to the nervous system, and most of these have been for brain tumors. In one example for a neurologic disease, a cell line genetically engineered to release ciliary neurotrophic factor (CNTF) was encapsulated, such that only proteins and smaller molecules could be exchanged with the extracellular fluid. These capsules were placed in the ventricles of patients with amyotrophic lateral sclerosis (ALS) to achieve sustained release of this growth factor into the CSF.[27] Although this procedure proved safe and was effective in maintaining increased levels of growth factor, it was ineffective at slowing the course of the disease. Another trial is planned to take skin fibroblasts from patients with Alzheimer's disease, engineer them to release nerve growth factor (NGF) and then transplant them into the brains of the same patients to try to prevent neuronal death.[28] For brain tumors, following resection of the main tumor mass, retrovirus producer cells have been implanted into the surrounding parenchyma known to contain infiltrative tumor cells. These producer cells release vectors bearing the gene for HSV-TK gene, which converts ganciclovir to a toxic nucleoside analogue, in an attempt to selectively infect and kill dividing tumor cells. This type of trial was carried out up to a phase III level with controls, but proved therapeutically ineffective.[1] Other smaller scale trials for tumors have included injection of adenovirus vectors bearing this same transgene, and of replication-conditional HSV and adenovirus vectors, which replicate selectively in dividing or p53-mutated tumor cells, respectively, with some anecdotal improvements reported.[29,30]

10.2 VEHICLES FOR GENE DELIVERY

10.2.1 VECTORS

Vector is a general term applied to the mode of packaging and delivery of genes to cells. It covers a wide range of modalities including application of naked DNA or DNA packaged in synthetic carriers or virus particles (virions). The nervous system is somewhat restrictive for gene delivery, as compared to other tissues, since direct access is limited to focal points of entry permitting only small volumes of inoculum. Since diffusion is limited in the nervous tissue (excluding the CSF or vasculature, see below), genes are typically delivered only to cells in a restricted area at the site of the injection. These and the detrimental effects of inflammatory responses in the nervous system, especially in the confines of the skull, have to a large part dictated the nature of the vectors used.

Although convenient in preparation and theoretically nonimmunogenic, DNA itself has proven relatively inefficient in delivery to the nervous system, such that relatively large amounts need to be used and can have some toxic effects,[31] albeit delivery via the CSF as liposomes appears quite effective.[32] Delivery has included direct injection, usually of DNA condensed with polycationic polypeptides, like polylysine and spermidine, or enveloped in cationic or anionic liposomes, or bombardment of DNA attached to gold particles via gene guns.[33,34] An additional limitation of bombardment is that the relatively large "bore" of the gun limits penetration into the brain and DNA passes only a few cell lengths. Synthetic carriers have been created to promote entry of DNA into cells. For example, in one configuration, DNA condensed with polylysine was coupled to a receptor ligand, transferrin, which induced receptor-mediated endocytosis selectively in cells with a high density of transferrin receptors.[35] The fusion protein from the virion of Sendai virus has been used to promote entry of liposomes into cells.[36] Further modifications of such synthetic carriers include specific internalizing and stabilizing molecules, such that the DNA is protected during transit to the cell nucleus. For example, incorporation of eukaryotic high mobility group proteins with nuclear localization signals into liposomes with DNA can promote access to the nucleus.[37,38] After gaining access to the cell nucleus, the foreign DNA will eventually be degraded (with circular DNA probably being somewhat more resistant to degradation as compared to linear DNA, but also less recombinogenic) unless it is imbued with the ability to either form a stable, extrachromosomal episome or to integrate into the host cell genome. Technology is developing for the creation of "designer" vectors, simulating the efficiency of gene delivery achieved by virus particles, and incorporating diverse elements from a number of viral and cellular sources so as to potentially allow selective entry into specific cell types and to control the fate of the transgenes.

For gene delivery to the nervous system, most investigators have used virus vectors, which have a high efficiency of gene delivery *in vivo*, through efficient cell/nuclear entry and control over the cellular fate of their DNA/RNA (Figure 10.2). In addition, some viruses have specific tropisms which can be channeled for gene delivery, notably the capsids of HSV-1 virions are transported retrogradely by an active process within neuronal processes allowing DNA to transit long distances within the nervous system.[39,40] However, most viruses are designed to take over and destroy the host cell, and the host organism has evolved tactics to try to thwart this activity, so use of viruses for benign delivery of genes is a challenge and requires extensive knowledge of the virus being harnessed. Most virus vectors are derived from human pathogens, which means that many individuals already have neutralizing antibodies to them, which can reduce effective gene delivery and elicit an immune response. This aspect turns out to be less of a problem in the nervous system, which is somewhat immune-compromised, as compared to the periphery, unless the vasculature is disrupted. Many viruses also express proteins that reduce antigen presentation by cells, which can be an advantage when expressing a therapeutic protein, such that the cells expressing it are not rejected by the immune system of the host. Such immune suppressive viral proteins have been removed from some vectors, however, (e.g., the E3 ORF from adenovirus vectors), thereby facilitating immune

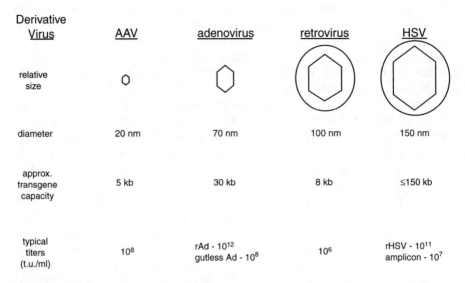

Derivative Virus	AAV	adenovirus	retrovirus	HSV
relative size				
diameter	20 nm	70 nm	100 nm	150 nm
approx. transgene capacity	5 kb	30 kb	8 kb	≤150 kb
typical titers (t.u./ml)	10^8	rAd - 10^{12} gutless Ad - 10^8	10^6	rHSV - 10^{11} amplicon - 10^7

FIGURE 10.2 Properties of virions used as vectors.

rejection of the transfected cell. Viruses can cause toxicity to cells via proteins in the virion itself, e.g., the fiber and penton proteins of adenovirus[41] and vhs and VP16 of HSV-1,[42] but this is limited to the number of particles injected and resolves as the input proteins are degraded. Additional toxicity can be caused by virus proteins encoded in the virus genome that are expressed in the host cell. Although many virus vectors have been engineered to be "gutless," such that they are stripped of all virus genes, and just use the virion for delivery of DNA/RNA (see below), two of the most commonly used vectors for the nervous system, recombinant adenovirus (rAd) and HSV-1 (rHSV) vectors, still retain virus genes. The typical replication-defective rAd vectors are deleted for E1a/b and E2 genes, which encode transcriptional activators for other virus genes. Thus expression of other virus genes is null-to-low; infection of host cells reveals upregulation of a number of genes associated with inflammatory and immune responses.[43] A number of versions of "nontoxic" replication-defective rHSV vectors have been generated. For example, deletion of three essential, immediate early genes, ICP4, ICP22, and ICP27, which encode viral transcriptional factors, reduces virus protein expression to a virtual null level.[44,45] Another approach to neutralizing virus gene expression in rHSV has been incorporation of a nontranscriptionally active form of VP16 into the virion and mutation in the gamma 34.5 genes, which reduce neurovirulence by confining viral replication largely to dividing cells.[46] Both these recombinant virus vectors are nonintegrating — adenovirus can exist stably in an as yet undefined latent state in some cell types,[47] while HSV can form a stable, episomal configuration in some neurons. For both these vectors, it has been difficult to achieve regulatable and stable transgene expression due to loss of viral DNA, methylation/nucleosomal configuration, and the dense matrix of viral enhancers and promoters in the virus backbone, which interfere with the transgene promoter. Still reports of stable expression in neurons do exist, notably

TABLE 10.2

"Safe" Vectors for the Nervous System

Type	Transgene Capacity (kb)	State in Nucleus
"Gutless" adenovirus	30	"Free"[a]
HSV amplicon	>50	"Free"[a]
AAV	4.5	Episomal and integrated
Lentivirus	8	Integrated (dividing and nondividing cells)
Retrovirus (MoMLV)	8	Integrated (dividing cells)

[a] Hybrid versions of these vectors are available which are episomal or can integrate into the host cell genome.[250]

with rHSV using the LAT promoter element, which is normally expressed during virus latency.[49,50] Although it may prove that vectors bearing virus genes can be sufficiently disarmed to prevent toxicity to neurons, safety concerns still remain, especially for use in the context of neurodegenerative conditions.

For neurologic and other long-term diseases, there is an increasing emphasis on the development of inherently safer vectors which employ virions to deliver genes efficiently, but which lack virus genes. In these vectors, only noncoding elements of the virus genome are retained as needed to mediate replication of the vector genome and packaging into virions, with an important aspect being that stocks are devoid of helper virus. Prototypes of these nontoxic vectors are HSV amplicon vectors, "gutless" or high-capacity adenovirus vectors, MoMLV retrovirus, lentivirus, and AAV vectors (Table 10.2). All these vectors are efficient at gene delivery to neurons and glia, but they differ in their transgene capacity, tropism, titers, and fate in the host cell nucleus. Both HSV amplicon and gutless adenovirus vectors have a relatively large transgene capacity (>50 kb and 30 kb, respectively), while that of retrovirus and AAV vectors is smaller (about 8 kb and 5 kb, respectively). Only lentivirus, AAV, and retrovirus can integrate efficiently into the host cell nucleus and hence assume a truly stable configuration, while the basic HSV amplicons and gutless adenovirus deliver "free" DNA to the nucleus.

For gene delivery to neural cells, a critical aspect is maintaining transgene expression for extended periods after stereotactic introduction. Transgene retention in cells has been achieved by incorporation of two elements of Epstein Barr Virus (EBV) — the latent origin of DNA replication (oriP) and the gene for the nuclear EBNA1 protein, which allow circular DNA to be replicated in synchrony with the host cells genome and distributed to daughter cells at mitosis.[51,52] There is no evidence, however, to suggest that these EBV elements serve to stabilize the foreign DNA in nondividing cells. In some cell types, the AAV ITR elements appear to be able to mediate formation of stable concatemers of DNA, which persist and express transgenes for extended periods.[53] Others have introduced relatively large DNA elements from mammalian chromosomes, such as centromeres, telomeres, and mammalian origins of DNA replication to create "mammalian artificial chromosomes,"[54,55] which can potentially be maintained as stable episomal elements in dividing and nondividing cells. Most of these mammalian elements are quite large (>12 kb), however, and require large-capacity cloning vehicles. Other DNA elements can promote integration of foreign DNA into the host cell genome, including for example: AAV ITR with or without the AAV *rep* gene;[56,57] loxP elements, which can be introduced into the genome and for which there are some pseudotypes in the mammalian genome, with the Cre gene;[58] and transposons.[59] Integration mediated by the AAV ITR without *rep* and by transposons appears to be random, while Rep proteins can promote specific integration of ITR-flanked sequences into the AAVS1 site on human chromosome 19q,[60] and Cre mediates integration at loxP sites. In general, site-specific integration is preferential to achieve predictability in the activity of a particular transgene promoter and to prevent accidental disruption of other cellular genes. DNA sequences can also integrate into the host cell genome by homologous recombination, e.g., for correction of mutant genes, but the efficiency of this event is very low.

A large number of promoter elements have been defined which allow selective expression of transgenes in particular cell types in the nervous system (Table 10.1), however, in many cases expression is disregulated due to elements in the vector genome or extinguished over time, primarily due to methylation[61] or condensation of the DNA into tight nucleosomal structures, which restrict access of transcriptional activators. The ability to maintain and regulate transgene expression in the host cell may ultimately depend on the ability of the vector to bring in a platform of DNA within which the transgene can operate. Such a platform might include, in addition to a mammalian or synthetic promoters and enhancers: insulators,[62] chromatin opening elements, and matrix or scaffold attachment regions,[63] to facilitate the reliable function of a promoter in a foreign domain. Some vectors are designed to express transgenes in the cytoplasm without requiring nuclear import. These include ones derived from other viruses including, vaccinia, Sindbis, and baculovirus. Self-promoting DNA transcription cassettes have been developed that encode prokaryotic RNA polymerases, which, in turn, can yield expression of transgenes utilizing the appropriate prokaryotic promoter.[64] Messages generated and expressed in the cytoplasm must be able to function without posttranscriptional modifications, which normally occur in the nucleus, including capping the 5′ terminal, splicing of introns, and polyadenylation.

10.2.2 Cells

In *ex vivo* therapy, cells are grown in culture, transfected with the desired transgene, and then administered to the host by stereotactic implantation or intravascular/intrathecal delivery. For these purposes, the cells employed can be either from primary cultures or from continous cell lines. Importantly, however, they should be nontumorigenic, easily obtainable, and free of endogenous pathogens.[65] Ideally, they should also be nonimmunogenic, if not syngeneic, to the host. The relative safety, efficacy, and versatility of the *ex vivo* gene delivery method has made it an attractive choice for experimental and clinical trials. In most paradigms, the grafted cells act

---

TABLE 10.3
Cells Used in *ex vivo* Gene Delivery Models

Primary:	Fibroblasts
	Endothelial cells
	Schwann cells
	Oligodendrocytes
	Neural stem cells
	Myoblasts
	Astrocytes
	Fetal neurons
	Adrenal chromaffin cells
Immortalized cell lines:	Fibroblast
	Neuroprecursor cells
	Temperature-sensitive neural cells

---

as biologic minipumps for the release of bioactive compounds, including neurotrophic factors (e.g., NGF, BDNF, bFGF[66]), neurotransmitters (e.g., dopamine and acetylcholine[3]), lysosomal enzymes,[67] and cytokines.[68] Such cells could also serve for degradation of toxic diffusible molecules which accumulate due to a metabolic deficiency state, e.g., to reduce levels of uric acid in the HPRT deficiency state (Lesch-Nyhan syndrome[69]); to promote neurite extension to a target, e.g., NGF in fimbria fornix lesion model;[70] or as a conduit for nerve regeneration, e.g., Schwann cells secreting growth factors placed in a tube near the cut end of nerve terminals.[71] Genetically modified hematopoietic precursor cells introduced into the bone marrow can also give rise to macrophages and lymphocytes which express transgenes and can migrate into the brain.[72,73]

In principle, many varied types of cells can be potential candidates for neural transplantation (Table 10.3). Some of the cell types used to date include: dermal fibroblasts,[74] endothelial cells,[3,75] Schwann cells,[76] myoblasts,[33,77] astrocytes,[78] oligodendrocytes,[79] fetal neurons,[80] adrenal chromaffin cells,[81] and neural progenitor cells (for review see Reference 82). With the advent of new vectors (AAV, HSV-AAV hybrid amplicons and lentivirus) with the capability of stably delivering transgenes to nondividing cell populations, there is no longer a requirement that donor cells be actively dividing for efficient gene transfer. The potential tumorgenicity of continuous cell lines has been partially overcome by transducing cells with genes for temperature-sensitive transforming proteins, such that they can be propagated (and hence are susceptible to retroviral transduction) in culture at 33°C, but stop dividing upon reaching rodent body temperature after transplantation (e.g., see Reference 83). Still many immortalized cell lines, e.g., mouse fibroblastic 3T3 cells, have proven nontumorigenic in the brains of animals[84] and can be further contained by encapsulation (see below).

Endothelial cells seem especially well suited for gene therapy since they are long-lived, multipotent cells and their phenotypic expression and functional capabilities can be manipulated by altering their local environment.[85] Like fibroblasts, endothelial cells are easily obtainable from peripheral tissues of animals and humans and can be readily expandable in culture. They have been genetically modified to achieve sustained, high levels of therapeutic transgene expression and can accommodate a broad spectrum of vectors.[86-89] Immortalized rat endothelial cells have been shown to migrate within subcutaneous glioma implants when injected directly into the tumor and to incorporate into tumor vasculature.[90] Myoblasts and fibroblasts[91] are also an attractive choice for cell-mediated therapy. They are easily propagated and can be screened and clonally selected in culture, and the former can be differentiated to a nonmitotic state by culture in low-serum-containing media.[92]

Potential replacement neurons and glia can be generated by using human or mouse neural progenitor cells obtained from fetal or adult mammals. These cells can be readily propagated in culture and can differentiate when grafted back into the nervous system of the host animal.[93-95] The great potential of these cells for gene delivery lies in their ability to migrate from the site of injection to areas of brain damage, e.g., ischemic damage[96] or tumors,[11] where they are most needed for therapeutic intervention. These progenitor-derived cell lines do not elicit a strong immune response when grafted back into the nervous system of nonsyngeneic experimental animals, due most likely to low immunological reactions in the brain, and low levels of expression of MHC-I and -II antigens on neural tissue.[97]

Although in some instances a stationary cell implant may be desirable, e.g., growth factor delivery to dopaminergic neurons in the substantia nigra,[98] some cells have the unique ability to migrate, and thus spread transgene products, within the brain. Such cells include oligodendrocytes, e.g., a cell line derived from them,[99] and neural progenitor cells.[100] This migratory capacity has been tested, for example, in treatment paradigms whereby immortalized C17.2 (mouse-derived) neural progenitors, engineered to express the pro-drug activating enzyme, cytosine deaminase, showed therapeutic efficacy in combination with 5-fluorocytosine when transplanted into experimental intracranial gliomas,[11] and were able to release lysosomal enzymes globally within the brain in a murine lysosomal deficiency state after introduction into the ventricles of newborn animals.[101]

Expression of foreign antigens by engrafted cells can lead to significant problems with immune rejection and inflammation. To address this issue, cells have been placed in semipermeable capsules with a pore size specifically designed to exclude antibodies, but allow the release of bioactive molecules.[102,103] Encapsulation has also proven useful for the transplantation of transformed cell lines, ensuring that their growth remains contained within the site of engraftment. Nonencapsulated packaging cell lines (derived from immortalized mouse fibroblasts), which release retrovirus vectors, have also been grafted directly into the brain,[4] but this method is made transitory by immunological rejection of cells due to the requisite expression of viral antigens, although the grafted cells do manage to deliver transgenes to a substantial number of surrounding tumor cells in experimental models.[4,104] A number of different cell types can be genetically modified for on-site vector production in the brain by employing amplicon vectors bearing components needed for retrovirus vector production[105] or by temporary arrest of replication-condition rHSV vectors.[12]

In order to determine the fate of donor/vehicle cells in the host animal, it is important to be able to distinguish them from endogenous cells. In some cases this can be done by using intrinsic properties of the donor cells. For example, male donor cells in a female host can be identified by *in situ* hybridization to repeat elements present in the Y chromosome,[106] and cross-species transfers can utilize species-specific antigenic markers, such as antibodies to human HLA histocompatibility proteins. Other means to mark cells prior to injection include uptake of nontoxic fluorescent dyes, such as carboxymethyl fluorescein diacetate (CMFDA; Molecular Probe) or propidium iodide, which bind tightly to protein and DNA, respectively, or incorporation of nucleoside analogues, such as BUdR, and ^{3}H-thymidine into the DNA which can be detected by immunocytochemistry[107] and autoradiography,[108] respectively. These labels show a decrease in intensity over time *in vivo* with gradual degradation of fluorescent molecules or dilution out with cell division.

Many new labeling techniques are becoming available which potentially allow tracking of groups of cells in the living animal, as well as later identification at the single cell level in tissue sections. For example, many cells will take up magnetic nanosphere particles which can be visualized by magnetic resonance imaging (MRI)[109,110] or compounds can be used which partition into cells and are labeled with high-energy, beta-emitting isotopes, like ^{111}-indium-oxine, that allow tracking with whole-body gamma imaging.[111] It is also possible to genetically label cells by transduction with marker genes, some of which can provide signals *in vivo* and in sections, including genes for green (GFP) and other hues of fluorescent proteins[112-114] that can be visualized using fluorescent microscopy (in sections) and two photon intravital microscopy *in vivo*;[115] the transferrin receptor, which binds labeled trans-ferrin-iron conjugates visualized by immunocytochemistry (sections) and MRI (*in vivo*);[116] luciferase metabolism of luciferin by light (sections) with CDC camera-enhanced (*in vivo*) microscopy;[117,118] proteases which can selectively liberate peptide-caged fluorescent fluors, by infrared microscopy (sections and *in vivo*[119]); and the dopamine type 2 receptor which can bind 3-(2′-[^{18}F]fluoroethyl) spiperone and can be visualized by positron emission spectroscopy (PET) (*in vivo*).[120]

10.3 ROUTES OF DELIVERY

In the CNS, the route of vector administration has been demonstrated to affect both transduction efficiency in the target region and spatial distribution of the delivered vector. Therefore, the extent of transgene expression, either in normal tissue or in brain tumors, depends dramatically on application routes and delivery modes.[121] While direct injection of vectors into tissue has been the simplest and most often used means to perform gene transfer in the CNS under experimental or clinical conditions, it also has several disadvantages. Among these are the inability to distribute vectors over large areas, and the trauma inflicted on the brain by direct injection.[26] On the other hand, there are other experimental modes of virus and nonvirus vector delivery, which, although not yet employed in clinical studies, hold promise for significant improvement of transgene delivery to the CNS. These are intravascular injection or infusion, and intrathecal or intraventricular injection of vectors, as well as use of cell dissemination (see above). Both routes of vector application make use of existing anatomical structures, such as intraventricular or subarachnoid space, and intravascular space. Therefore, vectors delivered by these routes are able to spread over considerable brain areas and may be physically targeted by modifying borders of these anatomical spaces by pharmacological or physical means.[122]

10.3.1 DIRECT INJECTION

The direct vector injection into the brain has been the first and since then the most applied technique in gene therapy studies. A broad range of vectors derived from adenovirus,[123-127] AAV,[132,133,70,134-136] herpes simplex virus,[137,9,138,45,7,140] retrovirus, including MoMLV and lentivirus,[141-144,130] and liposome-DNA complexes[32,145-147] have been successfully applied for gene delivery to normal neurons and glia (Color Figure 10.3)* and to neural derived tumors (Color Figure 10.4).*

Advantages of the direct injection technique are low systemic toxicity; high local vector concentration; local access to defined and specific brain areas, thereby diminishing the need for specific cell targeting; and distribution distances of several milimeters in the brain parenchyma.[17] For some neurologic diseases, like Parkinson's disease,[148-151] Huntington's disease,[152,153] and neuroendocrine deficiency states,[154] local gene expression may be sufficient to correct or ameliorate disease symptoms. In contrast, more widespread vector distribution would be essential for the achievement of therapeutic effects in the context of diffuse infiltrating neoplasms, neurodegenerative disorders (e.g., Azheimer's disease) or lysosomal storage diseases. The conventional technique of vector distribution by diffusion along a concentration gradient is mainly influenced by the counteracting pressure within the brain parenchyma or tumor interstitium. The increased interstitial pressure within neoplastic tissue serves to inhibit diffusion of the inoculum. Based on theoretical and experimental data, Jain et al.[155] calculated the time constants for diffusion of large proteins (MW around 150,000), which are smaller than viral vectors, in tumors: 1 h for a 100 μm distance, days for a 1 mm distance, and months for a 1 cm distance. This explains the very limited vector distribution of a few millimeters around the injection site[124,156,13] and the, so far discouraging, results in human trials of gene therapy for brain tumors.[157,158,1] In addition, the availability of receptor binding and entry and the relative size of the virions/vectors in relation to the interstitial space constricts gene delivery. Thus larger virions, e.g., adenovirus (about 70 nm diameter) and HSV (150 nm) diffuse over smaller regions than smaller virions, e.g., AAV (20 nm).

Convection-enhanced infusion is an alternative method for efficient and widespread delivery of macromolecules and particles to tumors or normal brain.[159,160] The application of a continuous positive pressure gradient by infusing small volumes (<0.5 μl/min) over an increased period of time (up to 24 h) allows improved particle distribution by bulk flow. Recently, Bankiewicz et al.[161] reported the successful use of convection-enhanced delivery to increase the tissue distribution of AAV vectors in a Parkinsonian monkey model system. They showed that AAV can be safely distributed over a whole anatomical target region, in this case the putamen and caudate nucleus. With further improvements of this technique, the zone of delivery

---

* Color figures follow page XXX.

from a single focal injection should be able to reach a cm wide distribution in the brain or tumor tissue.

Another way to increase foreign gene expression in a given area is the use of replication conditional viruses[124,162,163] or the transplantation of cells that have migratory properties[10,11,13] with or without viral vector production capabilities (see other sections). For example, the use of migratory active tumor or neuroprogenitor cells for on-site retrovirus production[6] should overcome limitations seen in the past with the use of stationary vector-producing fibroblasts.[164]

Direct brain injections in rodents require the use of a stereotactic device (David Kopf Instruments, CA; Stoelting, IL) and a stereotactic atlas[165,166] to target specific areas. Nevertheless, due to variations within strains of the same species, it is essential to verify the coordinates for specific strains within appropriate age and weight ranges. Differences due to sex are negligible.

After induction of anesthesia (mice: 4 μl/g mixture of 60% saline, 20% ketamine [100 mg/ml], and 20% xylazine [Rompun; 20 mg/ml]; rat: 200 μl/100 g of a mixture of 50% saline: 25% ketamine [100 mg/ml], 25% xylazine [Rompun; 20 mg/ml]), the skull is exposed with a midline saggital skin incision. According to the predefined stereotactic coordinates, the exact entry point can be marked with scratches on the skull surface using a metal caliper and the cranial sutures and the bregma as landmarks. The burr hole should be kept as small as possible so it can be used as a reference point for subsequent injections. The fixation of rats in the stereotactic apparatus with ear bars still allows some variation in the head position from procedure to procedure. To minimize breathing problems associated with ear bar insertion in mice, a special mouse bar is available (Kopf, CA), but the exact head positioning from procedure to procedure becomes even more difficult. Therefore, it is hard to hit the same target point with the second injection, even using the exact coordinates as previously. In addition, attention has to be paid to an exact axial head position. Even a minor change in angle leads to a considerable deviation of the needle tip at the target point several mm beneath the dura.

There is an inverse correlation between particle size and injection volume/injection speed. The larger the size of the injected particles, the smaller should be the infusion speed and injection volume per target point to avoid the creation of an intracranial sphere with disruption of the surrounding normal brain tissue. This can cause local brain damage with a consequent decrease in transduction efficiency in this area due to lack of oxygen and nutrients, with subsequent degradation of vectors and cells.

The following guidelines have been specifically worked out for stereotactic brain injections applying the bulk convection flow technique, but can also be used for brain injections with the diffusion technique.[167] With this method, it is essential to maintain a continuous pressure throughout the entire application time using an infusion pump (Stoelting). The volume of distribution (V_d) increases linearly with the infusion volume (V_i). The V_d/V_i ratio is relatively independent of the infusion speed and reaches values around 5.0 in normal brain tissue. There is an inverse correlation between infusion speed and the perfusion of the targeted brain area. At lower rates (0.1 and 0.5 μl/min) the infusate was almost entirely contained in the target, whereas at higher rates (1–5 μl/min), significant leakback was observed.[160]

To avoid reflux along the needle tract, the infusion speed should not exceed 0.5 µl/min and the needle itself should be as narrow as possible (e.g., 32 gauge). For all types of injections, at least 5 min should pass from the end of the infusion to needle retraction and retraction should be at a speed not greater than 1 mm/min. The burr hole is sealed with bone wax, and the extracranial surgical field is redisinfected. This is especially critical when injecting tumor cells, as inappropriately placed cells can produce extracranial tumor growth.

An alternative way to increase transduction rates is by injection of multiple small volume depots (<0.5 µl/area) throughout the tumor. As noted in Color Figure 10.4, the transduction rate with multiple injections is much higher in comparison to the single injection in a subcutaneous glioma model. However, this technique is more difficult to perform in an intracranial setting due to the potential for damage of the vasculature and resultant hemorrhage.

Injections can also be performed on animals pre- and postnatally. To define the exact target points, atlases of the developing mouse and rat brain are available.[168,169] Newborn mice are held in position and anethesized by being placed in a mold on ice for 5 min prior to the procedure. Intraparenchymal or intraventricular injections can be performed with pulled glass capillary pipettes connected to a nanoliter injection pump (Drummond Scientific Company, PA). This nanoliter injector offers volumes ranging from 2.3 nl to 69 nl. Translumination of the head using a translumina (Stoelting, IL), or if not available, a penlight can facilitate the correct needle position as the skull is translucent in very young pups. To deter the mother from killing the pups after the procedure, add a glove to the cage the day before; remove the mother and place her in a different cage for the duration of the procedure; carefully clean the injection site; and warm up the neonatal animals on a heating pad (Vetko, CO) covered by part of the nest bedding from the cage for up to 15 min. First return the pups, then the mother back to their cage. The best time for performing the procedure is early in the morning, because rodents are nocturnal animals and are less active during the day.

A similar technique can also be used for transuterine brain injections into fetal mice after surgical exposure. The earlier the procedure is performed the smaller the likelihood of preterm delivery. The bicornus uterus is exposed in anesthetized pregnant female rodents through a midline incision under sterile conditions and the use of a dissecting microscope. It is easy to recognize the fetal head through the uterine wall. The head can be fixed manually and after identification of the eyes for orientation purposes, intraparenchymal and intraventricular injections can be performed with an nl microinjector and the aid of translumination and vital dyes according to stereotactic coordinates.[170]

10.3.2 VASCULAR

10.3.2.1 Intravascular

Intravascular vector application employs the ubiquitously distributed network of arteries, veins, and capillaries present in the CNS. By virtue of its potential to reach all tissues and organs, the vascular system has been a popular route for cell and vector-mediated gene delivery, particularly due to the ability of intravenously and intraarterially administered cells to distribute themselves within the target (tissues, organs, and tumors) through a process termed transendothelial migration.[171] The endothelium itself is an excellent target organ for gene-based therapies for both vascular and nonvascular diseases due to its location at the interface between blood and tissue in the arterial, venous, and capillary circulation, with dense vascularization of normal brain and tumors. Endothelial cells, themselves, have important regulatory roles in angiogenesis, hemostasis, vascular reactivity, and responses to systemic immune and inflammatory stimuli.[172-175] Thus, for example, genetic modification of endothelial cells can facilitate the delivery of recombinant proteins into circulating blood or surrounding tissue.

Cell-mediated therapy is well suited for tumor vasculature, which is phenotypically distinct from normal blood vessels. Some molecular markers associated with tumor endothelium are absent or barely detectable in normal blood vessels,[176-179] including alpha$_v$ integrins,[180,181] and receptors for certain angiogenic growth factors.[182-184] Abundant evidence suggests that discrete features of tumor vasculature may allow selective tumor targeting.[185]

Tumor cells[186,187] and leukocytes[188,189] can preferentially home to specific organs, indicating that these tissues carry unique marker molecules accessible to circulating cells that determine their cell-to-cell interactions. Some organ-specific marker molecules on endothelial surfaces have been identified which can facilitate lymphocyte homing to lymphoid organs,[190] to tissues undergoing inflammation,[191-193] and to tumors.[194]

The process of transendothelial migration across blood vessels has been reported for intravenously administered/transplanted endothelial cells,[90,87,171] intraarterially injected myoblasts,[195,196] and intrasplenically transplanted hepatocytes.[197] Connective tissue fibroblasts have been demonstrated to be the origin of both pericytes and smooth muscle cells of blood vessels[198-200] and to participate as endothelial cells in the formation of new blood vessels during wound healing.[201]

Nonendothelial cells, such as mesenchymal cells[202] and tumor cells,[203] can integrate into the endothelial linings of newly formed blood vessels. Intraaortic injection of Tc-99 m radiolabeled lacZ-expressing myoblasts into mice produced clusters of lacZ-positive myofibers in limb muscle, lung, and the cortical layer of the kidney, presumably through active migration through the basal lamina.[195] Similarly, introduction of hepatocytes into the hepatic vascular bed has produced robust engraftment and survival, with entry and integration through disruption of the hepatic sinusoidal endothelium,[197] allowing extensive liver cell replacement in chronic disease.[204]

Numerous studies have shown that endothelial cells can undergo genetic manipulation and survive implantation to brain tumors,[90] normal brain,[205] and systemic microvessels following intravascular perfusion.[87,171] Immortalized rat endothelial cells migrate within subcutaneous glioma implants when injected directly into the tumor and incorporate into tumor vasculature.[90] Endothelial cells expressing recombinant marker proteins have shown stable implantation into peripheral arterial walls after vascular injection.[206,171,90,207] They also produce prolonged reporter gene expression in surrounding vasculature after transplantation in intracranial glioma models[13] and can be expected to proliferate in response to the specific mitogens often found within growing tumors.

Ojeifo et al.[207] showed that genetically modified human endothelial cells can be delivered intravenously to angiogenic sites and survive for extended periods of time. In an angiogenesis-directed model, he showed that lacZ-expressing HUVEC cells intravenously injected into nude mice targeted to experimental sites of angiogenesis (simulated wound healing) created by implantation of exogenous bFGF-secreting fibroblasts. Within a few minutes following injection, lacZ-HUVEC cells appeared in the lungs but were rapidly cleared; cells targeted to the angiogenic site prominently by 7 to 10 days and persisted in some cases for at least 4 weeks.[208] Immunohistochemical analysis suggested incorporation into host microvessels, with minimal presence in other tissues, including the lungs, liver, and kidneys.

Similarly, endothelial cells can survive for at least 4 weeks, proliferate, and integrate within host-derived tumor vessels after their implantation to subcutaneous gliomas.[90] In neonatal mice, endothelial cells stereotypically implanted into the caudate-putamen area became associated with the microvasculature in normal brain, and were found to express Glut1, a blood-brain barrier-associated endothelial cell glucose transporter.[205] Interestingly, the same cells implanted into adult brains only formed multicellular aggregates, highlighting the influence of environmental cues on endothelial phenotypic expression. Quinonero et al.[75] have also demonstrated integration of implanted endothelial cells into the host microvasculature of normal rat brain.

The following is given as an example of intravascular cell delivery (from Reference 208). Cells in culture, e.g., an immortalized line of murine endothelial cells,[209] are labeled in one or more of the following ways: stable transduction with GFP and drug selection of transduced cells over several weeks following infection with a retrovirus vector or transfection with a plasmid bearing GFP and neomycin transgene cassettes[113] (pEGFP-N3, Clontech Labs, CA); 24 h incubation with 25 μM CMFDA (Molecular Probes, Inc.) preceding the day of injection; or 90 min labeling of cells with [111]-indium-oxime (prepared fresh; 960 μCi per $1–3 \times 10^7$ cells in 1 ml), followed by centrifugation through a 40% Histopaque (Sigma) gradient and resuspension in phosphate-buffered-saline (PBS), pH 7.4. Cells (5×10^6 in 100 μl PBS per animal) are injected into the tail vein of mice anesthetized by intraperitoneal injection of ketamine/xylazine (see above). Scintigraphic images are acquired at varying times after injection in anesthetized mice using a small-field-of-view gamma camera (Sigma 410; Ohio Nuclear) driven by a Nuclear Mac (Scientific Imaging) with 20% windows over a 247-keV photopeak for acquisition of 100,000 counts. The biodistribution of radioactive label can also be quantitated upon sacrifice of the animals under heavy anesthesia (10 × anesthetizing levels) by dissecting out tissues of interest, weighing them, and determining levels of radioactivity with a gamma counter (incorporating an isotopic decay correction). Tissues can also be fixed, sectioned, and processed for direct fluorescence microscopic or immunocytochemistry.

10.3.2.2 Across the Blood-Brain Barrier

Intraarterial injection of vectors appears to have the potential for delivering vectors to a large proportion of brain and tumor cells without injuring normal structures or having other toxic consequences.[210-211] The normal blood-brain barrier (BBB) consists of endothelial cells bound together with tight junctions and wrapped by astrocytic processes, which severely limit the entry of substances into the interstitial space, while the tumor neovasculature has a more permeable barrier.[212,22,213] Although the blood-tumor barrier (BTB) is to some extent "leaky," it still limits delivery of high molecular weight substances to tumor tissue and to immediately adjacent, partially tumor-infiltrated areas of the brain.[214] In addition to the BTB, other factors can impair intravascular vector delivery. In order to infect the maximum number of cells, virus vectors must be delivered in sufficiently high titers and should not be inactivated by serum factors.[211] Areas and structures with higher interstitial fluid pressure may be less permissive to entry of macromolecules and particles.[155,215]

There are only a few studies performed with virus or nonvirus vectors injected intravascularly without modulation of the BTB or the BBB. Chauvet et al.[216] injected rAd vectors into the middle carotid artery (MCA) of a dog with a benign intracranial meningioma and were able to achieve a high percentage of transduced tumor cells without any concomitant toxic effects to the CNS. Delivery studies of vector across the unmodified BBB and the BTB have shown that there is only a small percentage of cells which can be transduced by this means[19,217] (Color Figure 10.5A).* Experiments with osmotic or pharmacologic barrier disruption[19,217,21,218,18,219] demonstrated significantly increased transduction rates of tumor cells after virus (rHSV or rAd) and nonvirus vector application. Several studies have focused on transient osmotic disruption of the BBB and the BTB, and this technique has been well characterized in animal models and in humans.[220,221,18,222] The mechanism of osmotic disruption of the barrier includes shrinkage of endothelial cells with subsequent opening of the capillary tight junctions, which is achieved by infusion of hypertonic solutions of sugar or salt into the arterial system.[223] Infusion of mannitol is most commonly used because of its relatively low toxicity and applicability to humans.[17] Mannitol offers the possibility of global delivery of drugs and virus vectors through the vasculature to a whole brain hemisphere, even in normal brain.[221,218] After intraarterial mannitol infusion into the MCA, glial cells were predominantly infected by rAd, while rHSV and iron oxide particles (MION) targeted neurons more efficiently.[17] The degree of BBB opening correlated with the relative transduction efficiency of neural cells.[220] Osmotic BBB disruption in combination with intraarterial administration of viral vectors potentially offers a method for global vector delivery to the brain.[18]

Vasoactive agents for modification of the BBB and BTB, such as leukotrienes,[224,225] bradykinin (BK) and its analog RMP-7,[21,226-231,19] histamine,[229,232] and calcium antagonists[230] appear to selectively increase permeability in capillaries in the CNS. BK, a nonapeptide hormone with peripheral vasodilatation effect, permeabilizes the vascular endothelium in brain capillaries at low concentrations (10 µg/kg/min).[22] BK exerts its effects by interaction with receptors on endothelial cells,[233] which mediate contraction of the cytoskeleton with subsequent temporary opening of the tight junctions[227,229,234] and may also increase the rate of pinocytosis/transcytosis in endothelial cells.[235] BK and RMP-7 also increase intracellular free calcium levels[227] and stimulate a nitric oxide-mediated pathway in tumor vasculature

---

* Color figures follow page XXX.

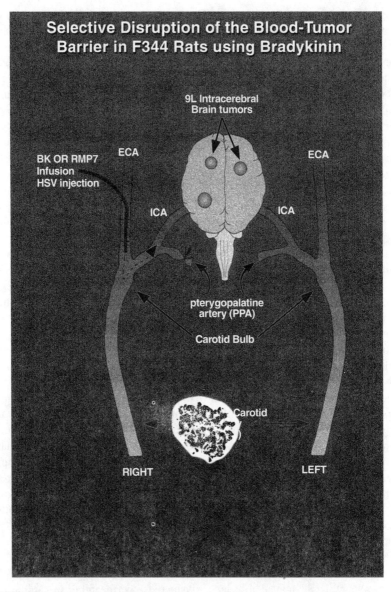

FIGURE 10.6 Intraarterial injection of vector in brain tumor model. Glioma tumors are established by injection of cells into three regions of the rat brain. After tumor growth, vector is injected into the carotid artery with and without bradykinin as a pharmacologic agent to temporarily open the blood-tumor barrier.

and/or in tumor cells themselves.[231] Barrier disruption by BK or RMP-7 can facilitate selective uptake of virus vectors administered through the carotid artery to single or multiple tumor foci (Color Figure 10.5B and and Figure 10.6). RMP-7 (Cereport, Alkermes Inc., MA) is an analogue of BK with a longer half life, greater potency *in vivo,* and specificity for the B_2 subclass of BK receptors, and without the systemic hypotensive effects of BK. RMP-7 has been shown to open the BTB to a greater extent than the BBB, allowing relatively specific delivery of therapeutic agents or vectors to brain tumors.

The technique of intraarterial delivery of vectors to the brain with or without barrier disruption has been developed and optimized mainly in rats. With some modifications, such as the use of the Seldinger technique for catheterization[236] instead of direct carotid puncture, it may be applicable to humans as well.

Briefly, animals are anesthetized by intraperitoneal injection of ketamine/xylazine (see above) and placed supine on a heating pad. Surgery is aided by an operating microscope. The right common carotid artery (CCA) is exposed through a 3 cm midline incision and subsequent dissection of neck tissues. After exposure and temporary clipping of the CCA, the external and internal carotid arteries (ECA and ICA) are dissected free. The thyroid artery and the pterygopalatine artery (PPA) are coagulated. The ICA is clipped, and the distal end of the ECA is ligated and incised proximal to the ligature. Polyurethane tubing (0.025 mm I.D., Braintree Sci., MA) connected to a 1 ml syringe and flushed with heparin (1000 U/ml), is inserted into the ECA and forwarded to the carotid bulb. CCA and ICA clips are removed in order to test the free passage of fluid and blood, and then the catheter is secured in place with 5-0 silk sutures and a drop of tissue adhesive (Vetbound©, 3M, MN). Bradykinin acetate (Sigma, MO) is dissolved in 0.9% NaCl at a concentration of 25 µg/ml, and the solution is kept on ice until immediately before infusion when it is warmed up to 37°C. BK is infused at a rate of 10 µg/kg/min for 10 min and immediately followed by bolus injection of vectors in 100 µl–1 ml of 0.9% saline. RMP-7 is prepared analogous to BK and infused at a rate of 0.1 µg/kg/min for 10 min. Vector injection may follow the infusion or may take place after 5 min of RMP-7 infusion with subsequent continuation of infusion. After finishing infusion and vector delivery, the proximal ECA is ligated and the catheter is removed. The soft tissues of the neck are approximated by sutures or tissue adhesive, and the skin is closed in one layer with 5-0 silk sutures.

10.3.2.3   Intrathecal and Intraventricular

Administration of vectors into the CSF spaces of the CNS may be used in the treatment of tumors, neurodegenerative disorders, and traumatic injuries.[237] After intraventricular or suboccipital intrathecal administration or rAd or rHSV vectors, a homogeneous distribution of transgene expression is found along the meninges and ependyma. Cortical transduction occurs as well, depending on the type of vector used. The route of vector clearance from the CSF, however, may result in antigen presentation and systemic immune responses).[238]

Recombinant virus vectors, e.g., rAd, rHSV, and AAV, have been used successfully to deliver genes via the intrathecal space to a variety of cell populations in the CNS. Neurodegenerative, disorders such as mucopolysaccharidosis,[239] or other types of lysosomal storage disease have been shown to improve significantly by intrathecal injection of vector into the CSF.[240-242]

Intrathecal delivery of vectors and therapeutic genes is an attractive approach because access to CSF is minimally invasive and distribution of virus vectors and cells may be facilitated by CSF circulation. Ram et al.[243] injected retrovirus producer cells into the leptomeningial space of tumor-bearing rats. Gene transfer was demonstrated in tumor foci growing in the cisterna magna, the injection site of the producer cells, although this procedure proved toxic in humans due to an inflammatory response.[158] Gene transfer into normal cells was also evaluated in rats and nonhuman primates without tumors after single and repeated intrathecal application of retrovirus producer cells.[244] Only choroid plexus cells, and no other normal CNS structures, showed transgene expression. Intrathecal application of rAd vectors has also been used for gene delivery. Bajocchi et al.[245] showed gene transfer into ependymal cells following direct injection into the ventricles. Ooboshi et al.[246,247] injected 1×10^9 pfu of rAd vectors intrathecally and demonstrated lacZ gene transfer into ependymal and leptomeningial cells, as well as into cerebral blood vessels. Marked toxicity was not observed in these studies. Vincent et al.[248] observed gene transfer into tumor cells along the entire neural axis after intrathecal administration of rAd vectors. After intrathecal application of replication-conditional rHSV vectors, Kramm et al.[238,15] showed extensive gene transfer into leptomeningial and parenchymal tumors. Additionally, ependymal and endothelial cells, as well as neurons projecting to the ventricles, showed marked transgene expression during the first two days after injection of HSV vectors (Color Figure 10.7).* Five and more days after vector application, normal cells no longer showed transgene expression. However, a considerable degree of animal toxicity was observed, probably due to an inflammatory reaction to the virus. Rosenfeld et al.[249] used an AAV vector to transduce medulloblastoma cells in a nude rat model of leptomeningial disease. After intrathecal application, tumor cells transduced with the marker gene lacZ were detected in the tumor mass, as well as in ependymal and subependymal areas, but not in normal brain parenchyma. No evidence of virus toxicity was noted during the course of the experiment.

The best way to perform intrathecal injection of vectors in the cisterna magna of rats is the stereotactic approach. Animals are anesthetized and fixated prone with a 45-degree head flexion in a stereotactic frame (D. Kopf Instruments, CA). A 22-gauge short-beveled needle (Beckton-Dickinson) is connected to a 80–100 µl microsyringe, placed in the stereotactic needle holder oriented 90 degrees to the skin, and forwarded exactly in the midline and in the space between the occiput and the arch of the atlas to a depth of approximately 4 mm from the skin level. CSF can be drawn and vectors are subsequently injected. Intrathecal placement of the needle can be confirmed by aspiration of CSF before and after the injection. Vectors can be injected in medium or in PBS containing 1 mM $MgCl_2$, 1 mM $CaCl_2$, and 1% sucrose. The volume used for injection may be as high as 100 µl, but smaller volumes are tolerated better, e.g., 10–20 µl, infused over a 2 min period.

10.4 DISCUSSION

Over the past 20 years, significant progress has been achieved in the evolving field of gene delivery to neural and other tissues in cell culture, in animals and even in

---

* Color figures follow page XXX.

TABLE 10.4
Ideals of Vector Design

- Efficient Gene Delivery (few particles per transduction, stable, high diffusion coefficient)
- Stable Nuclear State of the Transgene(s)
  (episomal, with replication in dividing cells, or site-specific integration)
- Large Transgene Capacity
- Accessible, Protected Configuration of Regulatory Elements
- Controlled Gene Expression (cell specific, regulatable)
- Intrinsically Nontoxic
- Low Immunogenicity
- Easy Construction and Rapid Generation
- No Contamination with Recombinant Virus
- High Titers

humans. Beginning with retrovirus-derived vectors, followed by rAd, AAV, rHSV, and HSV amplicon vectors, and lentivirus vectors, and more recently gutless Ad vectors, the characterization and application of these and synthetic vectors has led to a clearer picture about the strength and limitations of each vector system. As a logical consequence, efficacious parts of different viral systems have been combined and the second generation of viral vectors, the "hybrid vectors," e.g., Ad/EBV, HSV/EBV, HSV/EBV/RV, Ad/RV, Ad/AAV, and HSV/AAV,[250] has developed. In parallel, nonviral vector systems, starting with liposomes and later on molecular conjugates and polymers, have been designed and tested, ultimately leading to combinations of synthetic and viral derived elements.

All these modifications are directed toward one primary goal, the development of "safe and efficient vectors," which are nontoxic; nonimmunogenic; allow site-specific integration into the host genome and long-lasting transgene expression; infect both dividing and nondividing cells; and have a high transgene capacity (Table 10.4). In addition, to achieve optimal transgene delivery, particle titers and transduction efficiencies need to be as high as possible. The most recently generated vectors and vector combinations show more and more features of an "ideal vector." With the current technology for the design and development of vectors, most of the above requirements will be achievable in the near future. Of special interest is the search for sequences in the mammalian genome, which are related to loxP sites, and mammalian recombinases with similar function to the bacteriophage P1 recombinase Cre, as well as the ability of AAV LTR elements to integrate in the AAVS1 site on human chromosome 19 in the presence of transient Rep expression. Site-specific integration of transgene cassettes into the mammalian genome can provide a platform for exploring functional aspects of genomic sequences, which can be used to achieve regulatable, stable transgene expression.

With the rapid progress in the field, it has become more and more evident that improvements in vector delivery and cell-specific targeting are essential for the successful application of gene therapy paradigms in clinical settings. Physical limitations, such as differences in vector diameter and pore size of the parenchyma, and the electrical charge of virions or synthetic vectors, within the interstitial space of normal brain and tumor tissue are most likely responsible for the limited distribution and transduction seen with direct injection techniques. To solve this dilemma, two distinct research areas have emerged which focus either on decreasing the vector particle size or increasing co-ordinates of interstitial space. The covalent binding of proteins or small peptides to carrier molecules, e.g., the HIV tat protein, reduces particle size well below the physical restraints inherent in the interstitial space and results in an increase of interstitial particle distribution. With modifications in the electrical charges of virions or an expansion of the interstitial space, e.g., with the use of osmotically active substances and the application of bulk convection flow, a distribution of viral particles up to a centimeter range can be achieved. If wider distribution areas are desirable, improvements in vascular delivery methods, such as increased efficiency of BBB and BTB disruption techniques, and CSF delivery should allow a wider distribution and an increased number of vector particles within the brain. New research indicates that some cell types, such as neuroprecursor cells, can migrate widely in the brain and serve to distribute transgenes, for example, to experimental intracranial gliomas and ischemic lesions. Other cells, such as endothelial cells, can home to specific tissues or tumor through the vasculature. The combination of cell-mediated transport with onsite viral vector production, and thereby transgene amplification, may lead to local transgene concentrations high enough for adequate gene replacement therapy or efficient destruction of malignant cells. The mechanism behind the lesion-specific migration of neuroprecursor cells has not been elucidated. Presumably cytokine or chemokine gradients are responsible for this vectorial migration and the detection and creation of specific gradient profiles will enhance the efficacy of cell-mediated therapy paradigms.

The availability of more efficient delivery and distribution techniques will foster increasing efforts to target gene delivery to specific cell types. Two approaches are being explored to reach this goal: cell-specific transgene entry and cell-specific transgene expression. The discovery of cell- and tumor-specific antigens, allows targeting of vector entry. The detection of tumor-specific antigens has proven to be a difficult task, however, due to surface antigen masking used by tumor cells as a mechanism to escape recognition by the immune system. The technique of differential phage display using small peptide or single-chain antibody libraries allows the rapid screening of cell surface ligands in culture and *in vivo*, and the discovery of new cell-specific antigen-peptide binding partners is only a matter of time. Another approach is the characterization of surface molecules that are responsible for certain phenotypic features of cells, like the invasive character of malignant cells, e.g., matrix metalloproteinases, and their use as mediators of cell-specific binding. In addition, the use of cell-specific promoters or, in the case of malignant cells, cell-cycle-specific promoters, can increase the specificity of transgene expression. With further functional analysis of sequences derived through the Human Genome Project, new human cell-specific and function-specific promoters should become increasingly available.

Gene delivery to the nervous system is an expanding methodology with applications in basic neuroscience, brain dynamics, and gene therapy. Delivery without perturbation of existing structures and function is achievable, and beyond that it should be possible to target specific cells and set the range of intervention, as well as to control the levels of transgene expression throughout the life span of the cell/organism.

ACKNOWLEDGMENTS

We thank Ms. Suzanne McDavitt for skilled preparation of this manuscript; Ms. Deborah Schuback for skilled preparation of figures; and Dr. Joanna Bakowska and Dr. Christof Kramm for allowing us to use their unpublished figures. This work was supported by National Cancer Institute grant CA69246 (ABB, XOB), NINDS grant NS24279 (XOB), Deutsche Akademie der Naturforscher Leopoldina (NGR), and Deutsche Forschungsgemeinschaft (JAH).

REFERENCES

1. Rainov, N. G., A phase III clinical evaluation of Herpes Simplex Virus Type 1 thymidine kinase and ganciclovir gene therapy as an adjuvant to surgical resection and radiation in adults with previously untreated glioblastoma multiforme, *Hum. Gene Ther.*, 20, 2389, 2000.
2. Muldoon, L. L. et al., Delivery of therapeutic agents to brain and intracerebral tumors, in *Gene Therapy for Neurologic Disorders and Brain Tumors,* Chiocca, E. A. and Breakefield, X. O., Eds., Humana Press, Totowa, NJ, 1998, 295.
3. Gage, F. H. and Fisher, L. J., Intracerebral grafting: A tool for the neurobiologist, *Neuron*, 6, 1, 1991.
4. Short, M. P. et al., Gene delivery to glioma cells in rat brain by grafting of a retrovirus packaging cell line, *J. Neurosci. Res.*, 27, 427, 1990.
5. Miller, D. G., Adam, M. A., and Miller, A. D., Gene transfer by retrovirus vectors occurs only in cells that are actively replicating at the time of infection, *Mol. Cell Biol.*, 10, 4239, 1990.
6. Hampl, J. A. et al., Transduction of glioma cells into retrovirus vector producers with herpes simplex virus–Epstein Barr virus hybrid amplicon vectors in culture and in tumors, in preparation.
7. Martuza, R. L. et al., Experimental therapy of human glioma by means of a genetically engineered virus mutant, *Science*, 252, 854, 1991.
8. Goldsmith, K. T. et al., Trans complementation of an E1A-deleted adenovirus with codelivered E1A sequences to make recombinant adenoviral producer cells, *Hum. Gene Ther.*, 5, 1341, 1994.
9. Boviatsis, E. J. et al., Antitumor activity and reporter gene transfer into rat brain neoplasms inoculated with herpes simplex virus vectors defective in thymidine kinase or ribonucleotide reductase, *Gene Ther.*, 1, 323, 1994.
10. Gage, F. H., Stem cells of the central nervous system, *Curr. Opin. Neurobiol.*, 8, 671, 1998.
11. Aboody, K. S. et al., Foreign-gene-expressing mouse and human neural stem cells display extensive tropism for adult intracranial gliomas, *Proc. Natl. Acad. Sci. USA*, 97, 12846, 2000.
12. Herrlinger, U. et al., Neural precursor cells for delivery of replication-conditional HSV-1 vectors to intracerebral gliomas, *Mol. Ther.*, 1, 347, 2000.
13. Lal, B. et al., Endothelial cell implantation and survival within experimental gliomas, *Proc. Natl. Acad. Sci. USA*, 91, 9695, 1994.
14. Uteza, Y. et al., Intravitreous transplantation of encapsulated fibroblasts secreting the human fibroblast growth factor 2 delays photoreceptor cell degeneration, in Royal College of Surgeons rats, *Proc. Natl. Acad. Sci. USA*, 96, 3126, 1999.

15. Kramm, C. M. et al., Long-term survival in a rodent model of disseminated brain tumors by combined intrathecal delivery of herpes vectors and ganciclovir treatment, *Hum. Gene Ther.,* 7, 1989, 1996.

16. Snyder, E. Y., Taylor, R. M., and Wolfe, J. H., Neural progenitor cell engraftment corrects lysosomal storage throughout the MPS VII mouse brain, *Nature,* 374, 367, 1995.

17. Muldoon, L. L. et al., Comparison of intracerebral inoculation and osmotic blood-brain barrier disruption for delivery of adenovirus, herpesvirus and iron oxide nanoparticles to normal rat brain, *Am. J. Pathol.,* 147, 1840, 1995.

18. Nilaver, G. et al., Delivery of herpesvirus and adenovirus to nude rat intracerebral tumors after osmotic blood-brain barrier disruption, *Proc. Natl. Acad. Sci. USA,* 92, 9829, 1995.

19. Rainov, N. G. et al., Selective uptake of viral and monocrystalline particles delivered intra-arterially to experimental brain neoplasms, *Hum. Gene Ther.,* 6, 1543, 1995.

20. Rainov, N. G. et al., Long term survival in a rodent brain tumor model by bradykinin-enhanced intra-arterial delivery of a therapeutic herpes-simplex virus vector, *Cancer Gene Ther.,* 5, 158, 1998.

21. Barnett, F. H. et al., Selective delivery of herpes virus vectors to experimental brain tumors using RMP-7, *Cancer Gene Ther.,* 6, 14, 1999.

22. Inamura, T. and Black, K. L., Bradykinin selectively opens blood-tumor barrier in experimental brain tumors, *J. Cereb. Blood Flow Metab.,* 14, 862, 1994.

23. Miller, C. R. et al., Differential susceptibility of primary and established human glioma cells to adenovirus infection: Targeting via the epidermal growth factor receptor achieves fiber receptor-independent gene transfer, *Cancer Res.,* 58, 5738, 1998.

24. Jin, B. K. et al., Prolonged *in vivo* gene expression driven by a tyrosine hydroxylase promoter in a defective herpes simplex virus amplicon vector, *Hum. Gene Ther.,* 7, 2015, 1996.

25. Brenner, M. et al., GFAP promoter directs astrocyte-specific expression in transgenic mice, *J. Neurosci.,* 14, 1030, 1994.

25a. Martuza, R., personal communication.

26. Kramm, C. M. et al., Gene therapy for brain tumors, *Brain Pathol.,* 5, 345, 1995.

27. Zurn, A. D., Tseng, J., and Aebischer, P., Treatment of Parkinson's disease. Symptomatic cell therapies: Cells as biological minipumps, *Eur. Neurol.,* 36, 405, 1996.

28. Tuszynski, M. H. et al., Grafts of genetically modified Schwann cells to the spinal cord: Survival, axon growth, and myelination, *Cell Transplant,* 7, 187, 1998.

29. Markert, J. M. et al., Conditionally replicating herpes simplex virus mutant, G207 for the treatment of malignant glioma: Results of a phase I trial, *Gene Ther.,* 7, 867, 2000.

30. Sandmair, A. M., Vapalahi, M., and Yla-Herttuala, S., Adenovirus-mediated herpes simplex thymidine kinase gene therapy for brain tumors, *Adv. Exp. Med. Biol.,* 465, 163, 2000.

31. Brooks, A. I. et al., Reproducible and efficient murine CNS gene delivery using a microprocessor-controlled injector, *J. Neurosci. Meth.,* 80, 137, 1998.

32. Hagihara, Y. et al., Widespread gene transfection into the central nervous system of primates, *Gene Ther.,* 7, 759, 2000.

33. Jiao, S., Gurevich, W., and Wolff, J. A., Long-term correction of rat model of Parkinson's disease by gene therapy, *Nature,* 3621, 450, 1993.

34. Sato, H. et al., *In vivo* gene gun-mediated DNA delivery into rodent brain tissue, *Biochem. Biophys. Res. Comun.,* 270, 163, 2000.

35. Wagner, E. et al., Coupling of adenovirus to transferrin-polylysine/DNA complexes greatly enhances receptor-mediated gene delivery and expression of transfected genes, *Proc. Natl. Acad. Sci. USA,* 89, 6099, 1992.

36. Saeki, Y. et al., Development and characterization of cationic liposomes conjugated with HSJ (Sendai virus): Reciprocal effect of cationic lipid for *in vitro* and *in vivo* gene transfer, *Hum. Gene Ther.,* 8, 2133, 1997.

37. Aronsohn, A. I. and Hughes, J. A., Nuclear localization signal peptides enhance cationic liposome-mediated gene therapy, *J. Drug Target,* 5, 163, 1998.

38. Namiki, Y., Takahashi, T., and Ohno, T., Gene transduction for disseminated intraperitoneal tumor using cationic liposomes containing non-histone chromatin proteins: Cationic liposomal gene therapy of carcinomatosa, *Gene Ther.,* 5, 240, 1998.

39. Bearer, E. L. et al., Retrograde axonal transport of herpes simplex virus: Evidence for a single mechanism and a role for tegument, *Proc. Natl. Acad. Sci. USA,* 97, 8146, 2000.

40. Sodeik, B., Ebersold, M. W., and Helenius, A., Microtubule-mediated transport of incoming herpes simplex virus 1 capsids to the nucleus, *J. Cell Biol.,* 136, 1007, 1997.

41. Nemerow, G. R. and Stewart, P. L., Role of alpha(v) integrins in adenovirus cell entry and gene delivery, *Microbiol. Mol. Biol. Rev.,* 63, 725, 1999.

42. Breakefield, X. O. and DeLuca, N. A., Herpes simplex virus for gene delivery to neurons, *New Biol.,* 3, 203, 1991.

43. Benihoud, K. et al., The role of IL-6 in the inflammatory and humoral response to adenoviral vectors, *J. Gene Med.,* 2, 104, 2000.

44. Wu, N. et al., Prolonged gene expression and cell survival after infection by a herpes simplex virus mutant defective in the immediate-early genes encoding ICP4, ICP27, and ICP22, *J. Virol.,* 70, 6358, 1996.

45. Krisky, D. M. et al., Deletion of multiple immediate early genes from herpes simplex virus reduces cytotoxicity and improves gene transfer to neurons in culture, *Gene Ther.,* 5, 1593, 1998.

46. Palmer, J. A. et al., Development and optimization of herpes simplex virus vectors for multiple long-term gene delivery to the peripheral nervous system, *J. Virol.,* 74, 5604, 2000.

47. Ginsberg, H. S. and Prince, G. A., The molecular basis of adenovirus pathogenesis, *Infect. Agents Dis.,* 3, 1, 1994.

48. Millhouse, S. and Wigdahl, B., Molecular circuitry regulating herpes simplex virus type 1 latency in neurons, *J. Neurovirol.,* 6, 6, 2000.

49. Coffin, R. S. et al., The herpes simplex virus 2 kb latency associated transcript (LAT) leader sequence allows efficient expression of downstream proteins which is enhanced in neuronal cells: Possible function of LAT ORFs, *J. Gen. Virol.,* 79, 3019, 1998.

50. Soares, K. et al., cis-Acting elements involved in transcriptional regulation of the herpes simplex virus type 1 latency-associated promoter 1 (LAP1) *in vitro* and *in vivo, J. Virol.,* 70, 5384, 1996.

51. Yates, J. L., Warren, N., and Sugden, B., Stable replication of plasmids derived from Epstein–Barr virus in various mammalian cells, *Nature,* 313, 812, 1985.

52. Wang, S. and Vos, J.-M., A hybrid herpes virus infectious vector based on Epstein–Barr virus and herpes simplex virus type 1 for gene transfer into human cells *in vitro* and *in vivo, J. Virol.,* 70, 8422, 1996.

53. Peel, A. L. and Klein, R. L., Adeno-associated virus vectors: Activity and applications in the CNS, *J. Neurosci. Methods,* 98, 95, 2000.

54. Hamlin, J. L., Mammalian origins of replication, *Bioessays,* 14, 651, 1992.

55. Grimes, B. and Cooke, H., Engineering mammalian chromosomes, *Hum. Mol. Genet.,* 7, 1635, 1998.

56. Shelling, A. N. and Smith, M. G., Targeted integration of transfected and infected adeno-associated virus vectors containing the neomycin-resistance gene, *Gene Ther.,* 2, 1, 1994.

57. Balague, C., Kalla, M., and Zhang, W. W., Adeno-associated virus Rep78 protein and terminal repeats enhance integration of DNA sequences into the cellular genome, *J. Virol.,* 71, 3299, 1997.

58. Thyagarajan, B. et al., Mammalian genomes contain active recombinase recognition sites, *Gene,* 244, 47, 2000.

59. Zhang, L. et al., The Himar1 mariner transposase cloned in a recombinant adenovirus vector is functional in mammalian cells, *Nucl. Acids Res.,* 26, 3687, 1998.

60. Kotin, R. M., Linden, R. M., and Berns, K. I., Characterization of a preferred site on human chromosome 19q for integration of adeno-associated virus DNA by non-homologous recombination, *EMBO J.,* 11, 5071, 1992.

61. Jones, P. L. et al., Methylated DNA and MeCP2 recruit histone deacetylase to repress transcription, *Nat. Genet.,* 19, 187, 1998.

62. Rivella, S. et al., The cHS4 insulator increases the probability of retroviral expression at random chromosomal integration sites, *J. Virol.,* 74, 4679, 2000.

63. Kalos, M. and Fournier, R. E., Position-independent transgene expression mediated by boundary elements from the apolipoprotein B chromatin domain, *Mol. Cell Biol.,* 15, 198, 1995.

64. Brisson, M. et al., Subcellular trafficking of the cytoplasmic expression system, *Hum. Gene Ther.,* 10, 2601, 1999.

65. Isacson, O. and Breakefield, X. O., Benefits and risks of hosting animal cells in the human brain, *Nat. Med.,* 3, 964, 1997.

66. Hottinger, A. F. and Aebischer, P., Treatment of diseases of the central nervous system using encapsulated cells, *Adv. Tech. Stand. Neurosurg.,* 25, 3, 1999.

67. Ioannou, Y. A., Bishop, D. F., and Desnick, R. F., Overexpression of human alpha-galactosidase A results in its intracellular aggregation, crystallization in lysosomes, and selective secretion, *J. Cell. Biol.,* 119, 1137, 1992.

68. Nam, M. et al., Endothelial cell-based cytokine gene delivery inhibits 9L glioma growth *in vivo, Brain Res.,* 731, 161, 1996.

69. Jinnah, H. A. and Friedmann, T., Gene therapy and the brain, *Br. Med. Bull,* 51, 138, 1995.

70. Mandel, R. J. et al., Characterization of intrastriatal recombinant adeno-associated virus-mediated gene transfer of human tyrosine hydroxylase and human GTP-cyclo-hydrolase I in a rat model of Parkinson's disease, *J. Neurosci.,* 18, 4271, 1998.

71. Weidner, N. et al., Nerve growth factor-hypersecreting Schwann ell grafts augment and guide spinal cord axonal growth and remyelinate central nervous system axons in a phenotypically appropriate manner that correlates with expression of L1, *J. Comp. Neurol.,* 413, 495, 1999.

72. Ohashi, T. et al., Efficient transfer and sustained high expression of the human glu-cocerebrosidase gene in mice and their functional macrophages following transplan-tation of bone marrow transduced by a retroviral vector, *Proc. Natl. Acad. Sci USA,* 89, 11332, 1992.

73. Hoogerbrugge, P. M. et al., Donor-derived cells in the central nervous system of twitcher mice after bone marrow transplantation, *Science,* 239, 1035, 1988.

74. Wolff, J. A. et al., Grafting fibroblasts genetically modified to produce L-dopa in a rat model of Parkinson disease, *Proc. Natl. Acad. Sci. USA,* 86, 9011, 1989.

75. Quinonero, J. et al., Gene transfer to the central nervous system by transplantation of cerebral endothelial cells, *Gene Ther.*, 4, 111, 1997.

76. Guenard, V. et al., Syngeneic Schwann cells derived from adult nerves seeded in semi-permeable guidance channels enhance peripheral nerve regeneration, *J. Neurosci.*, 12, 3310, 1992.

77. Gussoni, E. et al., Normal dystrophin transcripts detected in Duchenne muscular dystrophy patients after myoblast transplantation, *Nature,* 356, 435, 1992.

78. Cunningham, L. et al., Nerve growth factor released by transgenic astrocytes enhances the function of adrenal chromaffin cell grafts in a rat model of Parkinson disease, *Brain Res.,* 658, 219, 1994.

79. Lachapelle, F. et al., Transplanted transgenically marked oligodendrocytes survive, migrate and myelinate in the normal mouse brain as they do in the shiverer mouse brain, *Eur. J. Neurosci.,* 6, 814, 1994.

80. Bjorklund, A., Dopaminergic transplants in experimental Parkinsonism: Cellular mechanisms of graft-induced functional recovery, *Curr. Opin. Neurobiol.,* 2, 683, 1992.

81. Decosterd, I. et al., Intrathecal implants of bovine chromaffin cells alleviate mechanical allodynia in a rat model of neuropathic pain, *Pain,* 76, 159, 1998.

82. Martinez-Serrano, A. and Bjorklund, A., Protection of the neostriatum against excitotoxic damage by neurotrophin-producing, genetically modifed neural stem cells, *J. Neurosci.,* 16, 4604–4646, 1996.

83. Trotter, J. et al., Lines of glial precursor cells immortalized with a temperature-sensitive oncogene give rise to astrocytes and oligodendrocytes following transplantation into demyelinated lesions in the central nervous system, *Glia,* 9, 25, 1993.

84. Rosenberg, M. B. et al., Grafting genetically modified cells to the damaged brain: Restorative effects of NGF expression, *Science,* 242, 1575, 1988.

85. Thorin, E. and Shreeve, S. M., Heterogeneity of vascular endothelial cells in normal and disease states, *Pharmacol. Ther.,* 78, 155, 1998.

86. Yao, S. N. et al., Expression of human factor IX in rat capillary endothelial cells: Toward somatic gene therapy for hemophilia B, *Proc. Natl. Acad. Sci. USA,* 88, 8101, 1991.

87. Dichek, D. A. et al., Seeding of intravascular stents with genetically engineered endothelial cells, *Circulation,* 80, 1347, 1989.

88. Zwiebel, J. A. et al., High-level recombinant gene expression in rabbit endothelial cells transduced by retroviral vectors, *Science,* 243, 220, 1989.

89. Zwiebel, J. A. et al., Recombinant gene expression in human umbilical vein endothelial cells transduced by retroviral vectors, *Biochem. Biophys. Res. Comm.,* 170, 209, 1990.

90. Conte, M. S. et al., Endothelial cell seeding fails to attenuate intimal thickening in balloon-injured rabbit arteries, *J. Vasc. Surg.,* 21, 413, 1995.

91. Edelstein, S. B. and Breakefield, X. O., Human fibroblast cultures, in *Physico-Chemical Methodoligies in Psychiatric Research,* Hanin, I. and Koslow, S., Eds., Raven Press, New York, 1980, 199.

92. Déglon, N. et al., Central nervous system delivery of recombinant ciliary neurotrophic factor by polymer encapsulated differentiated C2C12 myoblasts, *Hum. Gene Ther.,* 7, 2135, 1996.

93. Sah, D. W. Y., Ray, J., and Gage, F. H., Bipotent progenitor cell lines from the human CNS, *Nat. Biotech.,* 15, 574, 1997.

94. Snyder, E. Y. et al., Multipotent neural cell lines can engraft and participate in development of mouse cerebellum, *Cell,* 68, 33, 1992.

95. Snyder, E. Y., Grafting immortalized neurons to the CNS, *Curr. Opin. Neurobiol.*, 4, 742, 1994.

96. Snyder, E. Y. and Macklis, J. D., Multipotent neural progenitor or stem-like cells may be uniquely suited for therapy for some neurodegenerative conditions, *Clin. Neurosci.*, 3, 310, 1995.

97. Lampson, L. A., Molecular bases of the immune response to neural antigens, *Trends Neurosci.*, 10, 211, 1987.

98. Frim, D. M. et al., Implanted fibroblasts genetically engineered to produce brain-derived neurotrophic factor prevent 1-methyl-4-phenylpyridinium toxicity to dopaminergic neurons in the rat, *Proc. Natl. Acad. Sci. USA*, 91, 5104, 1994.

99. Louis, J.C. et al., CNTF protection of oligodendrocytes against natural and tumor necrosis factor-induced death, *Science*, 259, 689, 1993.

100. Snyder, E. Y., Neural stem-like cells: Developmental lessons with therapeutic potential, *Neuroscientist*, 4, 408, 1998.

101. Wolfe, J. H., Deshmane, S. L., and Fraser, N. W., Herpesvirus vector gene transfer and expression of beta-glucuronidase in the central nervous system of MPS VII mice, *Nat. Genet.*, 1, 379, 1992.

102. Hoffman, D. et al., Transplantation of a polymer encapsulated cell line genetically engineered to release NGF, *Exptl. Neurol.*, 122, 100, 1993.

103. Sagot, Y. et al., Polymer encapsulated cell lines genetically engineered to release ciliary neurotrophic factor can slow down progressive motor neuronopathy in the mouse, *Eur. J. Neurosci.*, 7, 1313, 1995.

104. Ram, Z. et al., *In situ* retroviral-mediated gene transfer for the treatment of brain tumors in rats, *Cancer Res.*, 53, 83, 1993.

105. Sena-Esteves, M. et al., Single step conversion of cells to retrovirus vector producers with herpes simplex virus-Epstein Barr virus hybrid amplicons, *J. Virol.*, 73, 10426, 1999.

106. Harvey, A. R. et al., Survival and migration of transplanted male glia in adult female mouse brains monitored by a Y-chromosome-specific probe, *Mol. Brain Res.*, 12, 339, 1992.

107. Bhide, P. G., Cell cycle kinetics in the embryonic mouse corpus striatum, *J. Comp. Neurol.*, 374, 506, 1996.

108. Schwartz, M. L., Rakic, P., and Goldman-Rakic, P. S., Early phenotype expression of cortical neurons: Evidence that a subclass of migrating neurons have callosal axons, *Proc. Natl. Acad. Sci. USA*, 88, 1354, 1991.

109. Lewin, M. et al., Tat peptide-derivatized magnetic nanoparticles allow *in vivo* tracking and recovery of progenitor cells, *Nat. Biotechnol.*, 18, 410, 2000.

110. Bulte, J. W. et al., Neurotransplantation of magnetically labeled oligodendrocyte progenitors: Magnetic resonance tracking of cell migration and myelination, *Proc. Natl. Acad. Sci. USA*, 96, 15256, 1999.

111. Schellingerhout, D. et al., Mapping the *in vivo* distribution of herpes symplex virons, *Hum. Gene Ther.*, 20, 1543, 1998.

112. Tsien, R. Y., The green fluorescent protein, *Annu. Rev. Biochem.*, 67, 509, 1998.

113. Aboody-Guterman, K. et al., Green fluorescent protein as a reporter for retrovirus and helper virus-free HSV-1 amplicon vector-mediated gene transfer into neural cells in culture and *in vivo*, *NeuroReport*, 8, 3801, 1997.

114. Hoffman, R. M., Orthotopic transplant mouse models with green fluorescent protein-expressing cancer cells to visualize metastasis and angiogeneisis, *Cancer Meta. Rev.*, 17, 271, 1999.

115. Potter, S. M. et al., Intravital imaging of green fluorescent protein using two-photon laser-scanning microscopy, *Gene,* 173, 25, 1996.

116. Weissleder, R. et al., *In vivo* magnetic resonance imaging of transgene expression, *Nat. Med.,* 6, 351, 2000.

117. Wood, K. V., Marker proteins for gene expression, *Curr. Opin. Biotechnol.,* 6, 50, 1995.

118. Contag, P. R. et al., Bioluminescent indicators in living mammals, *Nat. Med.,* 4, 245, 1998.

119. Weissleder, R. et al., *In vivo* imaging of tumors with protease-activated near-infrared fluorescent probes, *Nat. Biotech.,* 17, 375, 1999.

120. MacLaren, D. C. et al., Repetitive, non-invasive imaging of the dopamine D2 receptor as a reporter gene in living animals, *Gene Ther.,* 6, 785, 1999.

121. Zlokovic, B. V. and Apuzzo, M. L., Cellular and molecular neurosurgery: Pathways from concept to reality — Part II: Vector systems and delivery methodologies for gene therapy of the central nervous system, *Neurosurgery,* 40, 805, 1997.

122. Rainov, N. G., Breakefield, X. O., and Kramm, C. M., Routes of vector application for brain tumor gene therapy, *Gene Ther. Mol. Biol.,* 3, 1, 1999.

123. Badie, B. et al., Stereotactic delivery of a recombinant adenovirus into a C6 glioma cell line in a rat brain tumor model, *Neurosurgery,* 35, 910, 1994.

124. Boviatsis, E. J. et al., Gene transfer into experimental brain tumors mediated by adenovirus, herpes simplex virus, and retrovirus vectors, *Hum. Gene Ther.,* 5, 183, 1994.

125. Chen, S. H. et al., Gene therapy for brain tumors: Regression of experimental gliomas by adenovirus-mediated gene transfer *in vivo*, *Proc. Natl. Acad. Sci. USA,* 91, 3054, 1994.

126. Eck, S. L. et al., Treatment of advanced CNS malignancies with the recombinant adenovirus H5.010RSVTK: A phase I trial, *Hum. Gene Ther.,* 7, 1465, 1996.

127. Le Gal La Salle, G. et al., An adenovirus vector for gene transfer into neurons and glia in the brain, *Science,* 259, 988, 1993.

128. Morsy, M. A. et al., An adenoviral vector deleted for all viral coding sequences results in enhanced safety and extended expression of a leptin transgene, *Proc. Natl. Acad. Sci. USA,* 95, 7866, 1998.

129. Caillaud, C. et al., Adenoviral vector as a gene delivery system into cultured rat neuronal and glial cells, *Eur. J. Neurosci.,* 5, 1287, 1993.

130. Puumalainen, A. M. et al., Beta-galactosidase gene transfer to human malignant glioma *in vivo* using replication-deficient retroviruses and adenoviruses, *Hum. Gene Ther.,* 9, 1769, 1998.

131. Viola, J. J. et al., Adenovirally mediated gene transfer into experimental solid brain tumors and leptomeningial cancer cells, *J. Neurosurg.,* 82, 70, 1995.

132. Bartlett, J. S., Samulski, R. J., and McCown, T. J., Selective and rapid uptake of adeno-associated virus type 2 in brain, *Hum. Gene Ther.,* 9, 1181, 1998.

133. Lo, W. D. et al., Adeno-associated virus-mediated gene transfer to the brain: Duration and modulation of expression, *Hum. Gene Ther.,* 10, 201, 1999.

134. Mizuno, M. et al., Adeno-associated virus vector containing the herpes simplex virus thymidine kinase gene causes complete regression of intracerebrally implanted human gliomas in mice, in conjunction with ganciclovir administration, *Jpn. J. Cancer Res.,* 89, 76, 1998.

135. Okada, H. et al., Gene therapy against an experimental glioma using adeno-associated virus vectors, *Gene Ther.,* 3, 957, 1996.

136. Wu, P. et al., Adeno-associated virus vector-mediated transgene integration into neurons and other nondividing cell targets, *J. Virol.*, 72, 5919, 1998.

137. Chiocca, E. A. et al., Transfer and expression of the lacZ gene in rat brain neurons mediated by herpes simplex virus mutants, *New Biol.*, 2, 739, 1990.

138. Chambers, R. et al., Comparison of genetically engineered herpes simplex viruses for the treatment of brain tumors in a scid mouse model of human malignant glioma, *Proc. Natl. Acad. Sci. USA,* 92, 1411, 1995.

139. Kaplitt, M. G. et al., Long-term gene expression and phenotypic correction using adeno-associated virus vectors in the mammalian brain, *Nat. Genet.*, 8, 148, 1994.

140. Mineta, T., Rabkin, S. D., and Martuza, R. L., Treatment of malignant gliomas using ganciclovir-hypersensitive, ribonucleotide reductase-deficient herpes simplex viral mutant, *Cancer Res.*, 54, 3963, 1994.

141. Blömer, U. et al., Highly efficient and sustained gene transfer in adult neurons with a lentivirus vector, *J. Virol.*, 71, 6641, 1997.

142. Galipeau, J. et al., Vesicular stomatitis virus G pseudotyped retrovector mediates effective *in vivo* suicide gene delivery in experimental brain cancer, *Cancer Res.*, 59, 2384, 1999.

143. Naldini, L. et al., Efficient transfer, integration, and sustained long-term expression of the transgene in adult rat brains injected with a lentiviral vector, *Proc. Natl. Acad. Sci. USA,* 93, 11382, 1996.

144. Price, J., Turner, D., and Cepko, C., Lineage analysis in the vertebrate nervous system by retrovirusmediated gene transfer, *Proc. Natl. Acad. Sci. USA,* 84, 156, 1987.

145. Yagi, K. et al., Interferon-beta endogenously produced by intratumoral injection of cationic liposome-encapsulated gene: Cytocidal effect on glioma transplanted into nude mouse brain, *Biochem. Mol. Biol. Int.*, 32, 167, 1994.

146. Zerrouqi, A. et al., Liposomal delivery of the herpes simplex virus thymidine kinase gene in glioma: Improvement of cell sensitization to ganciclovir, *Cancer Gene Ther.*, 3, 385, 1996.

147. Zhu, J. et al., A continuous intracerebral gene delivery system for *in vivo* liposome-mediated gene therapy, *Gene Ther.*, 3, 472, 1996.

148. Bensadoun, J. C. et al., Lentiviral vectors as a gene delivery system in the mouse midbrain: Cellular and behavioral improvements in a 6-OHDA model of Parkinson's disease using GDNF, *Exp. Neurol.*, 164, 15, 2000.

149. During, M. J. et al., Long-term behavioral recovery in Parkinsonian rats by an HSV vector expressing tyrosine hydroxylase, *Science,* 266, 13990, 1994.

150. Leff, S. E. et al., Long-term restoration of striatal L-aromatic amino acid decarboxylase activity using recombinant adeno-associated viral vector gene transfer in a rodent model of Parkinson's disease, *Neuroscience,* 92, 185, 1999.

151. Lundberg, C. et al., Generation of DOPA-producing astrocytes by retroviral transduction of the human tyrosine hydroxylase gene: *In vitro* characterization and *in vivo* effects in the rat Parkinson model, *Exp. Neurol.*, 139, 39, 1996.

152. Schumacher, J. M. et al., Intracerebral implantation of nerve growth factor-producing fibroblasts protects striatum against neurotoxic levels of excitatory amino acids, *Neuroscience,* 45, 561, 1991.

153. Kordower, J. H. et al., Grafts of EGF-responsive neural stem cells derived from GFAP-hNGF transgenic mice: Trophic and tropic effects in a rodent model of Huntington's disease, *J. Comp. Neurol.*, 387, 96, 1997.

154. Geddes, B. J. et al., Long-term gene therapy in the CNS: Reversal of hypothalamic diabetes insipidus in the Brattleboro rat by using an adenovirus expressing arginine vasopressin, *Nat. Med.*, 3, 1402, 1997.

155. Jain, R. K., Transport of molecules in the tumor interstitium: A review, *Cancer Res.,* 47, 3039, 1987.

156. Rainov, N. G. et al., Retrovirus-mediated gene therapy of experimental brain neoplasms using the herpes simplex virus–thymidine kinase/ganciclovir paradigm, *Cancer Gene Ther.,* 3, 99, 1996.

157. Raffel, C. et al., Gene therapy for the treatment of recurrent pediatric malignant astrocytomas with *in vivo* tumor transduction with the herpes simplex thymidine kinase gene/ganciclovir system, *Hum. Gene Ther.,* 5, 863, 1994.

158. Ram, Z. et al., Therapy of malignant brain tumors by intratumoral implantation of retroviral vector-producing cells, *Nat. Med.,* 3, 1354, 1997.

159. Bobo, R. H. et al., Convection-enhanced delivery of macromolecules in the brain, *Proc. Natl. Acad. Sci. USA,* 91, 2076, 1994.

160. Morrison, P. F. et al., Focal delivery during direct infusion to brain: Role of flow rate, catheter diameter, and tissue mechanics, *Am. J. Physiol.,* 277, 1218, 1999.

161. Bankiewicz, K. S. et al., Convection-enhanced delivery of AAV vector in Parkinsonian monkeys: *In vivo* detection of gene expression and restoration of dopaminergic function using pro-drug approach, *Exp. Neurol.,* 164, 2, 2000.

162. Bischoff, J. R. et al., An adenovirus mutant that replicates selectively in p53-deficient human tumor cells, *Science,* 274, 373, 1996.

163. Yazaki, T. et al., Treatment of human malignant meningiomas by G207, a replication-competent multimutated herpes simplex virus 1, *Cancer Res.,* 55, 4752, 1995.

164. Harsh, G. R. et al., Thymidine kinase activation of ganciclovir in recurrent malignant gliomas: A gene-marking and neuropathologic study, *J. Neurosurg.,* 92, 804, 2000.

165. Franklin, K. B. J. and Paxinos, G., *The Mouse Brain in Stereotaxic Coordinates,* Academic Press, New York, 1997.

166. Paxinos, G. and Watson, C., *The Rat Brain in Stereotaxic Coordinates,* 4th ed., Academic Press, New York, 1998.

167. Chen, M. Y. et al., Variables affecting convection-enhanced delivery to the striatum: A systematic examination of rate of infusion, cannula size, infusate concentration, and tissue-cannula sealing time, *J. Neurosurg.,* 90, 315, 1999.

168. Kaufman, M. H., *The Atlas of Mouse Development,* Academic Press, New York, 1992.

169. Paxinos, G. et al., *Atlas of the Developing Rat Nervous System,* 2nd ed., Academic Press, New York, 1994.

170. Turkay, A., Saunders, T., and Kurachi, K., Intrauterine gene transfer: Gestational stage-specific gene delivery in mice, *Gene Ther.,* 6, 1685, 1990.

171. Messina, L. M. et al., Adhesion and incorporation of lacZ-transduced endothelial cells into the intact capillary wall in the rat, *Proc. Natl. Acad. Sci. USA,* 89, 12018, 1992.

172. Vane, J. R., Anggard, E. E., and Botting, R. M., Regulatory functions of the vascular endothelium, *N. Engl. J. Med.,* 323, 27, 1990.

173. Ward, P. A., Mechanisms of endothelial cell injury, *J. Lab. Clin. Med.,* 118, 421, 1991.

174. Pober, J. S., Cytokine-mediated activation of vascular endothelium. Physiology and pathology, *Am. J. Path.,* 133, 426, 1988.

175. Belloni, P. N. and Tressler, R. J., Microvascular endothelial cell heterogeneity: Interactions with leukocytes and tumor cells, *Cancer Mesta. Rev.,* 8, 353, 1990.

176. Folkman, J., Angiogenesis in cancer, vascular, rheumatoid and other diseases, *Nat. Med.,* 1, 27, 1995.

177. Folkman, J., Addressing tumor blood vessels, *Nat. Biotech,* 15, 510, 1997.

178. Risau, W. and Flamme, I., Vasculogenesis, *Annu. Rev. Cell. Biol.,* 11, 73, 1995.

179. Hanahan, D. and Folkman, J., Patterns and emerging mechanisms of the angiogenic switch during tumorigenesis, *Cell,* 86, 353, 1996.

180. Brooks, P. C. et al., Integrin alpha v beta 3 antagonists promote tumor regression by inducing apoptosis of angiogenic blood vessels, *Cell,* 79, 1157, 1994.

181. Hammes, H. P. et al., Subcutaneous injection of a cyclic peptide antagonist of vitronectin receptor-type integrins inhibits retinal neovascularization, *Nat. Med.,* 2, 529, 1996.

182. Martini-Baron, G. and Marme, D., VEGF-mediated tumor angiogenesis: A new target for cancer therapy, *Curr. Opin. Biotech.,* 6, 675, 1995.

183. Hanahan, D., Signaling vascular morphogenesis and maintenance, *Science,* 277, 48, 1997.

184. Risau, W., Mechanisms of angiogenesis, *Nature,* 386, 671, 1997.

185. Deneckamp, J. Endothelial cell proliferation is a novel approach to targeting tumor therapy, *Br. J. Cancer,* 45, 136, 1982.

186. Fidler, I. J. and Hart, I. R., Biological diversity in metastatic neoplasms: Origins and implications, *Science,* 217, 998–1003, 1982.

187. Johnson, R. C. et al., Endothelial cell membrane vesicles in the study of organ preferences of metastasis, *Cancer Res.,* 51, 394, 1991.

188. Springer, T. A., Traffic signals for lymphocyte circulation and leukocyte emigration: The multistep paradigm, *Cell,* 76, 301, 1994.

189. Salmi, M. et al., Selective endothelial binding of interleukin-2-dependent human T-cell derived from different tissues, *Proc. Natl. Acad. Sci. USA,* 89, 11436, 1992.

190. Rosen, S. D. and Bertozzi, C. R., The selectins and their ligands, *Curr. Opin. Cell Biol.,* 6, 663, 1994.

191. Bevilacqua, M. P. et al., Endothelial leukocyte adhesion molecule1: An inducible receptor for neutrophils related to complement regulatory proteins and lectins, *Science,* 243, 1160, 1989.

192. Siegelman, M. H., Rijn, M., and Weissman, I. L., Mouse lymph node hominy receptor cDNA clone encodes a glycoprotein revealing tandem interaction domains, *Science,* 243, 1165, 1989.

193. Cepek, K. L. et al., Adhesion between epithelial cells and T lymphocytes mediated by E-cadherin and the alpha E beta 7 integrin, *Nature,* 372, 190, 1994.

194. Johnson, R. C. et al., Lung endothelial dipeptidyl peptidase IV is an adhesion molecular for lung-metastatic rat breast and prostate carcinoma cells, *J. Cell Biol.,* 121, 1423, 1993.

195. Bresolin, N. et al., Intra-aortic injection of myoblasts in mdx mice: Genetic and technitium-99 m cell labeling and biodistribution, *Muscle Nerve,* 20, 757, 1997.

196. Neumayer, A. M., Di Gregoio, D. M., and Brown, R. H., Arterial delivery of myoblasts to skeletal muscle, *Neurology,* 42, 2258, 1992.

197. Gupta, S. et al., Entry and integration of transplanted hepatocytes in rat liver plates occur by disruption of hepatic sinusoidal endothelium, *Hepatology,* 29, 509, 1999.

198. Cliff, W. J., Observations on healing tissue: A combined light and electron microscopic investigation, *Philos. Trans. R. Soc. (London),* 246, 305, 1963.

199. Rhodin, J. A. G. and Fujita, H., Capillary growth in the mesentery of young normal rats. Intravital video and electron microscope analyses, *J. Submicrose Cytol. Pathol.,* 21, 1, 1989.

200. Nehls, V., Denzer, K., and Drenckhahn, D., Pericyte involvement in capillary sprouting during angiogenesis *in situ, Cell Tissue Res.,* 270, 469, 1992.

201. Kon, K. and Fujiwara, T., Transformation of fibroblasts into endothelial cells during angiogenesis, *Cell Tissue Res.,* 278, 625, 1994.

202. Aloisi, M., Giacomin, C., and Tessari, R., Growth of elementary blood vessels in diffusion chambers. I. Process of formation and conditioning factors, *Virchows Arch. B Cell. Pathol.,* 6, 350, 1970.

203. Hammersen, F., Endrich, B., and Messmer, K., The fine structure of tumor blood vessels. I. Participation of non-endothelial cells in tumor angiogenesis, *Int. J. Microcirc. Clin. Exp,* 4, 31, 1985.

204. Overturf, K. et al., Hepatocytes corrected by gene therapy are selected *in vivo* in a murine model of hereditary tyrosinemia type I, *Nat. Genet.,* 12, 266, 1996.

205. Johnston, P. et al., Delivery of human fibroblast growth factor-1 gene to brain by modified rat brain endothelial cells, *J. Neurochem.,* 67, 1643, 1996.

206. Nabel, E. G. et al., Recombinant gene expression *in vivo* within endothelial cells of the arterial wall, *Science,* 244, 1342, 1989.

207. Ojeifo, J. O. et al., Angiogenesis-directed implantation of genetically modified endothelial cells in mice, *Cancer Res.,* 55, 2240, 1995.

208. Brown, A.B. et al., Vascular targeting of therapeutic cells to tumors, submitted.

209. Arbiser, J. L. et al., Oncogenic H-ras stimulates tumor angiogenesis by two distinct pathways, *Proc. Natl. Acad. Sci. USA,* 94, 861, 1997.

210. Spear, M. A. et al., Targeting gene therapy vectors for CNS malignancies, *J. Neurovirol.,* 4, 133, 1998.

211. Muldoon, L. L. et al., Delivery of therapeutic genes to brain and intracerebral tumors, in *Gene Therapy for Neurological Disorders and Brain Tumors*, Chiocca, E. A. and Breakefield, X. O., Eds., Humana Press, Totowa, NJ, 1997, 128.

212. Cox, D. J., Pilkington, G. J., and Lantos, P. L., The fine structure of blood vessels in ethylnitrosurea-induced tumours of the rat nervous system: With special reference to the breakdown of the blood-brain-barrier, *Br. J. Exp. Pathol.,* 57, 419, 1976.

213. Yamada, K. et al., Quantitative autoradiographic measurements of blood-brain barrier permeability in the rat glioma model, *J. Neurosurg.,* 57, 394, 1982.

214. Groothuis, D. R. et al., Quantitative measurements of capillary transport in human brain tumors by computed tomography, *Ann. Neurol.,* 30, 581, 1991.

215. Jain, R. K., Barriers to drug delivery in solid tumors, *Sci. Am.,* 271, 58, 1994.

216. Chauvet, A. E. et al., Selective intraarterial gene delivery into a canine meningioma, *J. Neurosurg.,* 88, 870, 1998.

217. Rainov, N. G. et al., Intra-arterial delivery of adenovirus vectors and liposome-DNA complexes to experimental brain neoplasms, *Hum. Gene Ther.,* 10, 311, 1999.

218. Neuwelt, E. A., Pagel, M. A., and Dix, R. D., Delivery of ultraviolet inactivated ^{35}S-herpesvirus across an osmotically modified blood-brain barrier, *J. Neurosurg.,* 74, 475, 1991.

219. Schellingerhout, D. et al., Quantitation of HSV mass distribution in a rodent brain tumor model, *Gene Ther.,* 7, 1648, 2000.

220. Doran, S. E. et al., Gene expression from recombinant viral vectors in the CNS following blood-brain barrier disruption, *Neurosurgery,* 36, 965, 1995.

221. Neuwelt, E. A. and Hill, S. A., Chemotherapy administered in conjunction with osmotic blood-brain barrier modification in patients with brain metastases, *J. Neuro-Oncol.,* 4, 195, 1987.

222. Zünkeler, B. et al., Quantification and pharmacokinetics of blood-brain barrier disruption in humans, *J. Neurosurg.,* 85, 1056, 1996.

223. Rapoport, S. I. and Robinson, P. J., Tight-junctional modification as the basis of osmotic opening of the blood-brain barrier, *Ann. NY Acad. Sci,* 481, 250, 1986.

224. Black, K. L. and Chio, C. C., Increased opening of blood-tumor barrier by leukotriene C4 is dependent on size of molecules, *Neurol. Res.,* 14, 402, 1992.

225. Chio, C. C., Baba, T., and Black, K. L., Selective blood-tumor barrier disruption by leukotrienes, *J. Neurosurg.,* 77, 407, 1992.

226. Black, K. L. et al., Intracarotid infusion of RMP-7, a bradykinin analog and transport of gallium-68 ethylenediamine tetra-acetic acid into human gliomas, *J. Neurosurg.,* 86, 603, 1997.

227. Doctrow, S. R. et al., The bradykinin analog RMP-7 increases intracellular free calcium levels in rat brain microvascular endothelial cells, *J. Pharmacol. Exp. Ther.,* 271, 229, 1994.

228. Elliott, P. J. et al., Unlocking the blood-brain barrier: A role for RMP-7 in brain tumor therapy, *Exp. Neurol.,* 141, 214, 1996.

229. Inamura, T. et al., Intracarotid histamine infusion increases blood tumor permeability in RG2 glioma, *Neurol. Res.,* 16, 125, 1994.

230. Matsukado, K. et al., Selective increase in blood-tumor barrier permeability by calcium antagonists in transplanted rat brain tumors, *Acta Neurochir.,* 60, 403, 1994.

231. Nakano, S., Matsukado, K., and Black, K. L., Increased brain tumor microvessel permeability after intracarotid bradykinin infusion is mediated by nitric oxide, *Cancer Res.,* 56, 4027, 1996.

232. Nomura, T. et al., Effect of histamine on the blood-tumor barrier in transplanted rat brain tumors, *Acta Neurochir.,* 60, 400, 1994.

233. Hess, J. F. et al., Cloning and pharmacological characterization of a human bradykinin (BK-2) receptor, *Biochem. Biophys. Res. Comm.,* 184, 260, 1992.

234. Sanovich, E. et al., Pathway across blood-brain barrier opened by the bradykinin agonist, RMP-7, *Brain Res.,* 705, 125, 1995.

235. Raymond, J. J., Robertson, D. M., and Dinsdale, H. B., Pharmacological modification of bradykinin induced breakdown of the blood-brain barrier, *Can. J. Neurol. Sci.,* 13, 214, 1986.

236. Willson, J. K., A new technique for cerebral angiography: The variable stiffness guidewire, *Radiology,* 134, 427, 1980.

237. Driesse, M. J. et al., Distribution of recombinant adenovirus in the cerebrospinal fluid of nonhuman primates, *Hum. Gene Ther.,* 10, 2347, 1999.

238. Kramm, C. M. et al., Herpes vector-mediated delivery of marker genes to disseminated central nervous system tumors, *Hum. Gene Ther.,* 7, 291, 1996.

239. Elliger, S.S. et al., Elimination of lysosomal storage in brains of MPS VII mice treated by intrathecal administration of an adeno-associated virus vector, *Gene Ther.,* 6, 1175, 1999.

240. Ghodsi, A. et al., Extensive beta-glucuronidase activity in murine central nervous system after adenovirus-mediated gene transfer to brain, *Hum. Gene Ther.,* 9, 2340, 1998.

241. Ohashi, T. et al., Adenovirus-mediated gene transfer and expression of human beta-glucuronidase gene in the liver, spleen, and central nervous system in mucopolysaccharidosis type VII mice, *Proc. Natl. Acad. Sci. USA,* 94, 1287, 1997.

242. Sferra, T. J. et al., Recombinant adeno-associated virus-mediated correction of lysosomal storage within the central nervous system of the adult mucopolysaccharidosis type VII mouse, *Hum. Gene Ther.,* 11, 507, 2000.

243. Ram, Z. et al., Intrathecal gene therapy for malignant leptomeningial neoplasia, *Cancer Res.,* 54, 2141, 1994.

244. Oshiro, E. M. et al., Toxicity studies and distribution dynamics of retroviral vectors following intrathecal administration of retroviral vector-producer cells, *Cancer Gene Ther.,* 2, 87, 1995.

245. Bajocchi, G. et al., Direct *in vivo* gene transfer to ependymal cells in the central nervous system using recombinant adenovirus vectors, *Nat. Genet.,* 3, 229, 1993.

246. Ooboshi, H. et al., Adenovirus-mediated gene transfer *in vivo* to cerebral blood vessels and perivascular tissue, *Circ. Res.,* 77, 7, 1995.

247. Ooboshi, H., Rios, C. D., and Heistad, D. D., Novel methods for adenovirus-mediated gene transfer to blood vessels *in vivo*, *Mol. Cell Biochem.*, 172, 37, 1997.

248. Vincent, A. J. et al., Treatment of leptomeningial metastases in a rat model using a recombinant adenovirus containing the HSV-tk gene, *J. Neurosurg.*, 85, 648, 1996.

249. Rosenfeld, M. R. et al., Adeno-associated viral vector gene transfer into leptomeningial xenografts, *J. Neuro-Oncol.*, 34, 139, 1997.

250. Lam, P. Y. P. and Breakefield, X. O., Hybrid vector designs to control the delivery, fate and expression of transgenes, *J. Gene Med.*, 2, 395, 2000.

251. Schwartz, M. L. et al., Brain-specific enhancement of the mouse neurofilament heavy gene promoter *in vitro*, *J. Biol. Chem.*, 269, 13444, 1994.

252. Charron, G. et al., Multiple neuron-specific enhancers in the gene coding for the human neurofilament light chain, *J. Biol. Chem.*, 270, 30604, 1995.

253. Forss-Petter, S. et al., Transgenic mice expressing beta-galactosidase in mature neurons under neuron-specific enolase promoter control, *Neuron*, 5, 187, 1990.

254. Lemaire-Vieille, C. et al., Epithelial and endothelial expression of the green fluorescent protein reporter gene under the control of bovine prion protein (PrP) gene regulatory sequences in transgenic mice, *Proc. Natl. Acad. Sci. USA*, 97, 5422, 2000.

255. Auerbach, S. D. et al., Human amiloride-sensitive epithelial NA+ channel gamma subunit promoter: Functional analysis and identification of a polypurine–polypyrimidine tract with the potential for triplex DNA formation, *Biochem. J.*, 347, 105, 2000.

256. Maue, R. A. et al., Neuron-specific expression of the rat brain type II sodium channel gene is directed by upstream regulatory elements, *Neuron*, 4, 223, 1990.

257. Schimmell, J. J. et al., 4.5 kb of the rat tyrosine hydroxylase 5′ flanking sequence directs tissue specific expression during development and contains consensus sites for multiple transcription factors, *Mol. Brain Res.*, 74, 1, 1999.

258. Chan, R. M., Stewart, M. J., and Crabb, D. W., A direct repeat (DR-1) element in the first exon modulates transcription of the preproenkephalin A gene, *Mol. Brain Res.*, 45, 50, 1997.

259. McDonough, J. et al., Regulation of transcription in the neuronal nicotinic receptor subunit gene cluster by a neuron-selective enhancer and ETS domain factors, *J. Biol. Chem.*, 2000.

260. Naciff, J. M. et al., Identification and transgenic analysis of a murine promoter that targets cholinergic neuron expression, *J. Neurochem.*, 72, 17, 1999.

261. McLean, P. J. et al., A minimal promoter for the GABA(A) receptor alpha6-subunit gene controls tissue specificity, *J. Neurochem.*, 74, 1858, 2000.

262. Klein, M. et al., Cloning and characterization of promoter and 5′-UTR of the NMDA receptor subunit epsilon 2: Evidence for alternative splicing of 5′-non-coding exon, *Gene*, 208, 259, 1998.

263. Adam M. A. et al., Internal initiation of translation in retroviral vectors carrying picornavirus 5′ nontranslated regions, *J. Virol.*, 65, 4985–4990, 1991.

264. Miyao, Y. et al., Selective expression of foreign genes in glioma cells: Use of the mouse myelin basic protein gene promoter to direct toxic gene expression, *J. Neurosci. Res.*, 36, 472, 1993.

265. Mavria, G., Jager, U., and Porter, C. D., Generation of a high titre retroviral vector for endothelial cell-specific gene expression *in vivo*, *Gene Ther.*, 7, 368, 2000.

266. Kurihara, H. et al., Glioma/glioblastoma-specific adenoviral gene expression using the nestin gene regulator, *Gene Ther.*, 7, 686, 2000.

267. Fraefel, C. et al., Helper virus-free transfer of herpes simplex virus type 1 plasmid vectors into neural cells, *J. Virol.*, 70, 7190, 1996.

11 Neural Stem Cells and Brain Repair

Lorenz Studer and Ron McKay

CONTENTS

11.1 INTRODUCTION AND RATIONALE FOR THE USE OF BRAIN STEM CELLS

Understanding the genetic basis of brain function is a daunting task given the enormous complexity of the central nervous system (CNS). There are approximately one trillion cells in the human brain, with 10–100 billion neuronal cells displaying more than 50 distinct neurotransmitter phenotypes. Each of these neurons receives an average of over 1000 afferent synaptic inputs and will contact neighboring cells and organs via 1000 efferent synapses. Axonal arbors of single neuron in the cortico-spinal tract can travel through the whole length of the body and innervate targets at distances of more than one meter apart from the neuronal cell body. Understanding the function of such a complex structure requires knowledge about the basic elements that give rise to the extraordinary cell diversity in the brain. Developmental studies indicate that all the cells of the central nervous system are generated from a single layer of neuroepithelial cells. These neural plate cells are derived from the dorsal ectoderm of the embryo and induced by the notochord to undergo a well-defined set of morphological and molecular changes that lead to CNS induction, which in turn is followed by orchestrated waves of neural proliferation and differentiation.

The identification and *in vitro* isolation of stem cells from both the developing and adult mammalian CNS in vertebrates have provided powerful new tools to unravel multiple aspects of the genetic complexity of brain development and function. CNS stem cells serve as the universal building blocks that create cell diversity in the CNS via the generation of the correct numbers and types of neurons, astrocytes, and oligodendrocytes in each brain region. Defining the molecular machinery that controls stem cell maintenance, proliferation, and differentiation allows us to model brain development in a systematic, step-by-step fashion, starting from single stem cells to building complex neural circuits, both *in vitro* and *in vivo*.

Brain stem cells divide daily and efficiently for at least the first month of culture. The ability of CNS stem cells to grow at single-cell density allows for clonal analysis, the most stringent test for multipotency in stem cells. These studies[1-4] indicate that CNS stem cells clearly fulfill two major criteria established for other types of stem cells outside the CNS, namely self-renewal and multipotency. Asymmetric cell division, a criterion which is sometimes considered to be a general property of stem cells[5-9] and may actually occur within the developing neuroepithelium,[10,11] does not appear to be necessary in cultured CNS stem cells.[4]

In vitro stem cell proliferation can be used to expand the initial cell population sufficiently for large-scale biochemical studies. However, in some cases the feature of interest, i.e., the derivation of dopaminergic neurons, is not stable as CNS stem cells proliferate.[12,13] CNS stem cells can be derived from the earlier pluripotent embryonic stem (ES) cell.[14] The *in vitro* manipulation of ES cells further enhances our ability to provide unlimited numbers of brain cells *in vitro*. Such ES-derived CNS cells can be differentiated into highly specialized cells such as dopaminergic neurons,[15,16] or oligodendrocytes.[17,18] The use of CNS cells derived from mutant ES cells with specific genetic alterations can seamlessly combine the power of stem cell technology with the wealth of mouse genetics. There are a variety of assays to test the function of stem cell-derived cells ranging from molecular and histochemical characterization of the cells to functional biochemical or physiological assays such as measuring neurotransmitter release by high-pressure liquid chromatography[12,15,19] or synapse formation by electrophysiological studies.[15] Stem cells or stem-cell progeny can also be grafted back into various host environments including the embryonic and adult CNS to study functional integration *in vivo*. Finally, the identification of persisting endogenous stem cell populations in the adult brain offers the possibility of tracking and manipulating stem cells *in situ*.[20-27]

Mammalian CNS stem cells have been used as a biological assay system to test genetic vs. environmental influences on cell fate and function (for review, see Reference 28). These studies have challenged many long-held notions about plasticity and cell fate determination during development, as numerous cases now demonstrate that appropriate environmental cues dramatically alter the fate of CNS stem cells. For example, *in vitro* exposure of CNS stem cells to bone morphogenetic protein 2, 4, or 7 causes a rather unexpected transition from a cell with CNS stem cell properties to a cell exhibiting peripheral nervous system (PNS) stem cell markers that subsequently gives rise to PNS progeny including smooth muscle cells.[29-31] Cell fate switches of an even more dramatic scale were observed when injecting labeled CNS stem cells into non-CNS regions, resulting in CNS-derived muscle cells,[32,33] blood cells,[34] and cells with apparently pluripotent properties.[35]

The study of CNS stem cells not only provides important basic biological insights, but also allows us to manipulate or replace neural circuits in models of neurological disease. Brain stem cells are being developed as tools for cell and gene therapy by replacing specific types lost due to disease,[12,18,36] by compensating for specific genetic defects via the expression of a missing protein in grafted stem cells,[37] or by tracking down brain tumor cells *in vivo*.[38,39] The identification of persisting stem cells in the adult brain, as well as work showing contribution to the cell types of the CNS from stem cells outside the brain such as bone marrow stem cells,[40,41] offer potential new avenues for compensation of genetic deficits in the adult CNS.

Brain stem cell technology today has become an essential tool for many genetic approaches in modern neuroscience, ranging from the study of transcriptional control[42] to CNS proliferation, cell type specific–differentiation,[4] the formation of cell assemblies and functional networks, and the first preclinical attempts of brain repair.[12,18,36] The versatility of brain stem cells, combined with the power of genetics in the postgenomic age, should enhance our understanding of brain function and lead to effective brain repair strategies for neurological disorders.

11.2 METHODS

11.2.1 IDENTIFICATION, ISOLATION, AND *IN VITRO* CULTURE OF FETAL AND ADULT CNS STEM CELLS

Stem cells in the developing and adult brain can be identified by the expression of molecular markers such as the intermediate filament nestin,[43] the RNA-binding protein musashi,[44] as well as the expression of specific cell surface markers such as in human CNS stem cells CD133(+)/CD34(–).[45] The basic method of isolating neural stem cells *in vitro* is to dissect out a region of the fetal or adult brain that has been demonstrated to contain dividing cells *in vivo*. In the developing brain, most regions contain large numbers of stem cells, whereas in the adult brain the best results are achieved with tissue derived from the subventricular zone or the hippocampus.

The initial method for the isolation of CNS stem cells by Reynolds and Weiss[46] described the derivation of neurons and astrocytes from the adult mouse striatum. The cells were dissociated and subsequently propagated in serum-free medium containing high levels of epidermal growth factor (EGF). The clusters of free-floating proliferating neuroepithelial cells that form under these conditions were termed "neurospheres." Upon plating on an adhesive matrix such as poly-lysine and with-drawal of the mitogen, neurosphere-derived cells differentiate into neurons and glial cells. The basic "neurosphere" technology has been used in a very large number of studies. Some variations have been developed for specific applications such as substitution or supplementation of EGF with fibroblast growth factor (FGF),[47,48] the use of various medium supplements such as B27,[49] mechanical instead of enzymatic dissociation,[50] and the addition of leukemia inhibitory factor (LIF)[51] in some studies using human CNS stem cells.

An alternative strategy to neurosphere cultures was developed by several groups,[2-4,52] whereby CNS stem cells are cultured attached to a culture substrate such as fibronectin or laminin. Using this technique, single cells giving rise to individual clones of cells which can be traced easily by direct visual observation and time-lapse microscopy.[4,53] Variations of the method have been used for a wide range of application such as the *in vitro* generation of midbrain dopamine neurons,[12] the propagation of adult hippocampal precursors,[54] or the proliferation and differentiation of neural cells derived from mouse embryonic stem cells[15,18] (see below). We will briefly describe the isolation, propagation, and differentiation of CNS stem cells as monolayers. A more detailed step-by-step protocol has been published previously.[55]

11.2.1.1 Fetal Mouse or Rat CNS Stem Cell Monolayer Cultures

After sacrificing a time-pregnant rat by CO_2 asphyxiation, the embryos are removed as quickly as possible and placed in HBSS/HEPES buffer solution. The brain region of interest is dissected out using established anatomical landmarks and selected tissue pieces are transferred into a conical plastic tube and mechanically triturated. The gestational stage of the embryos should be selected such that the CNS region E-11 to E-12 of interest is proliferating, e.g., E-11 to E-12 for ventral mesencephalon, E-13 to E-14 for striatum, E-14 to E-15 for cortex, and E-15 to E-17 for hippocampus.

Depending on the quality of the dissection, the cells can be directly counted and plated or first spun at 200 g for 5 min and resuspended in culture medium, whereby debris and meninges can be easily removed by differential sedimentation. After determining the total cell number by trypan blue exclusion, medium levels are adjusted and the cells plated at a density of $10–30 \times 10^3$ cells/cm² on dishes precoated with polyornithine/fibronectin. Precoating is achieved by incubating plates sequentially in 15 µg/ml polyornithine (>2 h), followed by three washes in sterile PBS and incubation in 1 µg/ml fibronectin in PBS (>1 h). CNS stem cell growth is maintained in N2 medium supplemented with 10–20 ng/ml bFGF. In some studies, higher bFGF concentrations of up to 100 ng/ml have been used if cells were grown on laminin instead of fibronectin substrate. The N2 medium recommended for CNS stem cell culture[4] is based on the DMEM/F12 formulation and supplemented with glucose and bicarbonate as well as insulin, transferrin, selenite, progesterone, and putrescine (N2 components). Daily supplementation of the medium with bFGF is required to reliably prevent spontaneous cell differentiation. Medium is replaced every 2 days.

When the culture dish reaches about 60–70% cell confluence, passaging is required to prevent density-mediated growth arrest and differentiation. Cells are generally incubated in calcium/magnesium-free HBSS for 45 min followed by mechanical removal of the cells from the culture plate. Cells are washed off the plate with a 5 ml pipette followed by carefully scraping of the remaining attached cells with a cell lifter. The cells are collected in a conical plastic tube and spun down at 200 g for 5 min and resuspended in fresh N2 medium. At that stage, the passaged cell population consists of a relatively homogeneous population of multipotent precursor cells. These cells can be replated at various cell concentrations ranging from clonal density (1–20 cell/cm^2; cell lineage studies), to intermediate densities for continued cell expansion (1–20 × 10^3 cells/cm^2), to high-density cultures if region-specific neuronal differentiation is required (100–500 × 10^3 cells/cm^2). Cell differentiation is induced by withdrawal of the mitogen bFGF.

11.2.1.2 Monolayer Cultures of Adult Mouse or Rat CNS Stem Cells

Culture techniques for propagation and differentiation of stem cells derived from the adult brain closely match the conditions described above for fetal CNS stem cells. However, the initial dissection and dissociation of the cells requires a modified approach.

After sacrificing an adult mouse or rat with CO$_2$, the brain is removed *in toto* and cut into a series of coronal sections of about 1-mm thickness using a razor blade or an appropriate tissue chopper. The sections are collected in a dish containing tissue culture medium. Under the dissection microscope, the subventricular zone (SVZ) can be easily identified between the ventricular lining and the striatum. The SVZ is cut out with a tungsten needle or an equivalent instrument starting from the ventral tip of the lateral ventricle and ending at the corpus callosum. Similarly, the hippocampal formation can be identified and dissected out if hippocampal precursors are required. (See Paxinos and Watson[56] to identify the appropriate regions for dissection.)

The dissected tissue needs to be finely minced and subsequently incubated in 0.05% trypsin/0.02% EDTA + hyaluronidase for 30 min at 37°C in a shaking water bath. The digested fragments are washed by resuspension in N2 medium supplemented with kynurenic acid and DNase following centrifugation at 200 g for 5 min at 4°C. The tissue fragments are mechanically triturated and the supernatant containing suspended cells is transferred into a new tube. This procedure is repeated two or more times for optimum cell yield. Cells are then pelleted, resuspended in N2 medium, and plated at a density of 10–30 × 10^3 cells/cm^2 on culture dishes precoated with polyornithine/fibronectin (see above). The culture initially contains significant amounts of myelin and other small debris that might temporarily attach to the culture plate. However, consecutive medium changes during the culture period will dilute debris, and clusters of proliferating stem cells can be easily detected. Propagation, passaging, and differentiation procedures are identical to those described for fetal CNS stem cells (see above).

11.2.2 DERIVATION OF CNS STEM CELLS FROM MOUSE EMBRYONIC STEM CELLS

Embryonic stem cells have been at the basis of a revolution in mouse genetics. Genetic manipulations in ES cells allowed the generation of transgenic mice, knockout and knock-in mutants by gene targeting, and genetic screens based on random genetic mutations via gene trapping. Embryonic stem cells are pluripotent, i.e., able to generate all the different cell types of an organism. However, *in vitro* analysis of ES cell progeny has been difficult due to our inability to effectively harness the enormous differentiation potential of the cells.

Recently, strategies have been developed for the specific selection of distinct ES-derived cell classes such as blood cells,[57] cardiomyocytes,[58] or neural cells[59] based on cell-specific markers, i.e., antibody-mediated recognition of unique surface epitopes, or on promoter-driven expression of fluorescent reporters or antibiotics resistance genes. We have recently developed techniques that allow the efficient and specific generation of CNS cell from mouse stem cells without the need for any epitope-dependent cell-sorting step.[14,15,18] This method will be discussed in greater detail below. An alternative approach for the induction of CNS-specific differentiation of mouse ES cells is based on the use of specific stromal feeder layers hypothesized to provide a stromal-derived [CNS] inducing activity (SDIA).[16] Detailed methods are as follows.

The derivation of differentiated CNS progeny from mouse ES cells consists of five distinct steps (see Figure 11.1). In the first step (Stage I), ES cells are propagated in their undifferentiated state attached to culture dishes precoated with 0.1% gelatin in a standard ES medium containing 15% ES-qualified fetal bovine serum and supplemented with 1000–1500 U/ml LIF (LIF, leukemia inhibitory factor [ESGRO]). After 3–5 days of ES cell proliferation, the stem cells are trypsinized in 0.05% Trypsin/0.02% EDTA for 5 min at 37°C. Cell dissociation is stopped by adding serum-containing medium and the cell suspension is spun down in a tissue culture centrifuge at 200 g for 5 min. The cell pellet is subsequently resuspended in ES cell medium and cells are plated at about $20–40 \times 10^3$ cells/cm^2 on untreated Petri dishes (Stage II). Over the following 3–6 days, free-floating aggregates, so called embryoid bodies (EB), are being formed. At the end of Stage II, EB are collected and spun at low speed (100 g for 3 min) and resuspended in ES medium and plated onto culture dishes (Stage III). The following day, the medium is changed to a serum-free formulation supplemented with fibronectin at 5 µg/ml (ITSFn medium;[15] containing DMEM/F12 + glucose + bicarbonate + insulin, transferrin, selenite, and fibronectin).

After 5 to 8 days of growth in Stage III, cells start to express the neural stem cell marker nestin and are trypsinized and replated at $100–200 \times 10^3$ cells/cm^2 on polyornithine/laminin coated plates in N2 medium (see above) supplemented with 10 ng/ml bFGF and 1 µg/ml laminin (Stage IV). There are optional growth factors in Stage IV such as sonic hedgehog and FGF8 that can promote region-specific differentiation of ES-derived CNS cells.[15]

Stage V is induced by withdrawal of the mitogen bFGF and complementation of the medium with optional factors such as ascorbic or retinoic acid with subsequent differentiation of ES-derived CNS precursors into differentiated neuronal progeny. A modification of this protocol can be used for the efficient generation of astrocytic and oligodendrocytic progeny. Cells are passaged at the end of Stage IV and subsequently replated at $30–60 \times 10^3$ cells/cm^2 and proliferated in N2 medium supplemented with 10 ng/ml bFGF and 10 ng/ml EGF, passaged again, and further proliferated in N2 medium supplemented with 10 ng/ml bFGF + 10 ng/ml CNTF for astrocytic differentiation or 10 ng/ml bFGF + 10 ng/ml PDGF for oligodendrocytic differentiation. Mature astrocytic and oligodendroglial fates, respectively, are obtained upon subsequent withdrawal of the corresponding growth factors.

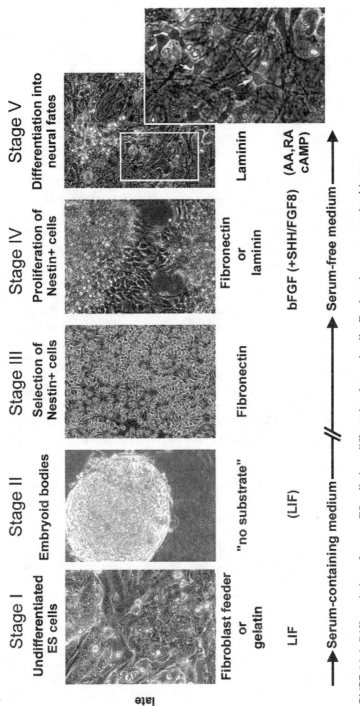

FIGURE 11.1 Differentiation of mouse ES cells into differentiated neural cells. Explanations are contained in text.

11.2.3 THE USE OF CNS STEM CELLS AS A TOOL FOR GENETIC SCREENS

CNS stem cells are particularly well suited as an *in vitro* assay system to identify factors that control neural fate. Though there are considerable differences among the various CNS stem cell techniques in the proportion of neurons to glial cells generated from stem cells, there is very little variation within a given technology. For example, the percentage of neurons derived from a single clone of proliferating CNS stem cells is constant independent of clone size, and every single clone gives rise to both neuronal and glial progeny.[4] Treatment of differentiating stem cells with platelet-derived growth factor (PDGF) results in the generation of higher proportions of cells expressing neuronal markers relative to glia.[4,60]

In contrast, treatment with the hormone triiodothyronine (T3) increases the generation of glia at the expense of the neuronal population.[4] Treatment of stem cells with leukemia inhibitory factor (LIF) or ciliary neurotrophic factor (CNTF) induces the generation of astrocytes at the expense of neurons and oligodendrocytes. The effect of ciliary neurotrophic factor on CNS stem cells is especially interesting since it acts as an instructive factor to direct the lineage choice of these cells.[4,61,62] However, developmental stage and regional identity of CNS stem cells are important factors that affect how an environmental signal is interpreted toward a specific cell fate response. For example, recent studies indicate that there might be a programmed sequence of neuron and glial cell production by multipotent cortical stem cells.[63,64] Signals that instructively drive astrocytic differentiation in both fetal and adult multipotent stem cells[4,61,62] do not act on multipotent cells derived from the early cortical anlage prior to the onset of gliogenesis.[29]

Neural stem cells can also serve as a powerful tool to unravel the signal transduction cascades guiding cell fate decisions downstream of the ligand-receptor interaction, such as the Jak-STAT pathway in astrocytic differentiation,[61,62] the role of neurogenin in promoting neuronal and inhibiting glial fates,[65] or the interaction of the STAT and Smad pathway via p300 in astrocytic differentiation.[66] The success of such studies is critically dependent upon a simple but efficient strategy of gene delivery into CNS stem cells. CNS stem cells have been transfected using a variety of transfection systems ranging from calcium phosphate technique,[61,65] Lipofectamine® (Gibco, Life Technologies),[67] or Effectene® (Quiagen).[68] However, transfection efficiencies are generally low. If higher transduction rates are required, viral vectors such as retroviral, adenoviral, adeno-associated virus (AAV)-based or lentiviral vectors are preferred. Nevertheless, for a quick routine assessment of a given construct in CNS stem cells, the use of transfection techniques is often adequate. Here we give an example of a transfection experiment using Lipofectamine Plus. Detailed methods are as follows.

CNS stem cells, isolated and grown as described above or Stage IV ES-derived CNS cells are passaged and plated at 2×10^3 cells/cm^2 on polyornithine/fibronectin-coated culture plates. Cells are subsequently proliferated in N2 plus 10 ng/ml bFGF and kept without penicillin/streptomycin for 3 days prior to transfection. The transfection is carried out at about 60% cell confluence for 3 h in 2 ml of medium containing 1–2 µg DNA per construct. The cells are then placed in fresh N2 medium plus 10 ng/ml bFGF and further proliferated or differentiated depending on the experimental question. These conditions may yield up to 10% transfection efficiency. The transfection is generally well tolerated by the stem cells within appropriate DNA concentrations. Higher concentrations lead to significant cell toxicity.

ES cells are particularly suited to systematically address the role of specific genetic pathways in brain development and function. In addition to exploiting the enormous libraries of targeted ES cell mutants for *in vitro* assays of neural induction, regional patterning, and cell differentiation, the differentiation technique described above allows assessing the role of specific genes in mutants that cause embryonic or postnatal lethality. ES cells have also been successfully used as a functional *in vitro* screen for gene discovery via gene trapping.[69-72]

CNS stem-cell technology has matured to the level where routine drug testing as well as *in vitro* genetic screens might become routine. CNS stem cells could ultimately replace the use of immortalized cell lines of questionable physiologic relevance for such functional tests. However, the *in vitro* differentiation technique remains an artificial system that needs careful evaluation of all the relevant parameters and confirmation via *in vivo* functional tests. For example, a recent study demonstrates the importance of adequate oxygen levels *in vitro* for CNS stem cell growth. Standard tissue culture incubator conditions are 5% CO_2 and 95% air, which exposes cells to a 20% O_2 environment. In mammalian brain, interstitial tissue O_2 levels range from about 1–5%.[73] When CNS stem cells are grown under such lowered oxygen levels (3 ± 2%), important basic biological parameters are changed. Proliferation of CNS stem cells is increased, cell death decreased, and neuronal subtype differentiation changed under lowered oxygen. Oxygen levels were found to directly affect gene expression levels indicating the importance of oxygen for the interpretation of *in vitro* gene expression screens. However, using elegant subtraction screens and appropriate control conditions, interesting data can be extracted from differential expression screens with CNS stem cells.[74] These studies are just the beginning of an exciting era in stem-cell research, which will lead to systematic analyses unraveling the signals that control stem-cell differentiation and determine cell number, type, and function in the CNS.

11.2.4 TRANSPLANTATION AND *IN VIVO* FUNCTIONAL ASSESSMENT OF CNS STEM CELLS

CNS stem cell transplantation plays an important role in studying basic biological questions such as stem cell plasticity and regional identity as well as in the assessment of the therapeutic potential of stem cells. Studies employing stem cell transplant approaches in the CNS encompass areas as diverse as developmental neurobiology and the study of neoplastic, metabolic, and degenerative brain disease. The transplantation studies can combine the use of stem cell technology with experimental surgery and state-of-the-art molecular biology and genetics.

CNS and ES-derived precursor and stem cells have been grafted in the developing[75-77] and adult[12] rodent brain in numerous experimental and therapeutical models. Several detailed methods for cell transplantation into the fetal, neonatal, and adult rodent brain have been published (for example, see Reference 17). In this chapter, we will focus on specific issues related to the transplantation of CNS stem or precursor cells and their progeny and provide specific techniques for the transplantation of stem-cell-derived dopamine neurons into the adult brain of a Parkinsonian rodent. Important strategic considerations in CNS stem-cell grafting are the stage of the cell population to be grafted, as well as the developmental stage of the host. For example, transplantation of undifferentiated precursor of stem cells into the developing brain leads to a seamless integration of the cells into the host brain, with both neuronal and glial components derived from the graft.[75,78,79] However, grafting of undifferentiated cells into the CNS results in exclusively glial differentiation in nonneurogenic regions of the adult brain,[80] but neuronal and glial differentiation when placed into adult neurogenic regions.[54,81] If efficient neuronal differentiation of grafted precursor cells is required in a region of the adult brain without efficient neurogenesis, i.e., in the striatum of Parkinsonian rats, stem cells may need to be predifferentiated *in vitro* prior to implantation.[12,82] Detailed methods are as follows.

For the functional assessment of dopamine neurons *in vivo*, an appropriate disease model is required. The system most commonly used is a unilateral 6-hydroxydopamine lesion of the medial forebrain bundle. The toxin is taken up specifically by dopaminergic terminals and retrogradely transported to the cell body, where it destroys the cell. Typically, a volume of 2.5 μl of the neurotoxin 6-hydroxydopamine bromide (6-OHDA) is injected at a concentration of 3.6 mg/μl at two sites along the MFB[83] using a Hamilton syringe attached to a stereotactic frame. The following stereotactic coordinates should allow for appropriate lesioning: AP –4.4 mm, ML –1.2 mm, and V –7.8 mm, toothbar set at –2.4 (ventral coordinates [V] refer to the level of the dura). An additional 3.0 μl were injected at: AP –4.0 mm, ML –0.8 mm, and V –8.0 mm, toothbar set at +3.4.

Precursor-derived dopamine neurons can be grafted either as cell suspension or as culture aggregates of up to 0.8 mm in diameter.[12] A cell suspension has the advantage that smaller injection tracks can be utilized with subsequently smaller host-derived immunological response and glial reaction. Cell aggregates on the other hand provide cell-to-cell signaling within differentiated progeny that helps to maintain phenotypic differentiation of the precursor cells and promotes cell survival in a host environment suboptimal for neuronal integration.

Cell suspensions are directly injected via a Hamilton syringe (23–26 G) or a fine glass capillary at a cell concentration of 50–100,000 cells/μl/min. A maximum of about 5 μl/site can be injected. Cell aggregates are loaded into a blunt spinal needle (18–22 G). Four to eight aggregates can be loaded into the needle that is attached to a holding device allowing independent movements of stylet and outer core. The cells are then deposited at AP +1.0 mm, ML –2.5 mm, and V –4.7 mm, toothbar set at –2.5 and held on place for 5 min before withdrawing the needle at a rate of 1 mm/min.

Behavioral testing is carried out in a rotometer system for the automated assessment of amphetamine-induced rotation behavior. Lesioned animals are tested at least twice before transplantation, and at least three times after transplantation within

2–3 week intervals. Animals that show a stable rotation response of >6 rotations/min upon the i.p. application of 2.5–5 mg/kg D-amphetamine are chosen for further grafting studies.

A successful graft should restore the rotation response completely. The behavioral tests are accompanied by histological assessment of graft survival. Standard histological and immunohistochemical procedures are used to identify the cell populations of interest. State-of-the-art genetic labels should be used for the unequivocal identification of the grafted cells within the host brain. This is especially important when grafting undifferentiated stem or precursor cell populations, as these cells tend to migrate extensively in the adult brain. Common labeling techniques include precursors derived from stable GFP transgenic lines, ROSA-26 derived precursors, or precursors stably transduced with a reporter system using viral vectors. If cross-species or male to female grafts are performed, DNA *in situ* hybridization for species or Y-chromosome-specific markers can be used. Alternatively, species-specific antibodies or transplantation of precursor cells labeled with BrdU, H^3-Thmydin, or lipophilic dyes prior to implantation have been used for graft detection. However, all these physical methods are prone to artifacts and should not be considered as a first choice technology but only if no alternative genetic labeling technique is possible. *In vivo* quantification of the grafted cells is carried using unbiased, stereological procedures such as the Cavalieri's estimator[84] for graft size and the optical fractionator to determine total numbers of grafted cells.

11.3 DISCUSSION

The biology of CNS stem cells is a rapidly evolving field. Figure 11.2 shows the interrelated fields that now comprise the broad domain of CNS stem cell research. It has been only about 10 years since the techniques were developed to reliably identify neural stem cells and to generate stem cell-derived neurons and astrocytes *in vitro*. More recently it has become possible to efficiently generate CNS cells from mouse embryonic stem cells. Today, these technologies allow the generation of unlimited numbers of many of the major cell types in the CNS. The next great challenges in CNS stem-cell biology are the further characterization of signaling pathways to generate any of the many different cell types in the CNS, and to develop alternative sources of CNS stem cells amenable to experimental and therapeutic applications.

There are many potential cell sources to derive CNS progeny. For example, recent data indicate that bone marrow derived cells might be able to generate cells with neural characteristics, both *in vitro*[85] and *in vivo*.[40,41] Another promising alternative approach is the use on nuclear transfer technology to generate ES-like cells[86] from adult somatic cells. Current studies[87] show that the initially laborious and inefficient generation of such nuclear transfer ES lines can be greatly optimized, and that nuclear transfer ES lines are indeed able to generate a complete animal. The *in vitro* differentiation potential can be directed toward the generation of very specific cell types, including the generation of functional dopaminergic neurons. ES lines generated via nuclear transfer on a routine basis could serve not only as an isogenic source for therapeutic cloning via cell transplantation, but also serve as an

Derivation of stem cells:	Pluripotent Cell	Organ-specific multipotent cells	Differentiated CNS cells	Therapeutic Application

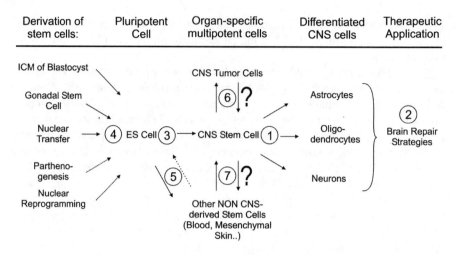

FIGURE 11.2 Interrelations of CNS stem cell research. Brain stem cell biology has evolved from a discipline focused on the conditions and signals that control (1) CNS stem cell identity, proliferation, and differentiation into many interrelated avenues of research that include studies on: (2) the application of stem cell progeny in brain repair; (3) the derivation and purification of ES-derived CNS stem cells; (4) the identification of novel potential sources for the derivation of stem cells; (5) transdifferentiation of stem cells, which challenge the plasticity of stem cells believed to be organ-specific; (6) the relationship between brain stem cells and brain tumor cells; and (7) non-CNS derived stem cells.

ideal assay system for drug testing and genetic screens individualized to a single patient or a specific class of patients. However, the ultimate goal in stem-cell technology might be to overcome the need for nuclear transfer and to reprogram an adult cell nucleus directly, via defined cytoplasmic factors that mimic conditions present in a oocyte immediately after nuclear transfer (this is when reprogramming of the transplanted nucleus occurs).

The dissection of molecular signals that control stem-cell differentiation and the use of CNS stem cells as a tool for drug testing and gene discovery will greatly profit from the advances in genomic sciences. With the complete draft of the human genome sequence available[88,89] and the genome sequence of many other species already completed[90-92] or expected to be completed in the near future, the progress in the elucidation of specific pathways in CNS stem-cell differentiation will speed up significantly. Large-scale genomic screens using CNS stem cells have been already initiated to define factors that control neuronal differentiation.[74]

In the near future we should be able to study interaction and cooperation of multiple signaling pathways in a complex fashion using functional genomics. However, basic stem-cell technology and a thorough understanding of the biological readout system is absolutely essential for both the generation and interpretation of the results. Transplantation studies will continue to be a hallmark technology to challenge stem-cell plasticity and to evaluate *in vivo* function. The rapid development of stem-cell technology and the powerful tools provided by genomic analyses will lead to an era where meaningful brain repair can be contemplated in the near future.

REFERENCES

1. Frederiksen, K., Jat, P. S., Valtz, N., Levy, D., and McKay, R., Immortalization of precursor cells from the mammalian CNS, *Neuron,* 1, 439–448, 1988.
2. Davis, A. A. and Temple, S., A self-renewing multipotential stem cell in embryonic rat cerebral cortex, *Nature,* 372, 263–266, 1994.
3. Kilpatrick, T. J. and Bartlett, P. F., Cloning and growth of multipotential neural precursors: Requirements for proliferation and differentiation, *Neuron,* 10, 255–265, 1993.
4. Johe, K. K., Hazel, T. G., Müller, T., Dugich-Djordjevic, M. M., and McKay, R. D. G., Single factors direct the differentiation of stem cells from the fetal and adult central nervous system, *Genes Dev.,* 10, 3129–3140, 1996.
5. Till, J. and McCulloch, E., A stochastic model of stem cell proliferation, based on the growth of colony-forming cells, *Proc. Natl. Acad. Sci. USA,* 51, 29–36, 1963.
6. Spangrude, G. J., Heimfeld, S., and Weissman, I. L., Purification and characterization of mouse hematopoietic stem cells, *Science,* 241, 58–62, 1988.
7. Baroffio, A., Dupin, E., and Le Douarin, N. M., Clone-forming ability and differentiation potential of migratory neural crest cells, *Proc. Natl. Acad. Sci. USA,* 85, 5325–5329, 1988.
8. Bronner-Fraser, M. and Fraser, S. E., Cell lineage analysis reveals multipotency of some avian neural crest cells, *Nature,* 335, 161–164, 1988.
9. Stemple, D. L. and Anderson, D. J., Isolation of a stem cell for neurons and glia from the mammalian neural crest, *Cell,* 71, 973–985, 1992.
10. Chenn, A. and McConnell, S. K., Cleavage orientation and the asymmetric inheritance of Notch1 immunoreactivity in mammalian neurogenesis, *Cell,* 82, 631–641, 1995.
11. Zhong, W. M., Feder, J. N., Jiang, M. M., Jan, L. Y., and Jan, Y. N., Asymmetric localization of a mammalian numb homolog during mouse cortical neurogenesis, *Neuron,* 17, 43–53, 1996.
12. Studer, L., Tabar, V., and McKay, R. D., Transplantation of expanded mesencephalic precursors leads to recovery in Parkinsonian rats, *Nature Neurosci.,* 1, 290–295, 1998.
13. Yan, J., Studer, L., and McKay, R. D. G., Ascorbic acid increases the yield of dopaminergic neurons derived from basic fibroblast growth factor expanded mesencephalic precursors, *J. Neurochem.,* 76, 307–311, 2001.
14. Okabe, S., Forsberg-Nilsson, K., Spiro, A. C., Segal, M., and McKay, R. D. G., Development of neuronal precursor cells and functional postmitotic neurons from embryonic stem cells *in vitro, Mech. Dev.,* 59, 89–102, 1996.
15. Lee, S.-H., Lumelsky, N., Studer, L., Auerbach, J. M., and McKay, R. D. G., Efficient generation of midbrain and hindbrain neurons from mouse embryonic stem cells, *Nat. Biotechnol.,* 18, 675–679, 2000.
16. Kawasaki, H., Mizuseki, K., Nishikawa, S., Kaneko, S., Kuwana, Y., Nakanishi, S., Nishikawa, S., and Sasai, Y., Induction of midbrain dopaminergic neurons from ES cells by stromal cell-derived inducing activity, *Neuron,* 28, 31–40, 2000.
17. Brüstle, O., Cunningham, M. G., Tabar, V., and Studer, L., Experimental transplantation into the embryonic, neonatal and adult mammalian brain, in *Current Protocols in Neuroscience,* McKay, R. D. and Gerfen, C. R., Eds., John Wiley & Sons, New York, 1997.
18. Brustle, O., Jones, K. N., Learish, R. D., Karram, K., Choudhary, K., Wiestler, O. D., Duncan, I. D., and McKay, R. G., Embryonic stem cell-derived glial precursors: A source of myelinating transplants, *Science,* 285, 754–756, 1999.

19. Wagner, J., Akerud, P., Castro, D. S., Holm, P. C., Canals, J. M., Snyder, E. Y., Perlmann, T., and Arenas, E., Induction of a midbrain dopaminergic phenotype in Nurr1-overexpressing neural stem cells by type 1 astrocytes, *Nat. Biotechnol.,* 17, 653–659, 1999.

20. Cameron, H. A. and McKay, R. D., Restoring production of hippocampal neurons in old age, *Nat. Neurosci.,* 2, 894–897, 1999.

21. Gould, E., Reeves, A. J., Fallah, M., Tanapat, P., Gross, C. G., and Fuchs, E., Hippocampal neurogenesis in adult Old World primates, *Proc. Natl. Acad. Sci. USA,* 96, 5263–5267, 1999.

22. Gould, E., Beylin, A., Tanapat, P., Reeves, A., and Shors, T. J., Learning enhances adult neurogenesis in the hippocampal formation, *Nat. Neurosci.,* 2, 260–265, 1999.

23. Gould, E., Reeves, A. J., Graziano, M. S., and Gross, C. G., Neurogenesis in the neocortex of adult primates, *Science,* 286, 548–552, 1999.

24. Van Praag, H., Kempermann, G., and Gage, F. H., Running increases cell proliferation and neurogenesis in the adult mouse dentate gyrus, *Nat. Neurosci.,* 2, 266–270, 1999.

25. Kempermann, G., Kuhn, H. G., and Gage, F. H., More hippocampal neurons in adult mice living in an enriched environment, *Nature,* 386, 493–495, 1997.

26. Eriksson, P. S., Perfilieva, E., Bjork-Eriksson, T., Alborn, A. M., Nordborg, C., Peterson, D. A., and Gage, F. H., Neurogenesis in the adult human hippocampus, *Nat. Med.,* 4, 1313–1317, 1998.

27. Craig, C. G., Tropepe, V., Morshead, C. M., Reynolds, B. A., Weiss, S., and van der Kooy, D., *In vivo* growth factor expansion of endogenous subependymal neural precursor cell populations in the adult mouse brain, *J. Neurosci.,* 16, 2649–2658, 1996.

28. Panchision, D., Hazel, T., and McKay, R., Plasticity and stem cells in the vertebrate nervous system, *Curr. Opin. Cell Biol.,* 10, 727–733, 1998.

29. Molne, M., Studer, L., Tabar, V., Ting, Y.-T., Eiden, M. V., and McKay, R. D., Early cortical precursors do not undergo LIF-mediated astrocytic differentiation. *J. Neurosci. Res.,* 59, 301–311, 2000.

30. Hazel, T. G., Panchision, D. M., Warinner, P., and McKay, R. D., Regional plasticity of multipotent precursors from the developing CNS, *Soc. Neurosci. Abstr.,* 23, 319, 1997.

31. Mujtaba, T., Mayer-Proschel, M., and Rao, M. S., A common neural progenitor for the CNS and PNS, *Dev. Biol.,* 200, 1–15, 1998.

32. Valtz, N. L., Hayes, T. E., Norregaard, T., Liu, S. M., and McKay, R. D., An embryonic origin for medulloblastoma, *New Biol.,* 3, 364–371, 1991.

33. Galli, R., Borello, U., Gritti, A., Minasi, M. G., Bjornson, C., Coletta, M., Mora, M., De Angelis, M. G., Fiocco, R., Cossu, G., and Vescovi, A. L., Skeletal myogenic potential of human and mouse neural stem cells, *Nat. Neurosci.,* 3, 986–991, 2000.

34. Bjornson, C. R., Rietze, R. L., Reynolds, B. A., Magli, M. C., and Vescovi, A. L., Turning brain into blood: A hematopoietic fate adopted by adult neural stem cells *in vivo, Science,* 283, 534–537, 1999.

35. Clarke, D. L., Johansson, C. B., Wilbertz, J., Veress, B., Nilsson, E., Karlstrom, H., Lendahl, U., and Frisen, J., Generalized potential of adult neural stem cells [see comments], *Science,* 288, 1660–1663, 2000.

36. McDonald, J. W., Liu, X. Z., Qu, Y., Liu, S., Mickey, S. K., Turetsky, D., Gottlieb, D. I., and Choi, D. W., Transplanted embryonic stem cells survive, differentiate and promote recovery in injured rat spinal cord, *Nat. Med.,* 5, 1410–1412, 1999.

37. Snyder, E. Y., Taylor, R. M., and Wolfe, J. H., Neural progenitor cell engraftment corrects lysosomal storage throughout the MPS VII mouse brain, *Nature,* 374, 367–370, 1995.

38. Benedetti, S., Pirola, B., Pollo, B., Magrassi, L., Bruzzone, M. G., Rigamonti, D., Galli, R., Selleri, S., Di Meco, F., De Fraja, C., Vescovi, A., Cattaneo, E., and Finocchiaro, G., Gene therapy of experimental brain tumors using neural progenitor cells [see comments], *Nat. Med.,* 6, 447–450, 2000.

39. Aboody, K. S., Brown, A., Rainov, N. G., Bower, K. A., Liu, S. X., Yang, W., Small, J. E., Herrlinger, U., Ourednik, V., Black, P. M., Breakefield, X. O., and Snyder, E. Y., Neural stem cells display extensive tropism for pathology in adult brain: Evidence from intracranial gliomas, *Proc. Natl. Acad. Sci. USA,* 97, 12846–12851, 2000.

40. Brazelton, T. R., Rossi, F. M. V., Keshet, G. I., and Blau, H. M., From marrow to brain: Expression of neuronal phenotypes in adult mice, *Science,* 290, 1775–1779, 2000.

41. Mezey, E., Chandross, K. J., Harta, G., Maki, R. A., and McKercher, S. R., Turning blood into brain: Cells bearing neuronal antigens generated *in vivo* from bone marrow, *Science,* 290, 1779–1782, 2000.

42. Josephson, R., Müller, T., Pickel, J., Okabe, S., Reynolds, K., Turner, P. A., Zimmer, A., and McKay, R. D., POU transcription factors control expression of CNS stem cell-specific genes, *Development,* 125, 3087–3100, 1998.

43. Lendahl, U., Zimmerman, L. B., and McKay, R. D., CNS stem cells express a new class of intermediate filament protein, *Cell,* 60, 585–595, 1990.

44. Sakakibara, S., Imai, T., Hamaguchi, K., Okabe, M., Aruga, J., Nakajima, K., Yasutomi, D., Nagata, T., Kurihara, Y., Uesugi, S., Miyata, T., Ogawa, M., Mikoshiba, K., and Okano, H., Mouse-Musashi-1, a neural RNA-binding protein highly enriched in the mammalian CNS stem cell, *Dev. Biol.,* 176, 230–242, 1996.

45. Uchida, N., Buck, D. W., He, D. P., Reitsma, M. J., Masek, M., Phan, T. V., Tsukamoto, A. S., Gage, F. H., and Weissman, I. L., Direct isolation of human central nervous system stem cells, *Proc. Natl. Acad. Sci. USA,* 97, 14720–14725, 2000.

46. Reynolds, B. A. and Weiss, S., Generation of neurons and astrocytes from isolated cells of the adult mammalian central nervous system, *Science,* 255, 1707–1710, 1992.

47. Gritti, A., Parati, E. A., Cova, L., Frolichsthal, P., Galii, R., Wanke, E., Faravelli, L., Morassutti, D. J., Roisen, F., Nickel, D. D., Vescovi, A. L., and Galli, R., Multipotential stem cells from the adult mouse brain proliferate and self-renew in response to basic fibroblast growth factor, *J. Neurosci.,* 16, 1091–1100, 1996.

48. Weiss, S., Dunne, C., Hewson, J., Wohl, C., Wheatley, M., Peterson, A. C., and Reynolds, B. A., Multipotent CNS stem cells are present in the adult mammalian spinal cord and ventricular neuroaxis, *J. Neurosci.,* 16, 7599–7609, 1996.

49. Svendsen, C. N., Fawcett, J. W., Bentlage, C., and Dunnett, S. B., Increased survival of rat EGF-generated CNS precursor cells using B27 supplemented medium, *Exp. Brain Res.,* 102, 407–414, 1995.

50. Svendsen, C. N., Ter Borg, M. G., Armstrong, R. E., Rosser, A. E., Chandran, S., Ostenfeld, T., and Caldwell, M. A., A new method for the rapid and long term growth of human neural precursor cells, *J. Neurosci. Methods,* 85, 141–152, 1998.

51. Galli, R., Pagano, S. F., Gritti, A., and Vescovi, A. L., Regulation of neuronal differentiation in human CNS stem cell progeny by leukemia inhibitory factor, *Dev. Neurosci.,* 22, 86–95, 2000.

52. Palmer, T. D., Ray, J., and Gage, F. H., FGF-2-responsive neuronal progenitors reside in proliferative and quiescent regions of the adult rodent brain, *Mol. Cell. Neurosci.,* 6, 474–486, 1995.

53. Qian, X., Goderie, S. K., Shen, Q., Stern, J. H., and Temple, S., Intrinsic programs of patterned cell lineages in isolated vertebrate CNS ventricular zone cells, *Development,* 125, 3143–3152, 1998.

54. Suhonen, J. O., Peterson, D. A., Ray, J., and Gage, F. H., Differentiation of adult hippocampus-derived progenitors into olfactory neurons *in vivo*, *Nature*, 383, 624–627, 1996.

55. Hazel, T., Culture of neuroepithelial stem cells, in *Current Protocols in Neuroscience*, McKay, R. D. and Gerfen, C. R., Eds., John Wiley & Sons, New York, 1997.

56. Paxinos, G. and Watson, C., *The Rat Brain in Stereotactic Coordinates*, Academic Press, New York, 2000.

57. Kennedy, M., Firpo, M., Chol, K., Wall, C., Robertson, S., Kabrun, N., and Keller, G., A common precursor for primitive erythropoiesis and definitive haematopoiesis, *Nature*, 386, 488–493, 1997.

58. Metzger, J. M., Lin, W. I., and Samuelson, L. C., Transition in cardiac contractile sensitivity to calcium during the *in vitro* differentiation of mouse embryonic stem cells, *J. Cell Biol.*, 126, 701–711, 1994.

59. Li, M., Pevny, L., Lovell-Badge, R., and Smith, A., Generation of purified neural precursors from embryonic stem cells by lineage selection, *Curr. Biol.*, 8, 971–974, 1998.

60. Williams, B. P., Park, J. K., Alberta, J. A., Muhlebach, S. G., Hwang, G. Y., Roberts, T. M., and Stiles, C. D., A PDGF-regulated immediate early gene response initiates neuronal differentiation in ventricular zone progenitor cells, *Neuron*, 18, 553–562, 1997.

61. Rajan, P. and McKay, R. D., Multiple routes to astrocytic differentiation in the CNS, *J. Neurosci.*, 18, 3620–3629, 1998.

62. Bonni, A., Sun, Y., Nadalvicens, M., Bhatt, A., Frank, D. A., Rozovsky, I., Stahl, N., Yancopoulos, G. D., and Greenberg, M. E., Regulation of gliogenesis in the central nervous system by the JAK-STAT signaling pathway, *Science*, 278, 477–483, 1997.

63. Burrows, R. C., Wancio, D., Levitt, P., and Lillien, L., Response diversity and the timing of progenitor cell maturation are regulated by developmental changes in EGFR expression in the cortex, *Neuron*, 19, 251–267, 1997.

64. Qian, X. M., Shen, Q., Goderie, S. K., He, W. L., Capela, A., Davis, A. A., and Temple, S., Timing of CNS cell generation: A programmed sequence of neuron and glial cell production from isolated murine cortical stem cells, *Neuron*, 28, 69–80, 2000.

65. Sun, Y., Nadal-Vicens, M., Misono, S., Lin, M. Z., Zubiaga, A., Hua, X., Fan, G., and Greenberg, M. E., Neurogenin promotes neurogenesis and inhibits glial differentiation by independent mechanisms, *Cell*, 104, 365–376, 2001.

66. Nakashima, K., Yanagisawa, M., Arakawa, H., Kimura, N., Hisatsune, T., Kawabata, M., Miyazono, K., and Taga, T., Synergistic signaling in fetal brain by STAT3-Smad1 complex bridged by p300, *Science*, 284, 479–482, 1999.

67. Roy, N. S., Wang, S., Jiang, L., Kang, J., Benraiss, A., Harrison-Restelli, C., Fraser, R. A. R., Couldwell, W. T., Kawaguchi, A., Okano, H., Nedergaard, M., and Goldman, S. A., *In vitro* neurogenesis by progenitor cells isolated from the adult human hippocampus, *Nat. Med.*, 6, 271–277, 2000.

68. Studer, L., Lee, S.-H., Panchision, D. M., Pickel, J., and McKay, R. D., Molecular control of dopaminergic differentiation in bFGF expanded midbrain precursors, *Soc. Neurosci. Abstr.*, p. 1344, 2000.

69. Baker, R. K., Haendel, M. A., Swanson, B. J., Shambaugh, J. C., Micales, B. K., and Lyons, G. E., *In vitro* preselection of gene-trapped embryonic stem cell clones for characterizing novel developmentally regulated genes in the mouse, *Dev. Biol.*, 185, 201–214, 1997.

70. Lyons, G. E., Swanson, B. J., Haendel, M. A., and Daniels, J., Gene trapping in embryonic stem cells *in vitro* to identify novel developmentally regulated genes in the mouse, *Methods Mol. Biol.*, 136, 297–307, 2000.

71. Salminen, M., Meyer, B. I., and Gruss, P., Efficient poly A trap approach allows the capture of genes specifically active in differentiated embryonic stem cells and in mouse embryos, *Dev. Dyn.,* 212, 326–333, 1998.

72. Thorey, I. S., Muth, K., Russ, A. P., Otte, J., Reffelmann, A., and von Melchner, H., Selective disruption of genes transiently induced in differentiating mouse embryonic stem cells by using gene trap mutagenesis and site-specific recombination, *Mol. Cell Biol.,* 18, 3081–3088, 1998.

73. Studer, L., Csete, M., Lee, S.-H., Kabbani, N., Walikonis, J., Wold, B., and McKay, R. D., Enhanced proliferation, survival and dopaminergic differentiation of CNS precursors in lowered oxygen, *J. Neurosci.,* 20, 7377–7383, 2000.

74. Geschwind, D. H., Ou, J., Easterday, M. C., Dougherty, J. D., Jackson, R. L., Chen, Z., Antoine, H., Terskikh, A., Weissman, I. L., Nelson, S. F., and Kornblum, H. I., A genetic analysis of neural progenitor differentiation, *Neuron,* 29, 325–339, 2001.

75. Brüstle, O., Maskos, U., and McKay, R. D. G., Host-guided migration allows targeted introduction of neurons into the embryonic brain, *Neuron,* 15, 1275–1285, 1995.

76. Brüstle, O., Choudhary, K., Karram, K., Huttner, A., Murray, K., Dubois-Dalcq, M., and McKay, R. G., Chimeric brains generated by intraventricular transplantation of fetal human brain cells into embryonic rats, *Nat. Biotechnol.,* 16, 1040–1044, 1998.

77. Brüstle, O., Spiro, A. C., Karram, K., Choudhary, K., Okabe, S., and McKay, R. G., *In vitro*-generated neural precursors participate in mammalian brain development, *Proc. Natl. Acad. Sci. USA,* 94, 14809–14814, 1997.

78. Fishell, G., Striatal precursors adopt cortical identities in response to local cues, *Development,* 121, 803–812, 1995.

79. Campbell, K., Olsson, M., and Björklund, A., Regional incorporation and site-specific differentiation of striatal precursors transplanted to the embryonic forebrain ventricle, *Neuron,* 15, 1259–1273, 1995.

80. Svendsen, C. N., Clarke, D. J., Rosser, A. E., and Dunnett, S. B., Survival and differentiation of rat and human epidermal growth factor-responsive precursor cells following grafting into the lesioned adult central nervous system, *Exp. Neurol.,* 137, 376–388, 1996.

81. Fricker, R. A., Carpenter, M. K., Winkler, C., Greco, C., Gates, M. A., and Bjorklund, A., Site-specific migration and neuronal differentiation of human neural progenitor cells after transplantation in the adult rat brain, *J. Neurosci.,* 19, 5990–6005, 1999.

82. Sinclair, S. R., Fawcett, J. W., and Dunnett, S. B., Dopamine cells in nigral grafts differentiate prior to implantation, *Eur. J. Neurosci.,* 11, 4341–4348, 1999.

83. Tabar, V. and Studer, L., Transplantation into the adult rodent brain. in *Current Protocols in Neuroscience,* McKay, R. D. and Gerfen, C. R., Eds., John Wiley & Sons, New York, 1997.

84. Gundersen, H. J. G., Bendtsen, T. F., Korbo, L., Marcussen, N., Moller, A., Nielsen, K., Nyengaard, J. R., Pakkenberg, B., Soerensen, F. B., Vesterby, A., and West, M. J., Some new, simple and efficient stereological methods and their use in pathological research and diagnosis, *APIMS,* 96, 379–394, 1988.

85. Woodbury, D., Schwarz, E. J., Prockop, D. J., and Black, I. B., Adult rat and human bone marrow stromal cells differentiate into neurons, *J. Neurosci. Res.,* 61, 364–370, 2000.

86. Munsie, M. J., Michalska, A. E., O'Brien, C. M., Trounson, A. O., Pera, M. F., and Mountford, P. S., Isolation of pluripotent embryonic stem cells from reprogrammed adult mouse somatic cell nuclei, *Curr. Biol.,* 10, 989–992, 2000.

87. Wakayama, T., Tabar, V., Rodriguez, I., Perry, A. C. F., Studer, L., and Mombaerts, P., Dopaminergic neurons from adult somatic cells via nuclear transfer, *Science,* in press.

88. Venter, J. C. et al., The sequence of the human genome, *Science,* 291, 1304, 2001.

89. Lander, E. S. et al., Initial sequencing and analysis of the human genome, *Nature,* 409, 860–921, 2001.

90. Tomb, J. F., White, O., Kerlavage, A. R., Clayton, R. A., Sutton, G. G., Fleischmann, R. D., Ketchum, K. A., Klenk, H. P., Gill, S., Dougherty, B. A., Nelson, K., Quackenbush, J., Zhou, L., Kirkness, E. F., Peterson, S., Loftus, B., Richardson, D., Dodson, R., Khalak, H. G., Glodek, A., McKenney, K., Fitzegerald, L. M., Lee, N., Adams, M. D., and Venter, J. C., The complete genome sequence of the gastric pathogen Helicobacter pylori, *Nature,* 388, 539–547, 1997.

91. Klenk, H. P., Clayton, R. A., Tomb, J. F., White, O., Nelson, K. E., Ketchum, K. A., Dodson, R. J., Gwinn, M., Hickey, E. K., Peterson, J. D., Richardson, D. L., Kerlavage, A. R., Graham, D. E., Kyrpides, N. C., Fleischmann, R. D., Quackenbush, J., Lee, N. H., Sutton, G. G., Gill, S., Kirkness, E. F., Dougherty, B. A., McKenney, K., Adams, M. D., Loftus, B., and Venter, J. C., The complete genome sequence of the hyperthermophilic, sulfate-reducing archaeon Archaeoglobus fulgidus, *Nature,* 390, 364–370, 1997.

92. Goffeau, A., Barrell, B. G., Bussey, H., Davis, R. W., Dujon, B., Feldmann, H., Galibert, F., Hoheisel, J. D., Jacq, C., Johnston, M., Louis, E. J., Mewes, H. W., Murakami, Y., Philippsen, P., Tettelin, H., and Oliver, S. G., Life with 6000 genes, *Science,* 274, 546, 1996.

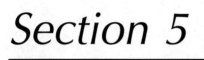

Section 5

12 Statistical Methods to Discover Susceptibility Genes for Nervous System Diseases

Jurg Ott

CONTENTS

12.1 INTRODUCTION

Close to 150 years ago, Gregor Mendel carried out experiments on the garden pea.[1] He was interested in characteristics (phenotypes) such as flower color and seed shape that show different well-defined characteristics. For example, flower color can be red or white. Specifically, he wanted to know how these phenotypes are expressed in offspring when they were different in the two parental plants. Through his breeding experiments he found what are now known as the Mendelian laws. He showed that "factors" (now called genes) determining these characteristics are passed from parents to offspring and do not change from one to the next generation. Also, some of them may be hidden (recessive) behind others (that are dominant).[2] The two main aspects of the Mendelian laws are segregation of genes through generations, and independent assortment of different genes.

Mendel worked on phenotypes that are highly heritable and that show "Mendelian" inheritance. How do we assess heritability? Essentially, heritability is measured by the extent that a trait (disease) runs in families, that is, by its familiality. There

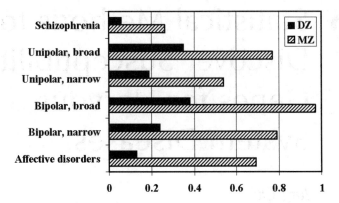

FIGURE 12.1 Concordance rates for psychiatric traits, based on other data.[3,ch. 16] The values for affective disorders are from older publications. For schizophrenia, results from various reports in Europe and the U.S. since 1964 have been combined. DZ = dizygotic twin; MZ = monozygotic twin.

are well-known traits that run in families yet are not genetically heritable, for example, measles. These may be transmitted through infection rather than genes. We usually assume, however, that such situations are recognized as nongenetic forms of transmission. Thus, familiality is still the yardstick by which the extent of heritability is measured. In particular, we establish concordance rates for various pairs of relatives, e.g., twins, cousins, etc. The concordance rate is the probability that the second member of the pair is affected given the first member is affected. Such rates must be established with special epidemiological methods because great care must be taken that ascertainment biases are avoided or properly taken into account. Heritability is often assessed through the difference in concordance rates for monozygotic (MZ) and dizygotic (DZ) twins, because MZ twins share 100% of their genes while DZ twins share on average only 50% of their genes. Figure 12.1 shows concordance rates for some psychiatric traits.[3] For example, the concordance rate for schizophrenia is 0.26 in MZ twins but only 0.06 in DZ twins. Clearly, these traits are highly heritable, i.e., they must be due to underlying gene(s).

Under specific mathematical models, heritability may be measured in terms of a proportion. A common mathematical model assumes that genetic and environmental and other factors (e.g., common family environment) act additively and independently. Heritability is then defined as the proportion of the variance of a phenotype that is explained by genetic contributions. For example, schizophrenia shows a heritability of 75–80%.

Many traits are heritable but their inheritance pattern is not Mendelian. Regarding human diseases, it is the rare ones that exhibit a Mendelian mode of inheritance. They cannot achieve a high frequency because selection works against them and keeps them at low frequencies. Common human traits, although heritable, generally show non-Mendelian inheritance patterns and are believed to be under the control of a number of susceptibility genes, only some of which are necessary for expression of the trait. This situation tends to shield them from strong selection.

12.2 CROSSING OVER AND GENETIC DISTANCE

There is a particular deviation from the Mendelian rules that we can profitably take advantage of. When two genes are in close physical proximity of each other, they are not inherited independently. Therefore, they show deviations from the Mendelian laws. Consider a genetic locus with two alleles numbered *1* and *2*. When a parent (a father, say) is heterozygous at that locus (has genotype *1/2*), then he will pass each of these two alleles with probability ½ to each offspring. This must also be the case for the grandparents of this parent, so that in a given offspring the chance is ½ that a given allele originated in the paternal grandmother or the paternal grandfather. However, consider a dominantly inherited disease with susceptibility allele *D*. Assume that the grandmother carried a *D* allele and had marker genotype *1/1*. She passed the *D* allele to her son (the father) who, in turn, passed it to an offspring. Grandmother, father, and offspring are then all affected with the trait. In this case, all affected offspring will receive the *D* and *1* alleles that both originated in the grandmother. The trait and marker loci are said to be genetically linked, because the *D* and *1* alleles passed by the grandmother appear to travel together from one generation to the next in a "linked" fashion. Marker locations are known on the human gene map but the location of the susceptibility gene may not be. Thus, with a dense map of markers, one may localize a trait susceptibility gene by testing each marker in turn whether it appears genetically linked to the trait locus.

Genetic linkage is not necessarily perfect. When a trait and a marker locus are at some distance from each other, a phenomenon called *crossing over* may occur in a parent such that an offspring receives an allele at the trait locus from one grandparent and an allele at the marker locus from the other grandparent. It is then the proportion of times that the two relevant alleles originated in different grandparents (the so-called recombination fraction) that reflects the distance between the two loci. Such crossing over may be observed in genetic experiments, but it was soon recognized that it corresponds to the cytogenetic observation that chromosome arms cross over each other in the course of meiosis. Because this happens more or less randomly along a chromosome, it occurs the more frequently between two loci the farther apart they are from each other. As outlined in the following paragraph, we are able to make use of this randomness to define a statistical concept called genetic distance.

The phenomenon of crossing over is the basis for statistical gene mapping. In the course of meiosis, crossovers between homologous chromosome strands lead to reciprocal exchanges such that in a resulting gamete, crossovers mark boundaries between genes originating from one grandparent and those originating from another grandparent.[4] On a gamete, the number of crossover points between two loci is defined as the genetic distance between them. For loci close together, the recombination fraction is essentially equal to genetic distance. For example, a recombination fraction of $\theta = 0.03$ corresponds to a genetic distance of 0.03 M (Morgans) or 3 cM (centiMorgans).

The concept of genetic distance has allowed the development of sophisticated statistical approaches for genetic mapping. Without actually seeing genetic loci, we can estimate the genetic distance between them, which is one of the fascinating aspects of genetic mapping. The basic idea is that pedigree data are collected, in which a particular heritable disease occurs, and family members are genotyped for a large number of genetic markers, which are known to reside along each of the human chromosomes.

Recombinations often cannot be observed unequivocally, for example, when a trait exhibits incomplete penetrance (not all individuals with a given genotype will express the phenotype) expression or when there are missing observations. Also, for complex traits, it is presumably multiple underlying susceptibility loci that can lead to disease so that an individual may be affected because of a trait locus at a position different from the one under study. For all these reasons, we must estimate recombination fractions and disease locus positions by a statistical estimation procedure called the maximum likelihood method. The likelihood is defined as the probability of occurrence of the data. We want to find that value of the recombination fraction for which the likelihood is largest, that is, that makes the data most plausible. Specifically, to localize the gene(s) underlying the disease on the human gene map, various positions of the hypothesized gene are assumed on a chromosome and for each assumed position, x, the likelihood, $L(x)$, is computed with the aid of a suitable computer program.[4] Varying values for x changes the resulting likelihood, $L(x)$. The value of x, at which the likelihood is highest represents the maximum likelihood estimate of the disease locus position. Results for all data are usually represented as a curve of lod scores, $Z(x) = \log_{10}[L(x)/L(x = \infty)]$, for various assumed disease locus positions, where the likelihood in the denominator refers to the situation that the disease locus is off the marker map (i.e., at position ∞ = infinity). Thus, the lod score is the logarithm of the odds for linkage of a disease gene to a given map position. Peaks of this lod score curve exceeding the value 3 are generally taken to indicate significant evidence for the position of a disease locus at that place.[5]

This brief description of gene mapping refers to genetic linkage analysis. A well-known alternative mapping method is based on linkage disequilibrium (association), which is the consequence of disease mutations in previous generations and the paucity of recombination between the disease locus and nearby marker loci. Therefore, alleles at such closely linked loci tend to appear associated over many generations after an initial disease mutation has occurred (see below).

12.3 MENDELIAN DISEASES

To estimate the position of a disease locus on a chromosome, one generally collects families in which the disease segregates. For example, Huntington disease is transmitted as a dominant trait with age-dependent penetrance. Close to 20 years ago, the gene underlying this trait was localized to a small region on human chromosome 4.[6] Pedigree lod scores had been obtained with the aid of the LIPED computer program,[7] as was done for most disease gene localizations in the 1970s and 1980s.

The likelihood for pedigree data is generally computed in a recursive fashion, i.e., calculations are done on small portions of the data at a time in such a way that the end result is the correct overall likelihood. Two successful recursion methods have been developed. The Elston-Stewart algorithm[8] recurses over individuals in a pedigree and allows for likelihood calculations on large pedigrees but only a small number of loci jointly, whereas the Lander-Green algorithm[9] recurses over loci and can generally accommodate all loci on a chromosome but can handle only small pedigrees. The former method has been implemented, for example, in the LIPED[7] and LINKAGE[10] programs, whereas the latter forms the basis for the GENEHUNTER[11] and ALLEGRO[12] programs.

Localization of a gene underlying a recessive or dominant trait is now relatively straightforward. However, there may still be complications that need to be mastered. For example, heterogeneity, when it is not properly allowed for, can mask the presence of a disease locus. Particularly for recessive traits with their typically small families, heterogeneity can be a major obstacle. Currently, the main challenge for statistical geneticists is finding genes underlying complex traits.

12.4   COMPLEX TRAITS

Many traits are heritable (run in families) and, thus, must at least partially be controlled by genes, yet they do not follow one of the known Mendelian modes of inheritance. They are generally referred to as complex traits. For example, schizo-phrenia has a heritability of at least 75%, i.e., a major portion of the phenotypic variance is due to genetic factors.

Linkage analysis methods are generally divided into two groups, parametric and nonparametric methods. In the former, a (Mendelian) mode of inheritance for the trait is defined and analysis is carried out as if the trait followed the specified inheritance model. Clearly, this means that the data are analyzed under "wrong" assumptions. However, many research articles have shown that two-point analysis is robust against many model misspecifications. For example, a recent article pro-posed a simple, fixed inheritance model, one for recessive and one for dominant inheritance, and demonstrated on the basis of simulated data that this approach is powerful in a wide range of complex inheritance models.[13]

Nonetheless, many researchers prefer approaches not requiring specification of an inheritance model for the trait (in human genetics, these are referred to parameter-free or nonparametric approaches). A well-known example is the affected sib pair (ASP) method. Ascertainment focuses on siblings affected with a trait, and their parents. The only connection to disease is through this kind of ascertainment. All calculations are done on the basis of marker inheritance. Specifically, for a given marker locus, the statistic of interest is whether a parent passes the same allele to the two offspring or not. If both receive the same allele from a parent, they are said to share this allele identical by descent (IBD). If they receive a different allele from a parent, there is no IBD sharing. According to the Mendelian rules, the two cases (IBD sharing — yes or no) occur with equal probabilities. However, if a suscepti-bility locus is in the vicinity of the given marker, one or the other of the parents may carry a disease allele in coupling with a particular marker allele, which is then more likely to be transmitted to both affected offspring. Therefore, allele sharing, significantly increased over the null value of 0.5, is considered evidence for the presence of a susceptibility locus.

As mentioned above, the presence of association (linkage disequilibrium) between a trait and a marker is generally also taken to indicate presence of a disease locus near that marker. A common data type for association studies rests on the case-control design. A number of individuals affected with a trait are ascertained, along with a number of control individuals not exhibiting the trait. Association is assumed present when the frequency distribution for marker genotypes or alleles is significantly different between cases and controls. Such case-control studies may be more powerful than linkage studies for the detection of weak trait loci.[14]

Major efforts are underway to develop genetic maps of single-nucleotide polymorphisms (SNP), that is, two-allelic PCR markers that are the most common source of human genetic sequence variation. Maps consisting of thousands of SNP will be used to localize genes underlying complex traits, for example, through case-control association studies or ASP linkage analyses. One of the shortcomings of most current approaches is that they look for one susceptibility gene after another, which disregards interactions among multiple susceptibility genes (this is also true for multipoint approaches, where "multi" refers to multiple marker loci, not multiple disease genes). Ideally, marker genotypes should be evaluated jointly as independent variables (inputs), where the dependent variable (output) might be binary (case-control data) or a multivariate quantitative trait such as the lod score obtained for each marker. The problem here is that the number of variables tends to be much higher than the number of observations. To solve this dilemma, a two-stage approach has been proposed in which a small subset of "relevant" marker loci is selected, and these are subsequently used for modeling of interactions.[15]

12.4.1 TWO-STAGE APPROACH TO GENOME SCREENS

The following discussion focuses on binary outcome data, for example, case-control data or two types of sib pairs, affected–affected and affected–unaffected, but the approaches outlined are applicable to a much wider range of study designs. The first stage will consist of selecting a small number of marker loci whose genotype or allele frequency distributions differ between the two outcome variables. In the second stage, various levels of interactions may be modeled, for example, through multiple logistic regression, neural network analysis,[16] or similar approaches. Below, a specific published approach to initial marker selection is outlined, Sequential Bootstrap Marker Selection.[15]

Consider a single marker and a relevant statistic for discrimination between cases and controls. For example, observations for the binary outcome variable (cases-controls) and the three marker genotypes may form a 2×3 contingency table, with χ^2 being the corresponding single-locus statistic. The basic idea behind this marker selection procedure is to form a univariate statistic, which is the sum of single-marker statistics over multiple (generally noncontiguous) marker loci. Other ways of combining information over marker loci are conceivable, but a sum seems a simple and straightforward approach. At this stage, interactions between markers are not taken into account except for those interactions that are implicit in the sum. Statistical significance of such a sum is evaluated by a sequential bootstrap procedure.[17-19]

Overall, the approach for marker selection is carried out as follows. Consider data of size n with disease phenotype, y, and K single-nucleotide markers with observed marker statistics, $(t_1, t_2, ..., t_K)$. Take B_1 bootstrap replicates of the same size from the original data. They represent variations of the observed data and will serve to eventually validate the selection procedure described below. From the original data and each of the B_1 bootstrap samples, take B_2 bootstrap samples under no association to calculate significance levels of the statistics.

Specifically, in the original data and each of its B_1 bootstrap replicates, order the single-locus statistics by size and form sums of K, $K - 1$, $K - 2$, etc. of these statistics, each time leaving out the smallest single-locus statistic. Evaluate the significance of these sums through the B_2 bootstrap samples and, in the original data and each of the B_1 replicates, repeat this process until the resulting sum is significant at the $\alpha = 0.05$ level. This leads to a preselection of a possibly different set of markers in the original data and each of the B_1 replicates. Now count the frequency, with which each marker is preselected over the $B_1 + 1$ data sets. A marker is selected for further analysis if this frequency is appreciable, for example, at least 50%.

12.4.2 EXAMPLE: RESTENOSIS IN HEART DISEASE

The marker selection procedure described above was carried out for a prospective case-control study on 779 heart disease patients, who had undergone angioplasty (opening of main artery). After 1 year, 342 of them (here called "cases") experienced a restenosis, that is, a renewed closing of arterial walls, while the remaining 437 individuals ("controls") did not experience restenosis. Based on previous work[20,63] candidate genes were hypothesized to be involved in the process of restenosis, and 1–2 SNP were genotyped in each gene, leading to a total of 89 genotyped SNP per individual. Of course, for this relatively small number of markers, a preselection procedure is not absolutely necessary, but researchers might have carried out a genome screen with thousands of markers, in which case some preselection would have been mandatory if marker genotypes are to be analyzed jointly.

The single-marker statistic of choice was χ^2 for the 2×3 contingency table mentioned above. For the selection procedure, bootstrap sample sizes of $B_1 = 499$ and $B_2 = 1000$ were used. The procedure eventually selected 11 out of the initial 89 markers as being relevant for the disease.[21]

In a second stage, logistic regression was used for modeling of the effects of the 11 markers. Linear and quadratic main effects per marker were introduced by defining two independent dummy variables, $(-1, 0, 1)$ and $(-\frac{1}{2}, 1, -\frac{1}{2})$, respectively, representing the 2 degrees of freedom of each 2×3 table. The first dummy variable reflects the allelic effects of a SNP marker, that is, the number of its 2 alleles, where the second dummy variable represents a quadratic deviation from the allelic effects. Thus, the two variables can accommodate dominance effects in the markers. For the restenosis data, significant linear and quadratic main effects, and significant two-way and three-way interactions were found.

12.5 SCAN STATISTICS FOR LOCALIZING SUSCEPTIBILITY GENES

In genome screens, lod scores are computed at each marker locus and local peaks of the resulting lod score curve are potentially indicative of the location of a susceptibility locus. The height of a local peak is taken as a measure for the degree of

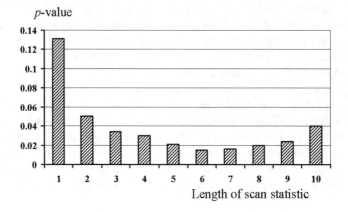

FIGURE 12.2 Results of the application of scan statistics to a genome screen for autism.

statistical significance at that position. Investigations have shown that, for equal peak height, true peaks tend to be wider than false peaks.[22] That is, a peak at a susceptibility locus is broader than a peak occurring as a random fluctuation. This property of true peaks can be taken advantage of by jointly considering lod scores at markers surrounding a local peak, rather than just the peak height by itself. One way of accomplishing this is by the use of scan statistics.[23] Consider a sequence of n consecutive markers, for example, markers 1 through n, or 2 through $(n + 1)$, or 3 through $(n + 2)$, etc. For each such sequence, compute the sum of the lod scores at the markers in the sequence. The largest sum over all possible such sequences is called the scan statistic, S_n, of length n. Clearly, scan statistics are somewhat reminiscent of Bayesian posterior probabilities, but scan statistics were introduced to human gene mapping in a pure likelihood ratio framework. Note also that summing lod scores over multiple markers is completely unrelated to haplotypes.

Applications of scan statistics to real data show that this approach tends to be more powerful than simply looking at maximum peak height. For example, in a genome screen for autism, 324 microsatellite markers were genotyped in two types of sib pairs, i.e., 86 affected–affected (AA) and 91 affected–unaffected (AU) sib pairs. Lod scores were computed for each sib pair and at each marker, the total lod was calculated as the sum of the lod scores over all sib pairs (lod scores for AU pairs generally were negative). Scan statistics were then applied to these total lod scores.[23] Significance levels associated with the scan statistic of a given length were approximated by permutation (randomization) tests. A graph of these p-values against lengths of scan statistics is shown in Figure 12.2 and clearly demonstrates the increased power of scan statistics over simple peak height, which is given by the scan statistic, S_1, of "length" 1.

Clearly, scan statistics do not address the specific properties of complex traits. That is, they do not consider multiple susceptibility loci jointly. However, they do add power to the evidence for linkage at local positions.

REFERENCES

1. Bowler, P.J., *The Mendelian Revolution*, Johns Hopkins University Press, Baltimore, 1989.
2. Strickberger, M.W., *Genetics*, 3rd ed., Macmillan, New York, 1985.
3. Vogel, F. and Motulsky, A.G., *Human Genet.*, 3rd ed., Springer, Berlin, 1997.
4. Ott, J., *Analysis of Human Genetic Linkage*, 3rd ed., Johns Hopkins University Press, Baltimore, 1999.
5. Morton, N.E., Sequential tests for the detection of linkage, *Am. J. Hum. Genet.*, 7, 277–318, 1955.
6. Gusella, J.F. et al., A polymorphic DNA marker genetically linked to Huntington's disease, *Nature*, 306, 234–238, 1983.
7. Ott, J., Estimation of the recombination fraction in human pedigrees: Efficient computation of the likelihood for human linkage studies, *Am. J. Hum. Genet.*, 26, 588–597, 1974.
8. Elston, R.C. and Stewart, J., A general model for the analysis of pedigree data, *Hum. Hered.*, 21, 523–542, 1971.
9. Lander, E.S. and Green, P., Construction of multilocus genetic maps in humans, *Proc. Natl. Acad. Sci. USA*, 84, 2363–2367, 1987.
10. Lathrop, G.M. et al., Strategies for multilocus linkage analysis in humans, *Proc. Natl. Acad. Sci. USA*, 81, 3443–3446, 1984.
11. Kruglyak, L., et al., Parametric and nonparametric linkage analysis: A unified multipoint approach, *Am. J. Hum. Genet.*, 58, 1347–1363, 1996.
12. Gudbjartsson, D.F., et al., Allegro, a new computer program for multipoint linkage analysis, *Nat. Genet.*, 25, 12–13, 2000.
13. Abreu, P.C., Greenberg, D.A., and Hodge, S.E., Direct power comparisons between simple LOD scores and NPL scores for linkage analysis in complex diseases, *Am. J. Hum. Genet.*, 65, 847–857, 1999.
14. Risch, N. and Merikangas, K., The future of genetic studies of complex human diseases, *Science*, 273, 1516–1517, 1996.
15. Hoh, J., et al., Selecting SNP in two-stage analysis of disease association data: A model-free approach, *Ann. Hum. Genet.*, 64, 413–417, 2000.
16. Lucek, P., et al., Multi-locus nonparametric linkage analysis of complex trait loci with neural networks, *Hum. Hered.*, 48, 275–284, 1998.
17. Efron, B., Computers and the theory of statistics: Thinking the unthinkable, *SIAM Review*, 21, 460–480, 1979.
18. Diaconis, P. and Efron, B., Computer-intensive methods in statistics, *Sci. Am.*, 248, 116–130, 1983.
19. Efron, B. and Tibshirani, R., Statistical data analysis in the computer age, *Science*, 253, 390–395, 1991.
20. Cheng, S., et al., A multilocus genotyping assay for candidate markers for cardiovascular disease risk, *Genome Res.*, 9, 936–949, 1999.
21. Zee, R.L. et al., Multi-locus interactions predict risk for post-PTCA restenosis: Approach to the genetic analysis of common complex disease, submitted, 2001.
22. Terwilliger, J.D. et al., True and false positive peaks in genomewide scans: Applications of length-biased sampling to linkage mapping, *Am. J. Hum. Genet.*, 61, 430–438, 1997.
23. Hoh, J. and Ott, J., Scan statistics to scan markers for susceptibility genes, *Proc. Natl. Acad. Sci. USA*, 97, 9615–9617, 2000.

13 Genetic Variation Analysis of Neuropsychiatric Traits

Michael E. Zwick, David J. Cutler, and Aravinda Chakravarti

CONTENTS

13.1 INTRODUCTION

The recent history and rapid progress of human genetics has largely consisted of the application of Mendelian genetics to identify rare allelic variants with large phenotypic effects. The rare variants that underlie "Mendelian traits" have a number of characteristics that have suggested the experimental approach employed to identify them. These rare variants are often deleterious missense mutations of very recent origin. Although rare in the entire human population, they are shared by the closely related members of a family. Within families, their genetics is usually simple — often only one or two different missense mutations in a single candidate gene lead to the same phenotype. Subsequent surveys of genetic variation among unrelated, phenotypically similar individuals often identify an extraordinary number of rare missense mutations in the same candidate gene.[1] In all of these cases, however, because the mutations are of recent origin, they are significantly associated with other pre-existing genetic variation in the surrounding genomic region. It is because of these many characteristics, even with few markers and incomplete genetic maps, that the mapping and identification of Mendelian traits has progressed so rapidly.[1,2]

But simple Mendelian traits are the exception, not the rule. The vast majority of human phenotypic differences, such as those influencing morphology, physiology, and behavior, are quantitative and complex in nature.[3] Complex traits exhibit a nearly normal distribution of phenotypes in human populations. Many *complex diseases* in humans may simply reflect the extremes of otherwise normal phenotypic traits.[4] Human complex traits, in general, have proven refractory to genetic analysis. Clearly, among the most difficult are human neuropsychiatric traits.[5-7] Traits, such as schizophrenia[8-13] or bipolar disorder,[14,15] must be complex in nature, for if they were fundamentally Mendelian, we would probably already understand their genetic basis.[5-7] With the recent completion of a reference human genome and genome-wide characterization of the patterns of human genetic variation,[16-18] human geneticists will now have accurate maps and many markers to employ in their genetic analysis of complex traits. In principle, it should now be possible to finally focus genetics on those traits that compose the vast majority of human phenotypic diversity.

But what has made the genetic analysis of complex traits so intractable? To a great extent, much of the difficulty must arise from differences in the patterns of genetic variation in complex and simple Mendelian traits. In contrast to those variants causing Mendelian traits, the genetic variants underlying complex traits may not consist solely of rare, deleterious missense mutations. The phenotypic effects of these genetic variants may be relatively small and not uniformly deleterious, while their interactions with other variants and the environment may be more complicated. They may have different times of origin and may or may not be in linkage disequilibrium with other variants in the same region, that themselves are of different ages. In addition, much of the difficulty must also arise from the lack of a human reference sequence and the fact that most of the techniques of human genetics, while being sufficient to identify the variants within a limited genomic region, fail to scale to whole genome experiments. However, as technological improvement allows the generation of vast quantities of variation data, human geneticists will likely require new models to both guide our future experiments and interpretation of the data. This is particularly true for studies that aim to understand the genetic basis of complex traits such as human neuropsychiatric disorders. The simple models that have proven so successful for Mendelian traits may not prove as successful for complex traits.

With this in mind, this chapter will address the following main areas:

1. What are the mean levels of human genetic variation throughout the genome? How do the levels vary across the human genome?
2. What genetic models might guide our analysis of the variation underlying neuropsychiatric traits?

While some technological limitations to the study of the genetics of complex traits are likely to remain, the vast majority will no longer exist with the completion of the human genome project. The greater challenge is an intellectual one — can we pose questions in a fashion that suggests the appropriate experiment that will lead to an answer. It is imperative that we clearly define the models and their predictions if we are to truly understand the genetic basis of complex traits.

TABLE 13.1
Genome-Wide Estimates of Human Nucleotide Diversity

Studies	Total Nucleotide Diversity ($\times 10^4$)
Halushka et al.[20]	8.3
Cargill et al.[21]	5.39
The International SNP Map Working Group[17]	7.50
Venter et al.[18]	8.90

TABLE 13.2
Comparison of X and Autosome Estimates of Nucleotide Diversity

Studies	Autosomal Nucleotide Diversity ($\times 10^4$)	X Chromosome Nucleotide Diversity ($\times 10^4$)
The International SNP Map Working Group[17]	7.65	4.69
Venter et al.[18]	8.94	6.54

13.2 PATTERNS OF HUMAN GENETIC VARIATION

With the initial sequencing and analysis of the human genome complete, what have we learned about the patterns and levels of DNA sequence variation? Single nucleotide polymorphisms (SNPs, pronounced snips) are both the most common and extensively surveyed form of genetic variation.[19] If we focus on SNPs, then four main conclusions can be drawn from recent analyses of human genetic variation. The first concerns the average level of variation typically measured as nucleotide diversity. Nucleotide diversity is the probability that the nucleotide at a randomly selected site in a randomly selected chromosome differs from the nucleotide found at the same site in another randomly selected chromosome. The average level of genetic variation in studies that have examined multiple genomic regions are remarkably consistent — ~8×10^{-4} (Table 13.1).[17,18,20,21] It seems unlikely that the average value will change significantly with future analysis.

The second main result is that levels of variation on the X and Y chromosomes are lower than those found on the autosomes (Table 13.2). This pattern of variation is expected because both sex chromosomes have a smaller effective population size than an autosome. As a consequence, the sex chromosomes are expected to both lose variation more rapidly and gain new variation more slowly than would a typical autosome. Furthermore, levels of variation on the sex chromosomes are expected to be influenced by natural selection, perhaps due to their hemizygosity, in a fashion different than that seen for a typical autosome.

The third main result concerns the variance in nucleotide diversity in different genomic regions. Both of the early studies that focused their attention to multiple genes noted that the most striking and substantial variation in nucleotide diversity was observed among different genes.[20,21] Confirming these observations at the level of the entire genome, the recent genome papers that examined the variation in nucleotide diversity in 100 to 200 kb bins[17,18] noted substantial variation among genomic regions. This observation likely results from local differences in mutation, recombination, and selection. When the final finished sequence is available for each chromosome, it should be possible to obtain a more precise quantification of the variation within and between specific chromosomes and chromosomal regions.

The fourth main result lies in the observation that missense mutations, those sites that alter amino acids in proteins, are individually rare. This observation was not unexpected, having been previously described in both *Drosophila*[22] and humans.[20,21,23] Furthermore, these types of studies have noted that coding regions have lower levels of nucleotide diversity than do noncoding regions. In fact, the Cargill et al.[21] nucleotide diversity estimate in 1999 is likely lower than other estimates because they specifically focused their attention on the coding regions of genes. The subsequent genome-wide surveys confirm all of these observations.[17,18] Furthermore, an analysis of the annotated 10,239 RefSeq genes[24] shows that between 0.1 and 0.2% of all identified SNPs appear to be missense mutations, with an even smaller percentage leading to nonconservative amino acid changes.[18] This obviously implies that between 99.8 and 99.9% of all SNPs fail to alter amino acids in a protein. Nevertheless, the SNP Consortium has shown that the public databases already contain at least one SNP within 93% of all genes, and that 98% of genes are within 5 kb of an identified SNP.[17]

13.3 COMPARATIVE GENOMICS — A PATH TO FUNCTION?

The critical issue we wish to focus on, however, arises from the observed great rarity of missense mutations and the fact that missense mutations typically underlie Mendelian traits. Thus, we have an *a priori* expectation that missense mutations are likely to have biological function by altering the primary sequence of the protein. But the application of this model for identifying the function of genetic variation that we observe is only relevant to the 0.1 to 0.2% of missense SNPs. What models should guide our search for function within the remaining 99.8 to 99.9% of SNPs that fail to alter amino acids in proteins? Many of these variants are likely to be neutral and lack functional effects, but surely not *all* of the remaining segregating sites lack biological function? Certainly some sites in noncoding and regulatory regions are likely to be important for understanding the genetic basis of complex traits.[23,25]

Thus we are faced with two critical questions. First, how might one find candidate sites underlying complex traits? Second, upon finding candidate sites, how might one demonstrate their function? The latter question might well be addressed through a variety of means. Clearly, refining or discovering new phenotypes, such as expression level, location, and timing as characterized through the use of expression microarrays,[26] will likely prove important. It also seems likely that comparative genomic diversity studies, that both identify genetic variation within humans and divergence between humans and other species, will prove increasingly important in assessing function of the vast majority of DNA sequence variation not leading to coding changes.[27] Segregating genetic variants might well influence all aspects of gene expression or alter splicing, with the ultimate effect of changing a phenotype of interest. But for the remainder of this chapter, we will focus on how one might find the candidate genetic variants underlying complex traits.

13.4 IDENTIFYING GENETIC VARIATION UNDERLYING COMPLEX TRAITS

As we discussed previously, the overall level of polymorphism in humans is quite small, with approximately 8 out of 10,000 nucleotides differing between two randomly chosen chromosomes. When new mutations do occur, they usually occur in a single gamete of a single individual. As a result, new mutations initially begin at a low frequency, with autosomal mutants starting at a frequency of approximately $1/10^{10}$ (assuming a human population size of five billion). A SNP at the meager frequency of 1% is, therefore, 10^8 times more common than a newly arising mutation. These facts imply there may be deep underlying differences between the patterns of genetic variation surrounding SNP of differing frequencies. These differing patterns of genetic variation may profoundly impact on a study's ability to find the genetic variation contributing to neuropsychiatric traits.

High frequency SNPs are, on average, older than low frequency SNPs. Although the precise age of any SNP is probably unknowable, the expected age in generations for a selectively neutral SNP is a simple function of its frequency, and is given by[28,29]

$$\frac{4Np\ln(p)}{1-p}$$

where p is the frequency of the SNP and N is the effective population size (effective population size is generally close to the harmonic mean population size over a long period of time). Figure 13.1 plots the expected age of a SNP as a function of its frequency (assuming an effective population size of 10,000 and a generation time of 20 years). Several features stand out. First, alleles of only modest frequency, say 5%, are, on average, quite old, over 100,000 years old. Increase that frequency by a factor of 10, and the average age only increases by around a factor of 5. Second, alleles with much lower frequency, say 0.01 to 0.1%, are much, much younger, approximately 1000 to 5000 years old (Figure 13.1). This has profound implications for the expected linkage disequilibrium surrounding SNPs.

When a SNP first enters the population, it occurs in a single genetic background, and it is in complete linkage disequilibrium (LD) with all other sites in the genome. In the absence of selection, this linkage disequilibrium decays over time. Figure 13.2 plots the amount of LD that ought to remain around a SNP with age equal to the expected age for a SNP of its frequency, for several genetic distances. It is clear that any site with frequency greater than 5% is unlikely to show LD with another site more than 1/10 of a CentiMorgan away. On the other hand, sites with much lower frequency (Figure 13.3) may often exhibit LD that extends over megabases (under the assumption that a CentiMorgan roughly corresponds to a megabase of physical distance).

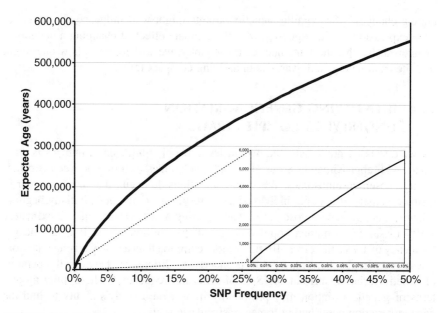

FIGURE 13.1 Mean age of neutral SNPs at intermediate frequency. Figure 13.1 plots the equation in the text for a mutant frequency between 0 and 50%. Human effective population size is assumed to be 10,000, and generation time is assumed to be 20 years. It should also be noted that the variance in the age is extremely large, on the order of the square of the mean. As a result, any particular SNP may be either much older or much younger than the mean age. Inset: Mean age of neutral SNP at low frequency. Low frequency SNPs are, on average, much younger than most of SNP likely to be contained in the public databases.

The public database, dbSNP, contains over 1.4 million SNPs.[17] This is a remarkable collection that will only improve in the coming months. Nevertheless, not all segregating sites will end up in dbSNP. There are two interrelated, but distinguishable, reasons for this. The first reason is simply technological. Not all genomic regions clone equally well[30] or exist between restriction sites of the proper size[31] and will therefore not be detectable by the two major SNP discovery efforts. This problem may be resolvable over time as technology changes. The second problem is harder. All the SNPs in dbSNP were discovered in relatively small population surveys. SNPs discovered by comparing BAC overlaps[30] were often detected in overlap depths of only size 2, and seldom larger than 10,[17] and the SNPs generated by reduced representation shutgun sequencing[31,32] of 24 individuals provide a maximum of 48 different chromosomes, but were often generated with coverage of only size 2. As a result, sites segregating in the general population simply may not be segregating in these small sample sizes. Figure 13.4 plots the probability that a site with a given frequency happens to be segregating in a sample of size 2 or of size 48. The remarkable feature of this graph is the relative poverty of sites with frequency below 1% or above 99%. Simply put, dbSNP is very unlikely to contain any SNP with very low or very high frequency. The only solution to this problem would require a much larger detection effort in much larger sample sizes.

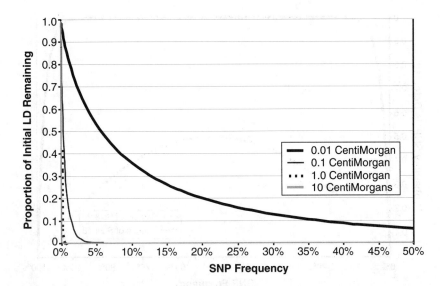

FIGURE 13.2 Decay of linkage disequilibrium (LD) around intermediate frequency SNPs. LD decays with rate $(1 - r)^t$, where r is a recombination fraction, and t is the length of time measured in generations.[33] To relate genetic distance to the recombination fraction, the Haldane map function was assumed, $r = (1 - e^{-2g})/2$,[34] where g is the genetic distance measured in Morgans. For these calculations, SNPs were assumed to have age exactly equal to their expected age.

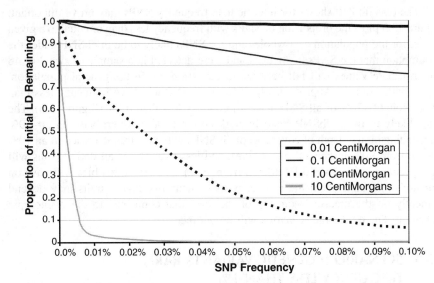

FIGURE 13.3 Decay of linkage disequilibrium around low-frequency SNPs. SNPs are assumed to have age identical to their expected age, and a Haldane map function is used (see Figure 13.2). Of course, since the age of a SNP is a stochastic variable, as is the decay of LD, the actual LD around a SNP may be much larger or smaller than this amount. Moreover, LD continues to be created by the introduction of new mutations.

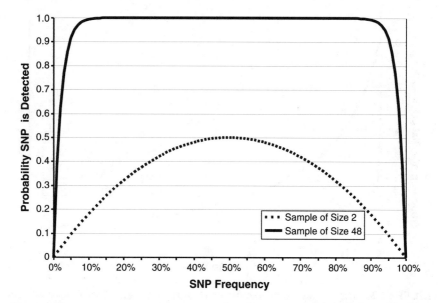

FIGURE 13.4 The probability that a site with a given frequency might be detected in samples of size 2 or 48. In general, the probability that a site with population frequency p is segregating in a sample of k chromosomes is $1 - p^k - (1 - p)^k$. If detection of segregating sites is perfect, Figure 13.4 gives the probability of detection. Of course, SNP detection is never perfect, and realized detection rates will be lower than those given.

The bias dbSNP shows for intermediate frequency SNPs may prove important. Figure 13.5 plots the proportion of SNPs with frequency less than or equal to a given value, for the population as a whole, and for SNPs likely to be in dbSNP under the assumption that the SNPs are neutral and were detected in a sample of size 2 or 48 (intermediate values will fall between these extremes). In the general population, most SNPs are rare. Approximately 90% of all SNP will have frequency less than 10%. Roughly 70% of all SNPs have a frequency less than 0.1% (Figure 13.6). The sites likely to be in dbSNP stand in stark contrast. Somewhere between 20–50% (depending on the depth of coverage) of all SNPs will have frequency less than 10% (down from 90% in general population), and less than 1% of all dbSNP sites will have frequency less than 0.1% (down from 70%). The effect this bias will have on our ability to understand the genetic basis of neuropsychiatric traits may depend critically on the precise frequency of the sites that contribute to the phenotypic variation. Three possible scenarios will be outlined.

13.5 SCENARIO I: GENETIC VARIANTS RARE; FREQUENCY LESS THAN 1%

Despite relatively limited technology, and few public genetic resources, the genetic variants causing Mendelian diseases have been mapped with extraordinary success.[1] Most of these variants are at low frequency, and therefore are likely to be in LD

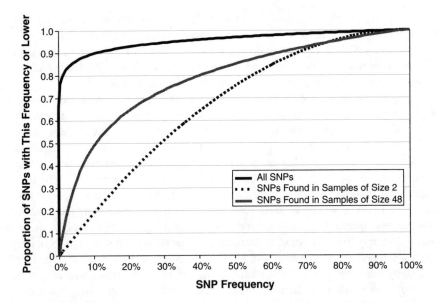

FIGURE 13.5 Effect of sample size on frequency of detected SNPs. In the general population, the proportion of neutral sites with frequency less than or equal to p is proportional to $2\log(2Np)$, where N is population size.[35] The proportion of sites in a sample of size k, with frequency less than or equal to p can be obtained from Equation 7 in Ewens.[36]

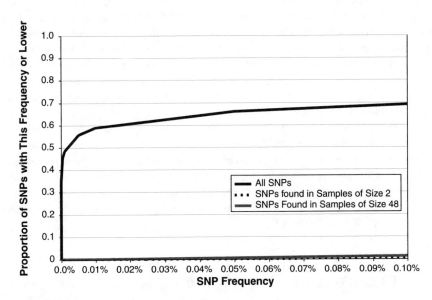

FIGURE 13.6 Only a very small fraction of rare SNPs will be found in the public database, despite the fact that most SNPs in the general population are rare.

with sites at relatively large distances. Hence, mapping efforts within families could be successful with relatively few, widely spaced markers. Complex disease traits, such as schizophrenia or bipolar disorder, may be caused by recently arising, low frequency mutations. However, if so, there must be many loci involved, each harboring allelic variants that might contribute to the disease process.

Imagine the following scenario. Consider a disease with population frequency of 1%. A recessive mutation at any of 100 loci or perhaps a co-dominant mutation at any of 10 loci might cause the disease. Further assume that any of these loci are sufficient to cause the disease. The sites causing the disease will have frequencies less than 1% in the general human population. These variants are extremely unlikely to be in dbSNP, but because their frequency is so low, they are likely to be in LD with sites that are in dbSNP. Significant LD is expected to extend over large physical distances. In order to map these variants, one can imagine employing the following experimental approach. First, one could mount a general population association study with a relatively limited number of markers, perhaps microsatellites, or as technology improves, a genome-wide set of SNP markers. In contrast to the number of probands in a study mapping a Mendelian trait, the number of individuals necessary to map a complex trait with this pattern of inheritance must be far larger. To a first approximation, if there are ten times as many loci, one might expect a typical study would require ten times as many probands. Of course, this approach will only discover a candidate region, and exact identification of the disease-causing site will require further direct sequencing (since the SNP is unlikely to be previously identified).

Alternatively, one could perform a family-based linkage study with an even wider genetic map, since LD within a family will be even greater than the population as a whole. Of course, since the population frequency of any of the disease causing sites is so low, no family is likely to harbor more than a few disease alleles. As a result the number of probands necessary for such a study may be much lower (since the number of loci contributing to the variation in any particular family will be lower). Such studies, even if they include very few markers, may be highly successful individually, due to the extensive LD surrounding the disease causing sites. Of course, finding the site itself will require additional sequencing, but a candidate region is likely to be found. *However, such studies will give the appearance of being unreproducible.* Simply put, the sites that cause the disease in one family are unlikely to cause the disease in another family. Further studies will fail to confirm linkage to the same region. This in no way implies that the original study was necessarily flawed. It is simply impossible for the same region to show linkage in large numbers of families because the frequencies of the individual mutations are so low. Common disease traits in different families will have different genetic causes, and the results of linkage studies will reflect this.

13.6 SCENARIO II: GENETIC VARIANTS COMMON; FREQUENCY GREATER THAN 10%

There are several important ways in which this problem is harder than the previous scenario, but there are certain advantages, as well. First, most of the sites that contribute to the disease are already in dbSNP, or are likely to be there soon. This may very well simplify things and may alleviate the need for further sequencing after linkage or association has been found. Unfortunately, linkage disequilibrium around these sites is likely to be extremely low. As a result, population-wide association studies may require genotyping many, or even all, of the sites in dbSNP. Technology to do so does not yet exist.

A more sensible current approach would be to search for linkage to a region within families (which will have much greater LD than the population as a whole). Since these sites are at high frequency, there is a much greater chance that linkage will be demonstrable in multiple families, and should therefore be sought. After linkage to a region has been repeatedly demonstrated, a population-wide association study can be considered. Genotyping of all the SNPs in dbSNP within the regions showing linkage might be feasible with current technology.

There is, though, one further complication. If a disease has a population frequency of say, 5%, and there are many sites at relatively high frequency that contribute to this disease, those sites must necessarily interact in some way to create the disease. No individual site is likely to be either necessary or sufficient. Some forms of genetic interaction (principally when the mutants have equal and additive effects) are relatively easy to model and understand.[3] Unfortunately, many forms of interaction have not yet been considered in detail. If loci interact in complicated ways, further conceptual advances will be needed before the genetic basis of these traits is likely to be understood.

13.7 SCENARIO III: GENETIC VARIANTS INTERMEDIATE; FREQUENCY GREATER THAN 1% BUT LESS THAN 10%

The final scenario is likely to be far more difficult than both of the previous scenarios for two main reasons. First, unlike the second scenario, the variants causing the disease are unlikely to be found in dbSNP. Second, in contrast to variants underlying Mendelian traits, intermediate frequency genetic variants may not even be in linkage disequilibrium with any site in dbSNP. Moreover, all of the difficulties of both of the previous scenarios may still apply. The only possible experimental approach to this scenario will require *de novo* SNP detection in individuals showing variation for the trait of interest. Once variation is detected, the problem will likely reduce to either scenario I or II, or some combination thereof.

Nevertheless, identifying genetic variants in this scenario is not hopeless! A sensible strategy might combine approaches from both of the first two scenarios. Begin with a large family-based linkage study (LD will be greatest within families) using a high density, publicly available SNP map. Set appropriately high, multiple test corrected *p*-values. If linkage is detected, do not attempt to replicate in other families, because replication is unlikely to be successful. Instead, move directly to a large, population-wide association study, using markers developed from samples showing variation for the trait of interest. Since a family-based linkage study is unlikely to provide resolution below megabase-sized regions, this approach may require large-scale resequencing.

13.8 CONCLUSIONS

The genetic basis of complex neuropsychiatric traits has not yet been elucidated.[5-7] In contrast, many rare Mendelian disorders have been fully characterized.[1] There are many reasons for this distinction. Some of the reasons have little to do with genetics. Diagnosis and phenotype characterization of neuropsychiatric traits is simply harder than diagnosis of, say, cystic fibrosis. Incorrect phenotypic description necessarily implies greater difficulty in understanding genetic causes. Such obstacles, though, are presumably solvable. There are strictly genetic obstacles as well.

There are several genetic features that allow relatively easy mapping of Mendelian traits. The sites contributing to variation in these traits are generally rare missense mutations that are in linkage disequilibrium with other surrounding sites. As a consequence, a researcher can both quickly identify a candidate region and then screen it for missense mutations. Subsequent studies of the Mendelian trait are often made easier because a small number of genetic loci (often only one) are usually involved. Hence, even when multiple mutations cause a Mendelian disease trait, these sites are usually located in the same genetic regions, so that multiple mapping studies find linkage to the same general region.

Complex neuropsychiatric traits may lack many of these features. First, it is nearly certain that diseases such as schizophrenia or bipolar disorder have multiple loci that contribute to the phenotype. As a result, study sizes required to detect any single locus must be much larger than for Mendelian traits. Moreover, the sites contributing to the phenotype might be much more frequent than for Mendelian traits, and consequently the LD around these sites might be much lower. This implies that much denser genetic maps may be required. Finally, the manner in which these sites interact may be complicated and require further refinement of our genetic models. None of these problems are insurmountable, but their existence has thus far slowed our understanding of neuropsychiatric traits.

The solution to these problems may depend on the exact genetic variants underlying these traits. If most genetic variants are rare (Scenario I), an approach similar to Mendelian traits can be used, albeit with much larger sample sizes. One important distinction remains: family-based linkage studies are expected to fail to reproduce. The genetic variants leading to the disease will simply be different in distinct families. As a result, understanding all the variants that contribute to the disease may require many independent linkage and subsequent association studies.

If most genetic variants are common (Scenario II), technology and public efforts may largely solve our problems. The public resources (dbSNP) are likely to contain most of the sites contributing to the disease. If technology ultimately permits genotyping all these SNPs cheaply and easily, a simple population association study with *all* markers is probably the best approach. If no such technology is available (and none is today), multiple family-based linkage studies, followed by association studies with all available markers in the region showing linkage, is probably the best approach. Some work may remain to understand the patterns of interaction between SNP, but the path to understanding is largely in view.

The most difficult situation involves SNPs at intermediate frequency (1–10%) with no truly rare sites (Scenario III). Such sites are quite old, relative to newly arising mutations, and likely to have little LD. They are also unlikely to be contained in dbSNP. If neuropsychiatric traits are ultimately caused by such allelic variants, large-scale, sample-specific resequencing will be necessary. If this technological hurdle can be overcome, even this most difficult of scenarios should fall to the general approaches discussed in Scenarios I and II.

Neuropsychiatric traits are complex, and as such, are more difficult to analyze than most single-gene Mendelian traits. Nevertheless, the genetics research program has hope. The number of required samples may be large and certain types of replication may be impossible to obtain. Further advances in our understanding of the patterns of interaction between variants may also be necessary. Nevertheless, there is every reason to suspect that even these most difficult of complex traits will ultimately be understood.

REFERENCES

1. Online Mendelian Inheritance in Man, OMIM™. McKusick–Nathans Institute for Genetic Medicine, Johns Hopkins University (Baltimore, MD) and National Center for Biotechnology Information, National Library of Medicine (Bethesda, MD), 2001, World Wide Web URL: http://www.ncbi.nlm.nih.gov/omim/.
2. Scriver, C. R., *The Metabolic & Molecular Bases of Inherited Disease*, McGraw-Hill/Medical Publishing Division, New York, 2001.
3. Falconer, D. S. and Mackay, T. F. C., *Introduction to Quantitative Genetics*, Longman, Burnt Mill, Harlow, Essex, England, 1996.
4. Collins, F. S., Shattuck Lecture — medical and societal consequences of the Human Genome Project, *N. Engl. J. Med.*, 341, 28–37, 1999.
5. Todd, R. D., Genetics of attention deficit/hyperactivity disorder: Are we ready for molecular genetic studies?, *Am. J. Med. Genet.*, 96, 241–243, 2000.
6. DeLisi, L. E., Craddock, N. J., Detera-Wadleigh, S., et al., Update on chromosomal locations for psychiatric disorders: Report of the interim meeting of chromosome workshop chairpersons from the VIIth World Congress of Psychiatric Genetics, Monterey, CA, Oct. 14–18, 1999, *Am. J. Med. Genet.*, 96, 434–449, 2000.
7. Baron, M., Genetics of schizophrenia and the new millennium: Progress and pitfalls, *Am. J. Hum. Genet.*, 68, 299–312, 2001.
8. Blouin, J. L., Dombroski, B. A., Nath, S. K. et al., Schizophrenia susceptibility loci on chromosomes 13q32 and 8p21, *Nat. Genet.*, 20, 70–73, 1998.
9. DeLisi, L. E., Shaw, S., Sherrington, R. et al., Failure to establish linkage on the X chromosome in 301 families with schizophrenia or schizoaffective disorder, *Am. J. Med. Genet.*, 96, 335–341, 2000.
10. Ekelund, J., Lichtermann, D., Hovatta, I. et al., Genome-wide scan for schizophrenia in the Finnish population: Evidence for a locus on chromosome 7q22, *Hum. Mol. Genet.*, 9, 1049–1057, 2000.
11. Brzustowicz, L. M., Hodgkinson, K. A., Chow, E. W. et al., Location of a major susceptibility locus for familial schizophrenia on chromosome 1q21–q22, *Science*, 288, 678–682, 2000.
12. Levinson, D. F., Holmans, P., Straub, R. E. et al., Multicenter linkage study of schizophrenia candidate regions on chromosomes 5q, 6q, 10p, and 13q: Schizophrenia linkage collaborative group III, *Am. J. Hum. Genet.*, 67, 652–663, 2000.

13. Schwab, S. G., Hallmayer, J., Albus, M. et al., A genome-wide autosomal screen for schizophrenia susceptibility loci in 71 families with affected siblings: Support for loci on chromosome 10p and 6, *Mol. Psychiatry,* 5, 638–649, 2000.

14. Visscher, P. M., Haley, C. S., Heath, S. C. et al., Detecting QTL for uni- and bipolar disorder using a variance component method, *Psychiatr. Genet.,* 9, 75–84, 1999.

15. Murphy, V. E., Mynett-Johnson, L. A., Claffey, E. et al., Search for bipolar disorder susceptibility loci: The application of a modified genome scan concentrating on gene-rich regions, *Am. J. Med. Genet.,* 96, 728–732, 2000.

16. International Human Genome Study Consortium, Initial sequencing and analysis of the human genome, *Nature,* 409, 860–921, 2001.

17. The International SNP Map Working Group, A map 'of human genome sequence variation containing 1.42 million single nucleotide polymorphisms, *Nature,* 409, 928–933, 2001.

18. Venter, J. C., Adams, M. D., Myers, E. W. et al., The sequence of the human genome, *Science,* 291, 1304–1351, 2001.

19. Collins, F. S., Guyer, M. S., and Chakravarti, A., Variations on a theme: Cataloging human DNA sequence variation, *Science,* 278, 1580–1581, 1997.

20. Halushka, M. K., Fan, J. B., Bentley, K. et al., Patterns of single-nucleotide polymorphisms in candidate genes for blood-pressure homeostasis, *Nat. Genet.,* 22, 239–247, 1999.

21. Cargill, M., Altshuler, D., Ireland, J. et al., Characterization of single-nucleotide polymorphisms in coding regions of human genes, *Nat. Genet.,* 22, 231–238, 1999.

22. Moriyama, E. N. and Powell, J. R., Intraspecific nuclear DNA variation in Drosophila, *Mol. Biol. Evol.,* 13, 261–277, 1996.

23. Zwick, M. E., Cutler, D. J., and Chakravarti, A., Patterns of genetic variation in Mendelian and complex traits, in *Annu. Rev. Genomics Hum. Genet.,* 1, 387–407, 2000.

24. Pruitt, K. D., Katz, K. S., Sicotte, H. et al., Introducing RefSeq and LocusLink: Curated human genome resources at the NCBI, *Trends Genet.,* 16, 44–47, 2000.

25. Mackay, T. F. C., Quantitative trait loci in *Drosophila, Nat. Reviews,* 2, 11–20, 2001.

26. Lipshutz, R. J., Fodor, S. P., Gingeras, T. R. et al., High density synthetic oligonucleotide arrays, *Nat. Genet.,* 21, 20–24, 1999.

27. Chakravarti, A., Population genetics — making sense out of sequence, *Nat. Genet.,* 21, 56–60, 1999.

28. Kimura, M. and Ohta, T., The age of a neutral mutant persisting in a finite population, *Genetics,* 75, 199–212, 1973.

29. Ewens, W. J., *Mathematical Population Genetics,* Springer-Verlag, Berlin, 1979, 165.

30. Taillon-Miller, P., Gu, Z., Li, Q. et al., Overlapping genomic sequences: A treasure trove of single-nucleotide polymorphisms, *Genome Res.,* 8, 748–754, 1998.

31. Altshuler, D., Pollara, V. J., Cowles, C. R. et al., An SNP map of the human genome generated by reduced representation shotgun sequencing, *Nature,* 407, 513–516, 2000.

32. Mullikin, J. C., Hunt, S. E., Cole, C. G. et al., An SNP map of human chromosome 22, *Nature,* 407, 516–520, 2000.

33. Crow, J. F. and Kimura, M., *An Introduction to Population Genetics Theory,* Burgess Publishing Company, Minneapolis, 1970, 48.

34. Weir, B. S., *Genetic Data Analysis II,* Sinauer Associates, Sunderland, MA, 1996, 231.

35. Sawyer, S. A. and Hartl, D. L., Population genetics of polymorphism and divergence, *Genetics,* 132, 1161–1176, 1992.

36. Ewens, W. J., A note on the sampling theory for infinite alleles and infinite sites models, *Theor. Pop. Biol.,* 6, 143–148, 1974.

Index

Index

A

AAV, *see* Adeno-associated virus
Aberrant pattern formation, 66
Actin probe sets, 161
Ad, *see* Adenovirus
Additive model, 32
Additivity, 6–7
Adeno-associated virus (AAV), gene delivery,
      230, 233, 237–238
  convection-enhanced, 242–243
  intrathecal, 249, 250
Adenovirus (Ad), 230, 234
Affected sib pair (ASP), 291, 294
Affymetrix GeneChip® technology
  advantages and drawbacks, 180
  image acquisition, 185
  noise assessment in initial experiments, 187
  procedural overview, 147
  reporting the most changed genes, 194
  sample quality control, 177
Age, 301, 302, *see also* Genetic variation analysis
ALS, *see* Amyotrophic lateral sclerosis
AMPA2 receptor, 198, *see also* Schizophrenia
Amplicon vectors, 237, *see also* Gene delivery
      methods; Herpes simplex virus
Amplification, 159–161, 183–184, *see also*
      Messenger RNA
Amygdala, 163, 164
Amyotrophic lateral sclerosis (ALS), 234
Analysis of variance (ANOVA), 12–14
Animals, genome-wide scan, 74–75
ANOVA, *see* Analysis of variance
Antibodies, 235
Anxiety, 119
Arabidopsis spp., 21
Array exclusion criteria, 196
Array printing, 183, 191
Arraying solutions, 183
Arraying surface, 183
Ascertainment, 31, 291
ASP, *see* Affect sib-pair
Assessment, in vivo, 275–277
Astrocytes, 274
Autism, 294
Autosomal nucleotide diversity, 299
Average difference, 150

Axonal markers, 94–95
Axon-guidance phenotypes, 105, *see also*
      Phenotypes

B

BAC, *see* Bacterial artificial chromosome
Backcross breeding scheme, 54, 71–72, 77,
      126–127
Background
  noise, 149, 150, 179
  staining, 106
  uneven, 190
Bacterial alkaline phosphatase (BAP), 212, 217, 219
Bacterial artificial chromosome (BAC)
  genetic variation analysis, 302
  positional cloning, 79–81
  rescue and applications of segregation/linkage
      analysis, 35, 37, 38
  transgenesis and gene targeting, 46
Bacteriophage clones, 35
BAP, *see* Bacterial alkaline phosphatase
BBB, *see* Blood–brain barrier
Behavior, visually- and auditory-driven, 67
β-geo reporter gene, 93–94
Bias, 192, 304
Biochemical pathways, 65–66
Biological changes, 187–189
Biological effect, 187
Biological functions, 195
Biotin labeling, 148, *see also* Labeling
Biotypes, 4
Bipolar disorder, 116
BK, *see* Bradykinin
Blood–brain barrier (BBB), 246–249, 252
Blood–tumor barrier (BTB), 247, 252
Bombardment, 235
Bradykinin (BK), 247–249
Brain, *see also* Individual entries
  gene delivery, *see* Gene delivery methods
  gene trap approach and development/wiring
    applications to actual data, 100–106
    data interpretation, 106–107
    detailed methods, 95–99
    possible methodologic variations, 99
    rationale, 92–95